Clinical Cardiac Electrophysiology:

Perioperative Considerations

Edited by
Carl Lynch III, MD, PhD
Professor of Anesthesiology
University of Virginia Health Sciences Center
Charlottesville, Virginia

With 19 contributors

J. B. LIPPINCOTT COMPANY
Philadelphia

Clinical Cardiac Electrophysiology:

Perioperative Considerations

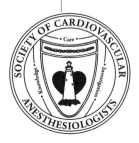

A Society of
Cardiovascular Anesthesiologists
Monograph

Acquisitions Editor: Mary K. Smith
Assistant Editor: Anne Geyer
Production Manager: Janet Greenwood
Production: Artech Graphics II, Inc.
Compositor: Artech Graphics II, Inc.
Printer/Binder: R.R. Donnelley & Sons

6   5   4   3   2   1

Library of Congress Cataloging-in-Publications Data

Clinical cardiac electrophysiology : perioperative considerations / edited by
    Carl Lynch III ; with 19 contributors.
        p. cm. — (A Society of Cardiovascular Anesthesiologists
    monograph)
        Includes bibliographical references and index.
        ISBN 0-397-51405-0
        1. Arrhythmia.   2. Heart conduction system.   3. Electrophysiology.
    4. Cardiovascular agents.   I. Lynch, Carl, 1949–   . II. Series.
        [DNLM: 1. Electrophysiology.   2. Heart—physiology.
    3. Heart Diseases—surgery.   4. Intraoperative Complications—diagnosis.
    WG 102 C6405 1994]
    RC685.A65C544   1944
    616.1'28—dc20
    DNLM/DLC
    for Library of Congress   94-2130
                                                                               CIP

The authors and publisher have exerted every effort to ensure that drug selection and dosage set forth in this text are in accord with current recommendations and practice at the time of publication. However, in view of ongoing research, changes in government regulations, and the constant flow of information relating to drug therapy and drug reactions, the reader is urged to check the package insert for each drug for any change in indications and dosage and for added warnings and precautions. This is particularly important when the recommended agent is a new or infrequently employed drug.

Publication Committee of the
Society of Cardiovascular Anesthesiologists

Edward Lowenstein, MD, Chairman
Boston, Massachusetts

David C. Warltier, MD, Vice Chairman
Milwaukee, Wisconsin

Thomas J.J. Blank, MD
New York, New York

William J. Greeley, MD
Durham, North C arolina

Carl Lynch III, MD, PhD
Charlottesville, Virginia

Contributors

Michael J. Barber, MD, PhD
Assistant Professor of Medicine
University of Virginia Health
  Sciences Center
Charlottesville, Virginia

Zeljko J. Bosnjak, PhD
Professor of Anesthesiology and
  Physiology
Director, Laboratory of Cellular
  Biology
Medical College of Wisconsin
Milwaukee, Wisconsin

Frederick A. Burrows, MD, FRCPC
Associate Professor
Department of Anesthesia
Harvard Medical School
Director, Cardiac Anesthesia Service
Children's Hospital
Boston, Massachusetts

John P. DiMarco, MD, PhD
Professor of Medicine
Director, Clinical Electrophysiology
  Laboratory
University of Virginia Health
  Sciences Center
Charlottesville, Virginia

Anne Hamilton Dougherty,
  MD, FACC
Associate Professor of Medicine
Associate Director of Cardiac
  Electrophysiology
University of Texas Medical School at
  Houston
Staff Cardiologist
Hermann Hospital
Houston, Texas

John D. Gallagher, MD
Associate Professor of
  Anesthesiology
Dartmouth Medical School
Hanover, New Hampshire
Attending Anesthesiologist
Dartmouth-Hitchcock Medical
  Center
Lebanon, New Hampshire

Robert M. Gow, MB, BS, FRACP
Assistant Professor
Department of Pediatrics
University of Toronto Faculty of
  Medicine
Head, Section of Electrophysiology
Division of Cardiology
Hospital for Sick Children
Toronto, Ontario

David E. Haines, MD
Associate Professor
Co-Director of Cardiac
  Electrophysiology
University of Virginia Health
  Sciences Center
Charlottesville, Virginia

Ld R. Herzog, MD
Department of Anesthesiology
University of Virginia Health
  Sciences Center
Charlottesville, Virginia

Sohail Jalal, MD
Assistant Professor of Medicine
University of Texas Medical School
  at Houston
Houston, Texas

Terry W. Latson, MD
Clinical Associate Professor
Department of Anesthesiology
University of Texas Southwestern
  Medical School
Dallas, Texas

Carl Lynch III, MD, PhD
Professor of Anesthesiology
University of Virginia Health
  Sciences Center
Attending Anesthesiologist
University of Virginia Hospital
Charlottesville, Virginia

Gerald V. Naccarelli, MD
Professor of Medicine
Director, Clinical Electrophysiology
University of Texas Medical School
  at Houston
Houston, Texas

Sunil Nath, MD
Instructor in Medicine
University of Virginia School of
  Medicine
Attending, Cardiac
  Electrophysiology
University of Virginia Health
  Sciences Center
Charlottesville, Virginia

Hue-Teh Shih, MD, MPH
Assistant Professor of Medicine
Co-Director, Interventional Cardiac
  Electrophysiology
Director, Molecular Cardiac
  Electrophysiology
University of Texas Medical School
  at Houston
Houston, Texas

William G. Stevenson, MD
Associate Professor of Medicine
Harvard Medical School
Co-Director of Clinical
  Electrophysiology
Brigham and Women's Hospital
Boston, Massachusetts

Lawrence A. Turner, MD
Associate Professor
Department of Anesthesiology
Medical College of Wisconsin
Assistant Director of Anesthesiology
Froedtert Memorial Lutheran
  Hospital
Milwaukee, Wisconsin

Deborah L. Wolbrette, MD
Assistant Professor of Medicine
Division of Cardiology
University of Texas Medical School
  at Houston
Staff Cardiologist
Hermann Hospital
Houston, Texas

James R. Zaidan, MD
Professor of Anesthesiology
Deputy Chairman for Education
Emory University School of
  Medicine
Chief of Anesthesiology
Grady Memorial Hospital
Atlanta, Georgia

Contents

Preface

Over the past few years, application of microelectrophysiologic and molecular biologic techniques (often in combination) has provided an increasingly detailed understanding of the membrane mechanisms that control cardiac electrical activity. Using agents that block specific currents, it has been possible to define the behavior of ion channels in isolated tissues or myocytes down to the movement of ions through an individual membrane channel protein. Injection of the DNA or messenger RNA for a particular channel into frog oocytes or cultured cell lines has permitted channels to be expressed and characterized in isolation. As precise understanding is gained about the structure, physiology, and pharmacology of channels, as well as their interaction in intact and pathologic myocardium, it will be possible to predict more effective therapeutic interventions. Currently, a complete and integrated understanding is prevented by the number, species specificity, and functional modulation of the various ion channels; by incomplete pathophysiologic models; by the complex architecture of the myocardium itself. For example, while drugs producing a longer blockade of sodium channels seemed theoretically beneficial, flecainide and encainide actually increased the incidence of sudden cardiac deaths that they were intended to decrease (the Cardiac Arrhythmia Suppression Trial).

Because our understanding of normal and pathologic cardiac electrical activity provides an incomplete explanation for certain events, and because our pharmacologic tools are not perfectly selective, much of our therapy is of necessity directed empirically. Fortunately, we have a spectrum of tools with which to treat the cardiac electrophysiological disorders that arise. A variety of drugs are efficacious; as their mechanisms are better defined, refinements in their use and synthesis of new analogues will improve the therapeutic-toxic ratio. Although not eliminating the underlying cause, programmable and increasingly sophisticated pacers and automatic defibrillators are being developed and implanted for effective treatment of arrhythmias when they occur. Employing sensors to monitor electrical behavior of the heart, these units are able to decide according to various algorithms the appropriate therapeutic intervention.

Considering the significant physiologic and sometimes psychologic stress that patients experience when undergoing surgery, an increased frequency of cardiac arrhythmias in the perioperative period is not surprising. Nevertheless, it is primarily those surgical patients who have preexisting cardiac disease, either caused by abnormalities at the molecular level (e.g., long QT syndrome, Duchenne muscular dystrophy) or secondary to structural or inflammatory lesions (e.g., coronary atherosclerosis, endocarditis), who are at greatest risk for serious sequelae from cardiac arrhythmias. From the common to the less common causes, one should be aware of the underlying problem and the appropriate pharmacologic and electrical therapies to provide appropriate treatment throughout the perioperative period, whether during cardiac or noncardiac surgery. This monograph attempts to provide a useful overview of cardiac arrhythmias from their cellular basis to the practical care of patients in whom they occur, from the channel pharmacology to newer monitoring modalities.

Thanks to the authors for providing excellent and thorough reviews of their subjects, with special thanks to David Haines, M.D., for his advice and collaboration in planning the associated symposium at the 1993 Meeting of the Society of Cardiovascular Anesthesiologists. Thanks also to Rhonda Taylor for secretarial assistance and to Anne Geyer and Mary K. Smith of J.B. Lippincott Company and Beth Oleska of Artech Graphics II for their help in overseeing production.

CARL LYNCH III, MD, PhD

Clinical Cardiac Electrophysiology:

Perioperative Considerations

Carl Lynch III

Cellular Electrophysiology of the Heart

1

The regular changes in the cardiac tissue membrane potential (action potentials) which mediate normal sinus rhythm, impulse conduction, and ultimately excitation-contraction coupling and the pumping of blood are determined by a well-ordered sequence of ion fluxes across the cell membrane. Specific ions flow when transmembrane proteins (channels) "open" and act as pores—ions flow through the interior of the proteins, passing into or out of the cell. In a normally well coordinated, self-regulating, periodic system, the ion fluxes of the heart determine the rhythmic pattern of the membrane potential. The function and response of the channels to membrane potential may also be modulated by other membrane receptors (e.g., muscarinic, adrenergic) either directly or via intracellular second messengers (e.g., cyclic AMP, inositol trisphosphate, Ca^{2+}). Cardiac arrhythmias are abnormalities in the regularly orchestrated bioelectrical behavior of the myocardium. Typical pharmacologic manipulation of arrhythmias centers around drugs which either 1) interact directly with channels, usually blocking or inhibiting them (e.g., lidocaine in the sodium channel; verapamil in the calcium channel) or 2) interact with receptors, modulating the channel's response (e.g., propranolol, by blockade of β-adrenergic receptors decreases Ca channel function). Additional therapy involves direct electrical interventions to establish or restore a normal rhythm (pacemaking, defibrillation). While much of our current therapy con-

Clinical Cardiac Electrophysiology: Perioperative Considerations
Edited by Carl Lynch III. J.B. Lippincott Company, Philadelphia, PA ©1994

tinues to be empirically derived, understanding of ion channel function from the molecular to the integrated cellular and tissue level has increased considerably since the first description of sodium, calcium, and potassium currents in heart. Our increasing knowledge of physiology and pharmacology permits a clearer mechanistic understanding of current therapy, as well as insight into potential new pharmacologic treatments. This chapter reviews the basic cellular electrophysiology and incorporates more recent discoveries, which more clearly define the basis for present and potential therapies.

MEMBRANE BIOPHYSICS—THE BASIS FOR CARDIAC ELECTRICAL BEHAVIOR

Ions, Membrane Potential, and Charge

As in all cells, the source of the membrane potential (V_m) of cardiac myocytes is the ion concentration difference, which is maintained across the cell membrane. Ionic concentration differences are primarily maintained by the impermeant nature of the lipid bilayer of the cell membrane—charged particles cannot pass into and through the highly hydrophobic lipid milieu of the bilayer. The ions (Na^+, K^+, Ca^{2+}, Cl^-) serve as stores of electrical energy by virtue of their concentration gradients, which establish a voltage defined by the Nernst equation (Table 1-1). The equilibrium potential (E_{ion}) for any ion is that voltage generated after ions have moved down their concentration gradient to a level at which a potential is created which completely opposes further ion flow. At equilibrium the chemical gradient is counterbalanced by a voltage gradient (defined by the Nernst equation) such that no current flows. However, when the membrane is impermeant to an ion, it cannot contribute to establishing the membrane potential.

At rest the only channels in the cell membrane that are open conduct almost exclusively K^+, so V_m is near E_K. By changing the permeability from predominantly one ion (K^+) to another (Na^+ and/or Ca^{2+}) with a different E, ion fluxes will occur and alter V_m. Both V_m and the change that it induces in membrane proteins, as well as the ion fluxes, can then serve as a medium for cellular signaling. Because the membrane bilayer and associated proteins are not totally impermeant to any ion and because ions periodically flow to transmit a signal, the ion gradients by themselves will ultimately run down, resulting in the loss of membrane potentials and inactivation of that means for cell signaling. The major mechanism for maintaining normal ionic gradients in the cell is the Na^+-K^+ ATPase, or sodium pump. Using the en-

TABLE 1-1. IONIC DETERMINANTS OF MEMBRANE POTENTIAL

Ion	Intracellular concentration (mM)	Extracellular concentration (mM)	Equilibrium Potential E_{ion} (mV)	Comment
K^+	100	3.5	−90	Establishes resting potential; however, conductance decreases with decreasing $[K^+]_o$, so that membrane potential will deviate from E_K as $[K^+]_o$ decreases
	120	5	−81	
		3.5	−95	
		5	−85	
		7	−77	
Na^+				
Quiescent	8	140	77	Normal, actively beating heart may have significantly higher $[Na^+]_i$
Beating	12		66	
Ca^{2+}				
Diastole	0.0001	1	122	The marked change in $[Ca^{2+}]_i$ contraction will influence I_{Ca} and I_{NaCa}
Systole	0.003		78	
Cl^-	18	100	−46	Normally contributes little to resting conductance

The equilibrium potential for each ion calculated according to the Nernst equation, where at 37°C:

$$E_{ion} = \frac{62\ mV}{n} \log \frac{[ion]_o}{[ion]_i}$$

where n is the charge on the ion and subscripts i and o designate intra- and extracellular (outside). 62 mV is calculated from $RT/F \times \ln 10$ (to convert from natural logarithms to $\log_{10}$); $R = 8.314$ V/coulomb/°K/mole, a farad (F) = 96,493 coulombs/mole of charge, and $T = 310$°K at 37°C. Intracellular concentrations are those based on estimates using intracellular ion selective electrodes[215–217]; values may vary depending on the tissue.[218]

ergy supplied by hydrolyzing ATP to ADP, three Na^+ are eliminated from the cell and two K^+ are pumped into the cell. This translocation is critical for taking care of "leaks" in the membrane, as well as reversing ion fluxes, which are employed in signaling. In addition to its using the Na^+ gradient to provide a rapid influx and depolarization, the energy generated by Na^+ entering down its gradient is also employed to perform work by membrane proteins such as the Na-Ca exchanger, which uses Na^+ influx to extrude Ca^{2+} from the cell.

It is important to remember that the ion flux required to establish the membrane potential depends on the ability of the membrane to act as an insulator and keep charges separate, a capability measured in charges per volt and termed "capacity." The capacity of biologic membranes is relatively constant at 1 microfarad (μF) per cm^2 of membrane surface (where a μF is 6.2×10^{12} charges/V), so that the capacity of a cell depends on its surface area. For a typical cardiac myocyte the total membrane surface is about ~7000μm^2 (7×10^{-5} cm^2) so the capacity is about 70 pF.[1] To generate a resting potential near E_K of –90 mV, about 40 million K^+ (or only about 1 in 40,000 of the total cell K^+) must leave the cell before the resting membrane potential is established at E_K and the flux ceases. While ion fluxes may be required to activate certain processes (e.g., Ca^{2+} for contraction), with regard to the membrane potential, a relatively small proportion of ions must cross the membrane to charge or discharge the cell membrane.

The Structure of Ion Channels

Although the amino acid sequence was defined and the three-dimensional structure was proposed long after the physiologic behavior of ion channels was well defined, it is nevertheless helpful to consider the structures which translocate ions and serve as the molecular basis of both normal and abnormal cardiac electrical behavior. Recent molecular biologic studies have permitted amino acid sequencing of a large number of the channel proteins themselves, revealing a similarity in structure among the major channel types, which may be ordered into superfamilies.[2–15] In order to rapidly conduct ions through the hydrophobic milieu of the bilayer, channels must span the bilayer. One superfamily of voltage-gated K^+ channels (see below: delayed rectifier and transient outward currents) is typically composed of approximately 600 amino acid subunits in which there are six sequence regions composed of hydrophobic amino acids.[9] These regions are thought to form α-helical transmembrane spanning regions, which may coalesce into a large transmembrane domain (Fig. 1-1). In order to conduct K^+, four of these units must combine to form the channel.

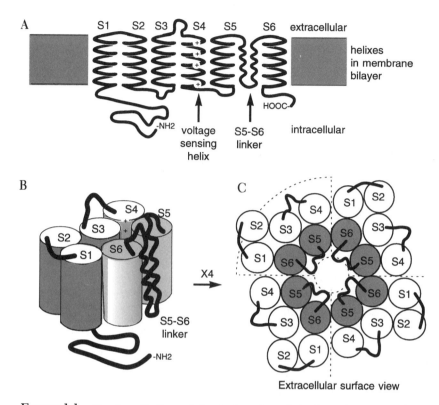

FIGURE 1-1. Structure of voltage-gated potassium channel. *A*, schematic representation of the proposed linear sequence (approximately 600 amino acids) of one subunit of a voltage-gated potassium channel as present in the membrane. The amino and carboxyl terminals are located in the intracellular compartment. Six hydrophobic transmembrane regions are identifiable which are thought to form α-helical transmembrane spans (S1–S6) as indicated. The fourth α-helical domain contains an arginine as almost every third amino acid and is felt to form the voltage-sensing region of the ion channel. The linking between helices 5 and 6 in the extracellular area folds back to line the pore of the channel. While voltage-gated potassium channels demonstrate this same structure with a six-transmembrane span, the subunits comprising the inward rectifier and the ATP-inactivated potassium channel appear to possess only two transmembrane spans. *B*, a three-dimensional representation of a potassium channel subunit in which the a-helical segments are indicated as cylinders. The exact packing order of S1–S4 is speculative. *C*, schematic representation showing the proposed packing of four potassium channel subunits to form the actual potassium channel, with ions passing through the central pore lined by the 5–6 linker sequences. For more complete descriptions consult Miller,[9] Hille,[77] Katz,[15] Tomaselli et al.,[14] and Kubo et al.[12]

For cation channels, the actual pore does not appear to be lined by transmembrane helices, but by an extracellular linking segment between helix spans 5 and 6, which contains acidic and polar amino acid residues. This segment folds back into the pore so that the polar or negatively charged sites become available for binding by ions as they pass through the channel.[14] This 5–6 linker region also is a major site for binding of drugs and other ions which block the channel.[16] The fourth transmembrane span contains a number of positively charged amino acids (typically arginine), which can respond to the membrane voltage field, acting as a voltage sensor and inducing a conformational change which will permit ions can flow through the central pore.[14] The combination of potassium channel subunits of differing types into a single four-subunit assembly markedly increases the possible number and the behavior of the assembled channels.[17]

In contrast to the potassium channels, the sodium and calcium channels are much larger (2000 to 2500 amino acids), with a sequence suggesting four potassium channel-like regions (six α-helical membrane spans) already linked together, creating four large transmembrane domains.[14,15] As with the potassium channel, the actual pore through which the ions pass is postulated to exist in the middle of the four hydrophobic domains, lined by the 5–6 linker regions. It is clear that the sodium and calcium channels are closely related in terms of structure,[18] and calcium channel characteristics can be conferred on the sodium channel by specific point mutations.[19] This similarity in structure may explain in part the fact that a number of drugs show only partial specificity for a particular channel. For example, both verapamil and nifedipine may block sodium channels as well as calcium channels, although not necessarily at clinical concentrations.[20-22] Likewise, local anesthetics as well as phenytoin can block calcium as well as sodium channels.[23-25]

In all channels studied, the intracellular regions possess serine or threonine amino acid residues, which may be phosphorylated by various regulatory kinase enzymes within the cell.[26] In the case of L-type calcium channels[27] and delayed potassium channels[28] phosphorylation results in enhanced voltage-dependent activation.

Ion Channel Gating

Most ion channels are not permanently open, but open (activate) in response to a voltage change across the membrane or binding of an agonist molecule. The channels then remain open for a particular period of time (on average), before they close (inactivate). Voltage-gated channels may remain open until the voltage change is reversed or may inactivate due to an ongoing conformational change in the molecule. As noted above, it is thought that the positively charged fourth α-helical span of each domain is the "voltage sensor" of the channel. When a voltage change is imposed across the membrane, the movement of this charged region of the channel protein can be measured.[29] Such a "gating charge" can be closely correlated with gating of the ionic currents through the specific channels.[30,31] Inactivation of the channel may in some cases be mediated by an intracellular segment (or "ball on a chain") which binds to the pore and "plugs" it, stopping ion flow.[9,17]

At rest, the membrane field is usually dominated by the bulk membrane potential established by the K^+ equilibrium potential (in which the loss of K^+ resulted in a negative charge inside the cell). However, the ion channels exist in a submicroscopic membrane environment, in which the electrical field the channel proteins "feel" is not exactly the measured membrane potential (the potential established between the bulk intra- and extracellular solution). Figure 1-2 shows a channel in a membrane with electrical charges (usually negative) present on the membrane surface (on phospholipids and on proteins), which alter the electrical field sensed by the membrane proteins, as well as attracting cations to the membrane surface.[32,33] When there are more ionic charges in solution, more of the negative surface charge is neutralized and the protein "feels" more of the measured potential. While this may seem a biophysical curiosity, it may explain certain clinical phenomena, such as restoration of cardiac conduction by Ca^{2+} administration when it is depressed by hyperkalemia (see below).

Ion Channel Conductance, Selectivity, and Density

By placing a polished microelectrode up against a cell and isolating a small patch (0.01 to 1 μm^2) of membrane (and also by incorporating channels into artificial membrane bilayers), the passage of ions through a single ion channel can be studied.[34] In this way it is possible to examine the microscopic function of a single protein complex. In order to pass through the pore, ions must first lose most or all of the surrounding water molecules (waters of hydration). This is only possible if they can move to an energetically more favorable location—so cations shed water molecules only because they are able to bind to the carboxylic amino acid groups which line the pore. The ions bind very, very briefly (2 to 10 ($2–10 \times 10^{-12}$ sec) psec) to these groups as they travel through the channel according to their electrochemical gradient. Some channels (notably K^+) may contain more than one ion at a time. The rate at which ions pass through the channel is the single channel conductance (γ), usually listed in picosiemans (pS, which is 10^{-12} A/V, or 10^{-12} coulombs/sec/V, where a coulomb is ~6.2×10^{18} charges) which is about 6 million ions/sec/V. The magnitude of ion flow may be better understood in terms of ions/msec: if a ion channel has a conductance of 8 pS and its ion has an electrical gradient of 130 mV (such as the sodium channel), and if the channel is open for 1 msec it will pass about 6000 ions (48 million ions/sec/V $\times$ 0.13 V $\times$ 0.001 sec). Certain

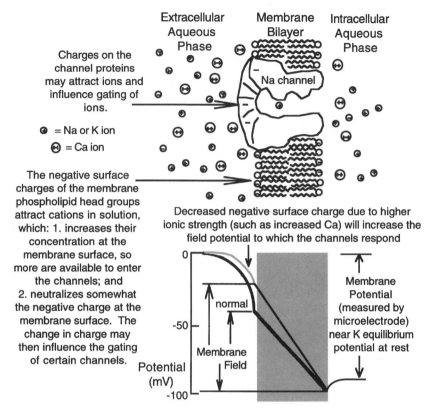

Charges on the channel proteins may attract ions and influence gating of ions.

⊙ = Na or K ion

⊖ = Ca ion

The negative surface charges of the membrane phospholipid head groups attract cations in solution, which: 1. increases their concentration at the membrane surface, so more are available to enter the channels; and 2. neutralizes somewhat the negative charge at the membrane surface. The change in charge may then influence the gating of certain channels.

Extracellular Aqueous Phase   Membrane Bilayer   Intracellular Aqueous Phase

Na channel

Decreased negative surface charge due to higher ionic strength (such as increased Ca) will increase the field potential to which the channels respond

Membrane Potential (measured by microelectrode) near K equilibrium potential at rest

normal

Membrane Field

Potential (mV)

0

-50

-100

FIGURE 1-2. A cross-sectional cartoon of the membrane bilayer with an incorporated sodium channel. The negative charges on the membrane phospholipids attract cations from solution, with Ca^{2+} (by virtue of its higher charge) being attracted more than Na^+ or K^+. The lower part of the figure depicts graphically the negative membrane surface charge, which alters the membrane potential gradient (electrical field) to which the channel protein actually responds. When there are more ions (especially divalent ions) in solution, the negative membrane charges are more neutralized (electrostatically or by direct binding) so that the electrical gradient across the membrane, to which the proteins respond, is increased. Although shown as a straight line, the actual potential gradient is nonlinear.

channels have very high conductances because ions pass through very quickly, while other channels pass ions far more slowly.

The selectivity of a channel describes its propensity to permit some ions to pass, but not others. Virtually no physiologically relevant channels will pass both cations and anions, probably because the acidic (negatively charged) amino acid residues required to attract cations will repel anions such as Cl⁻ and vice versa. Among the common cation channels (sodium, calcium, potassium), selectivity is imperfect so that other cations may occasionally pass, but each ion channel is named for the ion for which it is most selective. The calcium channel was initially thought to have a selectivity of Ca^{2+} over Na^+ of $\geq 10:1$.[35] It has subsequently been found that calcium channels have a more complex ionic permeation behavior, conducting Ba^{2+} and Sr^{2+} faster than Ca^{2+}, suggesting that Ca^{2+} actually binds more strongly to the channel, sometimes preventing or slowing passage of other ions.[36,37] In contrast, transition metal ions are typically inhibitory: Mn^{2+} passes very slowly, partially blocking the channel[38]; La^{3+} or Cd^{2+} block the channel at 1 to 100 µM.[39] Certain cardiac ion channels such as the pacemaker current are relatively nonselective and conduct Na^+ and K^+ almost equally well.

Channels occur in cell membranes with a certain density (number of channels per μm^2). In an opposite fashion from the K^+ efflux required to establish the resting membrane potential in a typical myocyte (surface area 7000 m^2), the Na^+ required to enter and depolarize the cell (from -90 to $+40$ mV) will be about 60 million ions. If these ions flow in 1 msec ($\sim 6 \times 10^{-9}$ A) as during a normal action potential, from the above calculation of sodium channel conductance (8 pS (8×10^{-12} A/V)) and assuming on average a voltage gradient of 60 mV driving Na^+ into the cell, the average current through each open channel will be about 5×10^{-13} A ($\sim 3,000$ ions/msec). Dividing reveals that about 20,000 open channels are required to depolarize the cell in 1 msec. For a typical myocyte,[1] there must be at least 3 sodium channels/μm^2. Since some channels may be inactive or not open rapidly and since the cell may be electrically coupled to other cells, the actual density for sodium channels has higher estimates (4 to 50 channels/μm^2), especially in certain species.[40] When such sodium currents are summed from the ~ 1 billion myocytes of a human heart, a substantial current will be generated. While the high Na^+ current needed for rapid depolarization and action potential conduction requires a high sodium channel density, a somewhat lower channel density of 0.3 to 5 channels/μm^2 is more typical for other myocyte ion channels, so that 1,500 to 25,000 units of a given channel are typically present in a cardiac myocyte.[41,42] However, the density may not be uniform over the entire cell surface.

CONDUCTION OF THE CARDIAC ACTION POTENTIAL—AN OVERVIEW

Figure 1-3 reviews the pattern of depolarization that occurs in the heart. The membrane potential changes in the heart can be viewed as a single depolarization, which sweeps through the whole heart, assuming various forms along the way. The cells of the heart are intimately connected and electrically coupled (by ion channels called "gap junctions") so that

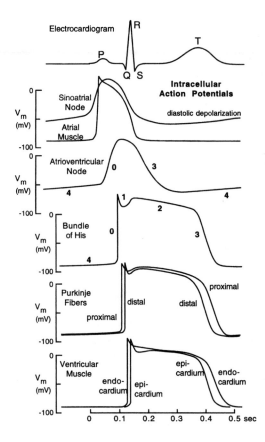

FIGURE 1-3. Conduction of the depolarization through the heart. The action potential configurations in the various regions of the heart and their temporal relationship are shown. The phases 0 through 4 of the action potential are indicated.

ionic currents pass among cells, spreading the depolarization. Depending upon the particular cardiac tissue, from a resting potential of from −90 to −60 mV, the tissue depolarizes to +30 to +10 mV.

The Sinoatrial (SA) Node

The sequence of depolarization is normally paced by a specialized region of tissue in the posterior wall of right atrium (near the superior vena cava), the SA node. Although difficult to identify grossly, this small group of microscopically distinct cells have few mitochondria, no sarcoplasmic reticulum (SR) or organized myofibrils, and have been termed "round cells."[43] When not undergoing an action potential (AP), the cells undergo a very gradual depolarization from ~−60 mV (the maximum diastolic potential at the end of repolarization) at a rate of 20 to 100 mV/sec. This process is the diastolic depolarization or pacemaker potential. When an internal membrane potential of ~–50 mV is reached, the depolarization rate increases to 1 to 2 V/sec, creating a slowly depolarizing AP. Although sodium channels may be present, they are inactivated by the low resting potential and slow diastolic depolarization and never observed physiologically. This nodal slow AP invades the zone of surrounding cells, which have more myofibrils and appear microscopically more as muscle.[43] These cells also spontaneously depolarize, but not as rapidly. As the cells become more musclelike in character, they have a more negative resting potential, more active sodium channels, and therefore a more rapid AP depolarization. Occasionally, islands of cells near the SA node may take over as dominant pacemaker with a slightly faster diastolic depolarization.

Atrial Muscle

Once the depolarization reaches atrial muscle, which has a –85 to –90 mV resting potential, the AP spreads rapidly, with a particularly rapid conduction pathway (Bachman's bundle) toward the left atrium. In comparison to ventricular muscle, the atrial AP is far more narrow, with a less prominent plateau, consistent with its far briefer contraction. However, the short AP duration in atrial tissue permits another action potential to occur much sooner, so that very high rates are achievable (i.e., atrial flutter).

Atrioventricular (AV) Node

As the borders of the atrial muscle are reached, the AP stops at the nonconducting fibrous tissue which forms the annulus for each atrioventricular valve. However, at a point at the junction of the atrioventricular septa, the arriving atrial AP passes through a transitional zone of cells (A–N zone), smaller than the normal atrial cells and separated from them by fibrous tissue. The AP is markedly slowed as it travels through the compact or N region of the node, composed of round cells similar anatomically and physiologically to the SA node cells. Located on the superior right ventricular edge of the ventricular septum, the AV node is similar to the SA node, except for having a slower rate of diastolic depolarization. The AP conduction velocity then increases as it travels through a second transitional zone (N–H region) to the bundle of His. By virtue of its slow diastolic depolarization, it is a backup system of impulse initiation, in case the SA node or atrial conduction fails. The delay between the atria and ventricular contraction permits better augmentation of ventricular filling. The node is also a frequency gate, so that if the atria flutter (~300 beats/min) or fibrillate, the ventricle will not be driven at a rate incompatible with effective pump function. This AV node represents the only electrical continuity and means of conduction between the atria and ventricles, except for individuals who have abnormal direct conduction pathways between the atria and ventricles (Wolff-Parkinson-White and Lown-Ganong-Levine syndromes).

The Bundle of His and Purkinje Network

A coordinated contraction of all the ventricular myocytes is required so that all of the energy of contraction can be used to expel blood, rather than have contracting muscle stretch out other muscle which has not yet begun to contract. Yet the myocytes that make up the bulk of the heart are specialized to develop tension, not to propagate a signal. Consequently, uniform initiation of contraction is achieved by specialized myocytes that form large diameter (50 to 80 μm) fibers with very sparse myofibrils or sarcoplasmic reticulum. The APs of this system are characterized by the most rapid rate of depolarization (400 to 800 V/sec) in heart. The bundle of His, which has a sufficiently high current density that it generates a discrete voltage signal, is a specialized group of large conducting fibers which emanates from the AV node and which divides into a right ventricular branch, an anterior left ventricular branch, and a large posterior left ventricular branch. Each branch forms a network of Purkinje fibers which spreads over the inner surface of the heart, rapidly carrying depolarization to the entire endothelial surface. These large diameter fibers have a conduction velocity of ~4 m/sec, which permits the rapid conduction of the AP from the AV node to the endocardial surface of the ventricular muscle in ~30 msec. The action potentials in the Purkinje system are also the longest in the heart, and the proximal fibers (nearest the bundle of His) are depolarized for a substantially longer time (40 to 80 msec) than those distal fibers that form electrical junctions with the myocardium. Consequently, the cardiac impulses cannot go retrograde (back toward the proximal, still depolarized, region) under normal conditions, protecting the heart from many erratic and premature beats that might be initiated in the distal regions. These fibers also have a very slow diastolic depolarization, which can provide pacemaking capability if the SA and AV nodes fail, albeit at very slow rates (10–20 beats/min).

The depolarization then sweeps through the ventricular wall in 80 to 100 msec, generating significant current, which is detected as the QRS complex of the electrocardiogram (ECG). This wave of depolarization is normally oriented in the endocardial-epicardial direction and is transmitted by the extracellular fluid to all parts of the body. During the plateau of the AP, virtually no current is flowing, and there is relative electrical silence. Epicardial action potentials are shorter than those in the endocardium; this causes a wave of repolarization that sweeps from the epicardium to the endocardium, which also causes a positive deflection of the ECG.

The sustained depolarization typical of myocardial tissue can be described by a depolarizing of Na^+ and/or Ca^{2+} influx, superimposed upon a resting potassium conductance (variably present) and a panoply of various potassium channels which orchestrate repolarization. It is the variable expression of these channels (tabulated in Table 1-2 and described below) in various regions of the heart (pacemaker sites, atria, conduction tissue, ventricle; see Table 1-3), which is responsible for the characteristic electrophysiologic behavior of each tissue. Modulation of these various membrane channels under the influence of various controlling molecules (G proteins) and second messengers can also dramatically alter membrane behavior.

DEPOLARIZING CURRENTS

The Sodium Channel

Sodium channels expressed in various tissues are primarily composed of a major α subunit of approximately 2000 amino acids. While the α subunit contains the main pore portion of the molecule, there may be up to two associated additional subunits (β_1 and β_2), depending on the particular excitable membrane source. Like many membrane proteins, N-glycosylation of the α subunit appears to be required for normal cellular processing, combination with the β subunits, and normal channel function.[18] The sodium channel of cardiac muscle, like that of skeletal muscle, appears to be a complex of α and β_1 subunits. The α subunit from heart (designated h1) is one of at least six different sodium channel α subunits that have been cloned[44] and has the distinction of being resistant to blockade by tetrodotoxin, a distinct feature of the cardiac Na^+ conductance.[45]

Of critical importance to cardiac muscle function is the fact that there are enough working sodium channels so that a large depolarizing current can spread very rapidly to adjacent cells and tissue, so that APs are conducted rapidly throughout the mass of muscle. This is accomplished by the abundance of these channels in the sarcolemma of the atrial and ventricular muscle and especially in the Purkinje system. The major factor affecting the availability of these channels is the

TABLE 1-2. CARDIAC ION CHANNELS

Channel		Activation Kinetics	Inactivation Kinetics	Blocking Agents
Sodium ("fast")	I_{Na}	Very fast	Fast, except for a "slow" fraction which reopen	Local anesthetics and phenytoin (use-dependent), tetrodotoxin (weak)
Calcium ("slow")				
L-type (or high voltage activated)	$I_{Ca,L}$	Fast	Slow, large Ca^{2+}-dependent component	DHPs, benzothiazepines, phenylalkyl-amines, Cd^{2+}
T-type	$I_{Ca,T}$	Fast	Moderate	Amiloride, tetramethrin, Ni^{2+}, Cd^{2+}
Potassium				
Inward rectifier	I_{K1}	Instantaneous	Mg^{2+} block of outward current	Cs^+, Ba^{2+}(100 μM), thiopental, 0K^+
Plateau or background	I_{Kp}	Instantaneous or very fast	None?	Ba^{2+}(1 mM)
Delayed rectifier	I_{Ks}	Slow	None	Tetraethylammonium
	I_{Kr}	Moderate	Slight	d-Sotalol, dofetilide, E4031, 1 μM La^{3+}
	I_{RAK}/I_{HK2}	Very rapid activation	None	Quinidine (HK2)
Transient outward	I_{to1}	Fast	Moderate	4-AP (1 mM), bupivacaine, quinidine
	$I_{K(Ca)}, I_{to2?}$	Fast	Moderate	4-AP (10 mM), tedisamil
ACh, adenosine, PAF activated	$I_{K(ACh,Ado)}$	Instantaneous when activated via G protein (G_i)	Mg^{2+} block of outward current	
ATP (-inactivated)	$I_{K(ATP)}$	Instantaneous when active	Mg^{2+} block of outward current	Glybenclimide, amiodarone, quinidine
Na-activated	$I_{K(Na)}$	Instantaneous when active		
Fatty-acid activated	I_{KAA}, I_{KPC}	Instantaneous when active		
Pacemaker (non-specific)	I_f	Slow	Slow	1–2 mM Cs^+
Chloride				
Transient outward	$I_{Cl(Ca)}, I_{to2?}$	Fast	Fast	SITS, DIDS
CFTR	$I_{Cl,cAMP}$	Instantaneous when active	None	DNDS
Stretch-activated				SITS, 9-AC, DNDS

4-AP, 4-aminopyridine; CFTR, cystic fibrosis transmembrane regulator; DIDS, 4,4'-dithiocyanatostilbene; DNDS, 4,4'-dinitrostilbene-2,2'-disulfonic acid; PAF, platelet-activating factor; SITS, 4-acetomido-4'-isothiocyanatostilbene-2,2'-disulfonic acid; 9-AC, anthracene-9-carboxylic acid.

TABLE 1-3. DISTRIBUTION AND EFFECTS OF ION CURRENTS IN VARIOUS TISSUES OF THE HEART

Current	SA node	Atrial Muscle	AV Node	Bundle of His-Purkinje Fibers	Ventricular Muscle Endocardium	Ventricular Muscle Epicardium
Sodium channel (I_{Na})	Inactive, if present	Prominent	Inactive	Prominent	Prominent	Prominent
Calcium channel (I_{Ca})	Contributes to DD and provides AP	Yes	Contributes to DD and provides slow AP	Yes	Yes	Yes
Pacemaker (I_f)	DD	No (?)	DD	DD	Present, but inactive	Present, but inactive
Inward rectifier (I_{K1})	No	Yes, modest	No	Modest	Yes	Yes
Plateau current ($I_{K,p}$)	?	?	?	?	Yes	Yes
Delayed rectifier (I_K)	Yes, turn off contributes to DD	Yes, may be modest	Yes, turn off contributes to DD	Yes	Yes	Yes
Transient outward (I_{to})	No	Large, shortens APD	No	Yes, causes prominent "notch" in AP	None or minimal	Yes, contributes to shorter APD
ACh-induced ($I_{K(ACh)}$, $I_{K(Ado)}$)	Yes, markedly decreases rate of DD	Yes	Yes, markedly decreases rate of DD	Yes, markedly decreases rate of DD		Minimal, if any
ATP-gated channel ($I_{K(ATP)}$)	Yes, decreases rate of DD	? (probably present)	Probably similar to SA node	? (probably present)	Yes, activation decreases APD	Yes, activation decreases APD

The above summary is a general pattern that may be dependent upon the particular species examined. Compiled from various sources.[1,111,136,138,223,224] membrane voltage. Since normal cardiac muscle is at −80 to −90 mV, it is within 15 to 25 mV of threshold (−70 to −65 mV), that is, the potential at which sufficient sodium channels open so that the entering ions cause a regenerative depolarization. Although the channels activate with a time constant of less than 1 msec, most of them also inactivate rapidly at a negative V_m close to threshold. The relationship between V_m and the number of available sodium channels is given by the steady-state inactivation (or h_∞) curve (h was the term Hodgkin and Huxley used for the inactivation variable in their original mathematical formulation describing sodium currents). Figure 1-4 shows the steady-state (hence the infinity subscript) relation between Vm and the fraction of available (noninactivated) sodium channels. Negative to −80 mV, a large fraction of sodium channels are noninactivated; rapid depolarization positive to −70 mV results in a regenerative opening and an AP. When the potential stays positive to −65 mV, more than 75% of the channels are inactivated, and there are usually too few to open to generate a propagated AP. In the region from −80 to −65 mV, 65 to 35% of the channels may be available, which may be enough for an abnormally slow conducted AP. The rate of depolarization (dV/dt) is roughly proportional to the number of working sodium channels.

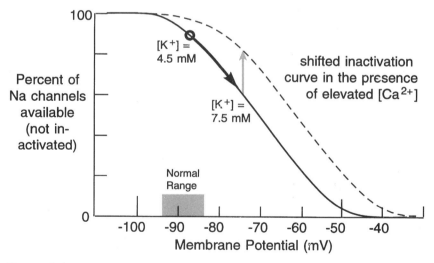

FIGURE 1-4. The steady-state inactivation of the sodium channel. The *solid curve* indicates the usual fraction of sodium channels available for opening as a function of membrane potential. With depolarization, such as that caused by hyperkalemia (indicated by the *arrow*), fewer channels will be available and conduction will be slowed (increased K^+ conductance may also contribute). At any given membrane potential, increased extracellular Ca^{2+} makes more sodium channels available for activation (*dashed* inactivation curve), which will tend to restore conduction. These actions are consistent with neutralization of the membrane surface charge, which influences sodium channel gating (see Fig. 1-2).

Obviously, changes in the extracellular K+ ($[K+]_o$) will alter the resting potential of the myocyte, which changes the density of available (noninactivated) sodium channels responsible for the conduction the AP in Purkinje fibers and muscle. As indicated in Figure 1-4, when myocardial cells are partially depolarized, as occurs with hyperkalemia, fewer channels are available (not inactivated) so that conduction is impaired. This is readily apparent clinically as the widened QRS complex representing slowed conduction through the ventricle. As shown in Figure 1-2 the sodium channel protein is really responding to the membrane (electrical) field, which will differ somewhat from the actual membrane potential and depend upon the membrane surface charges present. By increasing the extracellular Ca^{2+} concentration, negative membrane surface charges will be neutralized and fewer sodium channels will be inactivated at any given potential (*dashed curve*, Fig. 1-4). This phenomenon accounts for the improved conduction observed when Ca^{2+} is administered in hyperkalemic states.

The extreme of the voltage-inactivating effect is employed clinically in the application of *cardioplegia*, in which solutions containing 20 to 30 mM K+ depolarize the heart to −50 to −35 mV. In this potential range almost all sodium channels will be inactivated and normal cannot occur, resulting in plegia ("paralysis") of the heart. Fortunately, this potential is above the mechanical threshold, that is, Ca^{2+} is not released from the SR to activate contractions, so that energy is not consumed generating tension. However, a very small fraction of cardiac sodium channels appear not to inactivate rapidly, providing sustained current even during the plateau of the AP.[46,47] This subpopulation of channels provides a small "window" current, which persists for the duration of depolarization and contributes to the plateau phase of the Purkinje fiber AP.[46,48] One of the actions of pure sodium channel blocking agents (tetrodotoxin and certain local anesthetics) is to decrease the plateau duration, presumably by blocking the sodium channel window current which persists in the plateau, although this effect is modest in ventricular muscle.

CALCIUM CHANNELS

The slow inward current of cardiac tissue has been clearly defined as the influx through channels specific for divalent cations (Ca^{2+} and also Ba^{2+} and Sr^{2+}). While there are at least two types of calcium channels in heart, the L-type ($I_{Ca,L}$, long lasting, high threshold, high voltage-activating positive to −40 mV) channel is by far the most common and important myocardial calcium channel.[49-52] Through this channel flows the Ca^{2+} current that maintains the internal Ca^{2+} stores necessary for contractions. It is modulated by β-adrenergic stimulation[27,53] and is also the channel sensitive to the commonly used "calcium antagonists," the 1,4-dihydropyridines (DHPs, e.g., nifedipine), the phenylalkylamines (e.g., verapamil), and the benzothiazepines (e.g., diltiazem).[36] In close homology with the sodium channel, the L-type channel consists of a major pore-forming α_1 subunit composed of 2171 amino acids as defined by cloning and expression.[4] It is closely related (66% homology) to the skeletal muscle DHP receptor α_1 subunit, which gates Ca^{2+} release in skeletal muscle. The α_1 subunit usually has four associated subunits (α_2, β, γ, and δ), at least some of which appear to modulate ionic conductance.[4,18] Also like the sodium channel, asymmetric membrane charge movements are detectable with applied voltage changes (gating charge) and appear to reflect voltage-induced molecular rearrangements of the calcium channel voltage sensor associated with opening of the channel to Ca^{2+} flux.[30,31]

Inactivation of the channel appears to reflect multiple processes. Although there is a component of voltage-dependent inactivation of L-type channels, a major mechanism that closes the channel appears to be a rise in $[Ca^{2+}]_i$, possibly the Ca^{2+} that accumulates at the interior channel mouth.[54,55] Increased intracellular Ca^{2+} also appears to enhance currents, apparently increasing the number of activated channels.[55] Clustering of calcium channels may influence the accumulation of local $[Ca^{2+}]_i$, and thus the regional variation in density of calcium channels may also play a modulatory role.[56]

Alterations in Ca^{2+} entry via the calcium channel appear to have complex effects upon cardiac APs, which are related in large part to the secondary effects of altered intracellular Ca^{2+} ($[Ca^{2+}]_i$). When Ca^{2+} entry was altered by varying extracellular Ca^{2+} ($[Ca^{2+}]_o$), the increased $[Ca^{2+}]_o$ resulted in an increased AP plateau and shortened duration, while decreased $[Ca^{2+}]_o$ resulted in a decreased AP plateau and increased AP duration.[57]

The DHPs bind to particular modulatory sites on the cardiac calcium channel and depending upon the exact drug structure, they may either inhibit channel opening or shift the channel to a long opening mode (Bay K 8644).[58] Specific drug effects are also dependent on the membrane potential, which can markedly influence channel structure and function.[58] For example, partial depolarization, as occurs in cardiac muscle with hyperkalemia or ischemia (or as is normally present in vascular smooth muscle) markedly increases the binding of and inhibition by nifedipine. This behavior may explain the particular benefit of these agents in primarily depressing vascular smooth muscle or benefiting ischemia. This potential dependent binding represents a variation of the modulated receptor hypothesis first proposed for the local anesthetics' binding by sodium channels by Hille,[59] and Hondeghem and Katzung.[60] In contrast, the frequency-dependent binding of verapamil is reminiscent of the frequency-dependent block of sodium channels by the local anesthetics.[36,61,62] The binding to the DHPs also modulates the binding of the phenylalkylamines or benzothiazepines by altering their distinctly different sites on the molecule. As with decreased $[Ca^{2+}]_o$, blockade of the calcium channel has variable effects on the AP. In Purkinje fibers, although the AP plateau may be depressed, the AP duration may be enhanced.[63] This effect may be related to Ca^{2+}-dependent control of re-polarizing currents (see below).

Although the L-type channel is more common, the more transient calcium channel (T-type, low threshold, activating when the membrane potential becomes positive to -65 mV) identified as $I_{Ca,T}$ has been identi-fied in all tissues where it has been sought.[49,50,64–67] While it accounts for only 10% of I_{Ca} ventricular and atrial tissue,[49,67] it may account for as much as 40% of current in Purkinje fibers,[50] in which it may contribute to the pacemaker depolarization as in sinoatrial tissue. While $I_{Ca,T}$ ap-pears to contribute to the latter half of the SA nodal depolarization, and its blockade (by 40 µM Ni^{2+}) can slow spontaneous nodal firing, $I_{Ca,L}$ ap-pears to contribute the majority of depolarizing current in nodal cells.[66,68] Its role in a variety of myocardial cells is not fully understood.

The calcium channels are not only a site for pharmacologic inter-vention by specific Ca^{2+} entry blockers; it is also clear that these chan-nels are susceptible to inhibition by volatile anesthetics. Since the ini-tial suggestion that halothane depresses Ca^{2+} currents,[69] it has become clear that the volatile anesthetics specifically depress Ca^{2+} currents of myocardial tissue.[70–75]

THE POTASSIUM CURRENTS

Outward K^+ currents are the primary neutralizing flux that counteract the inward Na^+ and Ca^{2+} currents. Over the past decade it has become evident that K^+ efflux occurs at various points in the cardiac AP cycle by ion passage via a wide variety of different potassium channels, rep-resenting at least two different ion channel superfamilies.[5,7,9,12,13] These channels are responsible not only for the pattern of repolarization in various regions of the heart, but are also responsible for modulating ex-citability via membrane conductance (G_m) and refractory periods. While similarities and differences exist among species which must be defined, the large number of different channels clearly represent distinct recep-tors with the potential for specific pharmacologic intervention.

The Resting Potassium Conductance:
the Inward Rectifier (G_{K1})

Although its physiologic behavior has been defined for a many years, the channel responsible for the myocardial resting conductance is in many ways the most interesting in terms of structure, function, and distribution. In comparison with the amino acid sequence and proposed structure of the subunits of voltage-activated potassium channels (with six transmembrane spans) (Fig. 1-1),[5,7,9,13] the recently defined amino acid sequence of the inwardly rectifying potassium channel differs sharply.[12] The protein is composed of 428 amino acids with hydrophobic regions compatible with only two α-helical transmembrane spans. However, the linking region between these spans appears to form a pore-lining segment similar to that proposed for the larger six span domains present in other types of channels.

Anomalous and Rectifying Behavior

This resting potassium conductance of myocardium (G_{K1}) is responsible for maintaining the resting potential based on its equilibrium potential. But G_{K1} has two peculiar and very important characteristics that are primarily responsible for membrane electrical behavior of the heart. First, the channel conductance is proportional to $[K^+]_o$ or actually to the square root of $[K^+]_o$.[76] That is, the greater $[K^+]_o$ is, the more rapidly ions can pass through the channel, but when $[K^+]_o$ is low, ion flux is reduced. This peculiar behavior may have two benefits: K^+ loss may be reduced when $[K^+]_o$ is low; also, when $[K^+]_o$ is higher and the sodium pump is activated, the negative current generated (3 Na^+ are pumped out for 2 K^+ pumped in) will not hyperpolarize the cell too much.[77] A consequence is the myocardial membrane potential dependence on $[K^+]_o$, which is shown in Figure 1-5A. The agreement with the predicted Nernstian potential for K^+ (E_K) is excellent until $[K^+]_o$ becomes very low (<3 mM), where the cell fails to hyperpolarize further. When the $[K^+]_o$ gets very low, G_{K1} shuts down and even very slight Na^+ leaks prevent further hyperpolarization or may even depolarize the cell, which occurs in ≥1 mM $[K^+]_o$.[78] Consequently, with extreme hypokalemia, even though the E_K is very negative, dysrhythmias may become common because there is little K^+ conductance to stabilize the membrane. Because the ability to pass current increases with $[K^+]_o$, even against a larger concentration gradient, this behavior of G_{K1} has been called "anomalous."

Second, many ion channels pass ions equally well in either direction, according to the potential applied and transmembrane ion concentrations. While G_{K1} readily permits external K^+ to enter from outside the cell, with increasingly positive (>10 to 20 mV) electrical gradients applied inside the cell (such as an action potential), less K^+ can leave the cell—the channel acts as a diode rectifier; hence the name "inward rectifier." This effect is shown in the relation of current to voltage (the "I-V curve") of Figure 1-5*B*. Intracellular Mg^{2+} is the species responsible for this blockade of outward K^+ current.[79-82] This rectifying action has a very important result. When cardiac muscle is depolarized to the plateau potential (0 to +25 mV) by entry of Na^+ and Ca^{2+}, K^+ cannot escape from the cell to eliminate the positive

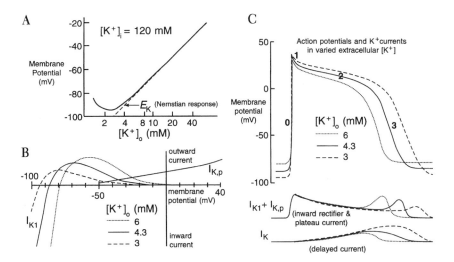

FIGURE 1-5. Electrophysiologic effects of the inward rectifier (G_{K1}) A, the calculated Nernstian equilibrium potential for K^+ (E_K) and the membrane potential (V_m) recorded in myocardial tissue are shown. The intracellular K^+ ($[K^+]_i$) is based on measurements in Purkinje fibers and ventricular muscle.[215-217] Because of the dependence of the resting potassium conductance (G_{K1}, inward rectifier) on extracellular K^+ ($[K^+]_o$), as $[K^+]_o$ declines the membrane potential deviates increasingly from the E_K as G_{K1} decreases. At very low $[K^+]_o$ cell may actually depolarize.[218] B, the current voltage (I-V) relation of G_{K1}. When increased potassium is present externally, there is a shift toward depolarizing potential; however, the conductance increases so that a greater depolarizing current will be required to decrease the membrane potential. With decreased $[K^+]_o$, the lower conductance means that less current will generate a larger depolarization. The outward current observed with positive potentials appears to be due to a separate plateau or background conductance ($I_{K,p}$).[85] C, effects of varied $[K^+]_o$ on the action potential configuration. As the delayed current is activated, with increased $[K^+]_o$, G_{K1} is enhanced so that outward currents through it results in an earlier repolarization and termination of the plateau. The opposite effect is observed in low $[K^+]_o$ when G_{K1} is reduced, and the delayed current must contribute more outward current, which requires a longer period of time-dependent activation. In either case, a component of repolarizing outward current is present because of $I_{K,p}$ which is relatively insensitive to $[K^+]_o$. Inward currents are not shown. For a discussion and mathematical description, respectively, see Surawicz[219] and Luo and Rudy.[83]

charge. If this channel was not blocked by Mg^{2+} to prevent outward K^+ flux, K^+ would rapidly leave the cell and repolarize it, so that the plateau would be small or nonexistent. This failure to eliminate K^+ down its electrochemical gradient potentiates the depolarization caused by Na^+ entering through the sodium channels and is also in large part responsible for maintaining the plateau. As shown in Figure 1-5B, after the cell becomes positive to –65 to 70 mV, the greater positive potentials actually cause less current to flow (the "negative conductance" region).

Effects of Altered Extracellular $[K^+]_o$

Effects of hypo- and hyperkalemia are shown in Figure 1-5B. When $[K^+]_o$ is low (2.5 mM), conductance (G_{K1}) is low (membrane resistance is high) so only a modest amount of current (flowing from adjacent tissue) leads to a large depolarization from the resting potential (–95 mV) to near the AP threshold (–70 mV). In contrast, when $[K^+]_o$ is high, the membrane conductance is high (membrane resistance is low) because of the inherent behavior of G_{K1}, so that positive charges (as K^+) escape from the cell more rapidly. Although the membrane is more depolarized, so that it is closer to the action potential threshold, the cell is not likely to be more excitable because current will more readily pass through the membrane and therefore not depolarize it. While a smaller depolarization is required to get to threshold (–65 mV) from the resting potential (–75 mV), similar or more current (either from adjacent tissue or from an electrode) may be necessary to depolarize the cell as when $[K^+]_o$ is low. A major effect of hyperkalemia is slowed conduction, because of the need for increased depolarizing current to reach threshold, and a decrease in AP duration as more outward K^+ current repolarizes the cells. When external K^+ is moderately low (2 to 3.5 mM) and the cells are more hyperpolarized, more sodium channels are available, which results in improved AP conduction. Although a greater depolarization is required, the G_{K1} is decreased so that any current (from adjacent cells or cell damage) causes more depolarization toward threshold. The result may be greater excitability and a tendency for extra depolarizations (ectopy).

The peculiar behavior and $[K^+]_o$ dependence of G_{K1} also profoundly influence repolarization. Once the repolarization process has been initiated by the delayed current (I_K, see below) and V_m becomes negative to –30 mV, G_{K1} increases progressively, permitting an increase in K^+ efflux and enhancing the rate of repolarization. In fact, G_{K1} is the major repolarizing current during the later half of repolarization.[83] Consequently, increases or decreases in conductance with hyper- or hypokalemia, respectively, will decrease or increase AP duration as indicated in Figure 1-5C.[83] Likewise, specific blockade of G_{K1} by 50 to 100 μM Ba^{2+1} or by certain barbiturates (e.g., thiopental)[84] will prolong the AP duration by extending the later phase.

The Background (or Plateau) K⁺ Current

Single channel studies clearly demonstrate that G_{K1} turns off completely with depolarization (as shown in Fig. 1-5B) in the presence of physiologic intracellular Mg^{2+}[79-82] and will therefore be unavailable to provide any countercurrent to the calcium current and sodium current during the AP plateau. Yet an outward current (as shown in Fig. 1-5B) is evident on whole cell recordings even at positive potentials.[76] This current has been observed in isolation and is termed the "background or plateau current" ($I_{K,p}$).[85] This background conductance is blocked by only 1 mM Ba^{2+}, but is unaffected by low or 0 $[K^+]_o$ which inactivates G_{K1}. This small amplitude current activates very rapidly on depolarization[86] and can provide a small amount of outward current that will contribute toward repolarization of the action potential during the plateau[85] (Fig. 1-5C).

The Delayed Rectifier Current

As is present in nerve and skeletal muscle, cardiac tissue displays a time- and voltage-dependent K⁺ current, which is responsible for repolarizing cells.[87] Since its activation was slow,[87] and because this channel was thought to pass outward current, as opposed to the inward rectification of G_{K1}, this repolarizing current has been frequently termed the "delayed rectifier" (also "delayed current"). Initially labeled I_{x1} in studies by Noble and Tsien,[88,89] the current (conductance) is typically indicated as simply I_K (G_K).[90] Because of the dependence of G_{K1} on $[K^+]_o$, the influence of G_K on repolarization will vary with $[K^+]_o$. As indicated in Figure 1-5C, when $[K^+]_o$ is low, G_K plays a more dominant role, while in higher $[K^+]_o$, G_{K1} contributes more to repolarization.

Subtypes of the Delayed Current

The delayed K⁺ current is kinetically complex, sometimes showing a fast and a slow component of activation.[89-91] Recently, two distinct components have been separated. In guinea pig myocytes, a rapidly activating, inwardly rectifying current can be specifically blocked by certain class III antiarrhythmic agents (*d*-sotalol, E-4031, dofetilide) as well as by 1 µM La^{3+}, which has been designated $I_{K,r}$.[92–94] This current is similar in kinetics to an I_K present in the atrial node[95] and recently identified in rabbit ventricular myocytes, where an E-4031-sensitive $I_{K,r}$ appears to be the only delayed current present.[42] When $I_{K,r}$ is blocked, the plateau of the AP is significantly prolonged although the phase 3 repolarization rate is largely unaltered.[42,92] Following blockade of $I_{K,r}$, a nonrectifying delayed current that activates more slowly remains ($I_{K,s}$).

I_K plays a critical role in initiating phase 3 of repolarization. The kinetically slower I_K channel ($I_{K,s}$) activity appears to be modulated by a number of pathways including intracellular Ca^{2+},[96,97] the cyclic AMP-dependent protein kinase,[28,98] and protein kinase C.[28,96,99,100] Ca^{2+} is not required for I_K, but rather enhances the current. This Ca^{2+}-dependent control component permits the increased contractile state of the myocardium, associated with an increase in rate and increased intracellular Ca^{2+}, to increase channel opening, which will then decrease AP duration. The decreased AP duration will then cause earlier relaxation and a longer diastolic interval, permitting more appropriate cardiac metabolic balance, while the delayed current (I_K) appears to provide a long-term modulatory role for Ca^{2+} in altering the ventricular AP duration, $I_{K(Ca)}$, which appears to play a prominent role in controlling the initial phase 1 repolarization.

Finally, in rat atrium a very rapidly activating, noninactivating, 4-aminopyridine-sensitive current has been documented (I_{RAK})[101] and is probably responsible for the very fast repolarization observed in tissue from this species. The channel has also been cloned and expressed, showing great similarity to a neuronal rat K$^+$ channel BK2.[6] The kinetic behavior and conductance are very similar to those of a quinidine-sensitive potassium channel cloned from human atrium (HK2)[102–104] as well as a channel observed in canine ventricular epicardium.[105]

Transient Outward Current and Early Repolarization

Additional outward potassium currents, which are variably present in different cardiac tissues, are the transient outward currents (I_{to}; also early outward, I_{eo}).[63,106–108] These rapidly activating and inactivating currents markedly alter or abbreviate the AP configuration in various cardiac myocytes, influencing excitability and contractility. The identification as K$^+$ currents is strongly suggested by their blockade or inhibition by 4-aminopyridine[109–113] and internal quaternary ammonium ion.[114] The current appears in both Ca^{2+}-dependent[63,113,115,116] and -independent forms,[1,111,117] the latter having many features in common with the neuronal transient K$^+$ current I_A.[1,111] Such a current has recently been cloned and expressed from human heart.[118] A transient outward cardiac current was initially identified as a Cl$^-$ current,[119] and this concept has re-emerged to describe a component of transient inward current.[120,121]

Pacemaker Tissue

In SA nodal tissue, a transient current has been documented that contributes to repolarizations; however, it differs somewhat from the other I_{to} in being somewhat less selective for K$^+$.[122]

Atrial Muscle

Two transient outward currents have been documented in atrial muscle from a variety of tissues, the most prominent being the Ca^{2+}-independent form, which is largely responsible for the dramatically abbreviated plateau of the atrial AP.[123,124] A variation within the atrial muscle, with more I_{to} present in the epicardium than in the endocardium, has also been described.[117] The role and requirement of the Ca^{2+}-dependent I_{to} ($I_{K(Ca)}$) are unclear.

The His-Purkinje System

A prominent feature of the Purkinje fiber AP is the phase 1 notch, indicated in Figure 1-3. Unlike atrial muscle in which the repolarization process initiated by I_{to} persists and results in a shortened AP duration, in Purkinje fibers the AP plateau is restored for an extended period, in part due to the decreased G_{K1}.[91] Although there is the certainty of an I_{to} in Purkinje fibers, its exact character seems to be species dependent. In cow and calf Purkinje fibers, a Ca^{2+}-activated current, which is via a very high conductance channel conductance,[63,116] is clearly demonstrable, while in the sheep conduction system, I_{to} appears to be Ca^{2+}-independent.[107,116] These differences emphasize that pharmacologic intervention toward a single K$^+$ current will require species-specific data.

The Ventricular Epicardium

The presence of I_{to} in ventricular cells varies with species, being the major repolarizing current in rat (although Ca^{2+} activation seems controversial),[108,113] but absent in guinea pig.[100] Although a 4-aminopyridine-sensitive current has been demonstrable in ventricular myocytes,[125] it has became evident more recently that there is differential transmural distribution. Approaching the epicardial surface, cells have shorter APs and a phase 1 "notch" not observed in endocardial myocytes,[126-128] but similar to those of Purkinje fibers which possess I_{to}. Although the similar orientation of the R and T vectors on the ECG required a more brief epicardial versus endocardial AP duration, the cellular basis for this phenomenon became documented only when the presence of a prominent I_{to} was found in epicardial, but not in endocardial, canine ventricular cells.[112] Application of 4-aminopyridine eliminates the phase 1 notch in the epicardial AP and increases AP duration, but has minimal effect on endocardial AP duration. The effect of I_{to} in ventricular tissue is shown in Figure 1-6, where the presence of the early outward current in the epicardium leads to the phase 1 notch and a subsequent decrease in AP duration. In addition to explaining the ECG T-wave direction, these currents may have relevance in various disease states.

THE CURRENTS OF THE PACEMAKER POTENTIAL

The prominent diastolic depolarization (DD) of pacemaker tissue is a result of a complex combination of changing ionic permeabilities. Although the primary responsible current is the pacemaker current (I_f), the ionic flux through the delayed potassium and calcium channels also contributes.

The Pacemaker Current, I_f

Although initially defined as an inactivating potassium current (I_{K2}),[88,129] subsequent studies clearly demonstrated that an inward current was responsible for the DD.[130-134] The current was found to be an ionic pathway activated by membrane potentials negative to −40 to −50 mV and sensitive to changes in both $[Na^+]_o$ and $[K^+]_o$. The channel is relatively nonspecific, permitting both depolarizing Na^+ influx as well as K^+ efflux. However, near diastolic potentials in nodal tissue (−50 to −60 mV), the channel permits a depolarizing inward current (mostly Na^+ influx). The channel is inhibited and enhanced by muscarinic[134] and adrenergic receptor activation,[135] respectively, thereby modulating heart rate.

The Coordinated Pacemaker Currents

The Absence of G_{K1} in the Pacemaker Tissue

The magnitude of the G_{K1} in resting ventricular and atrial muscle constitutes a considerable stabilizing influence to keep the tissue near E_K. The massive influx of Na^+ current is required to generate the depolar-

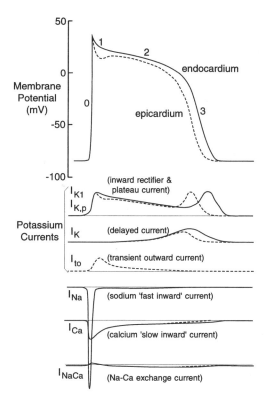

FIGURE 1-6. Contribution of the transient outward current (I_{to}) to the action potential is demonstrated. In epicardial ventricular muscle and also in the Purkinje fiber, a transient outward current occurs immediately after depolarization and is responsible for the larger phase 1 repolarization. Some persistent I_{to} then results in an abbreviated plateau and an earlier repolarization mediated initially by I_K and subsequently by I_{K1}. The inward currents, including that through the Na-Ca exchange (I_{NaCa}), are also indicated.

ization. In contrast, the pacemaking nodes of the heart have virtually no detectable resting G_{K1}-type conductance.[136] This makes it possible for the very modest currents generated by the pacemaker channels to depolarize these cells to the activation threshold potential of calcium channels. In contrast, the His-Purkinje system possesses sufficient resting G_{K1} that spontaneous depolarizations occur at a much slower rate than observed in the nodal tissue, but less than is present in ventricular muscle,[91] so that spontaneous depolarizations are possible. Pacemaker currents can be observed in ventricular tissue[137]; however, because of the sustained high level of G_{K1}, such currents probably have modest physiologic significance with spontaneous pacemaking rarely present.

The Role of the Other Currents

While the role and necessity of a depolarizing inward current have been emphasized,[138] other currents in all likelihood contribute to DD. The deactivation of the delayed current (I_K, probably the slow component) and the activation of calcium currents appear to contribute.

When 40 μM Ni^{2+} is applied to block $I_{Ca,T}$, the later half of the DD is remarkably prolonged, suggesting that $I_{Ca,T}$ in addition to I_f contributes depolarizing current to the DD.[66] Since the $I_{Ca,L}$ is not activated negative to −40 mV, it appears to contribute to the phase 0 of the AP, but not to DD. Figure 1-7 shows schematically the currents that appear to account for the cyclic nature of the cardiac pacemaker.

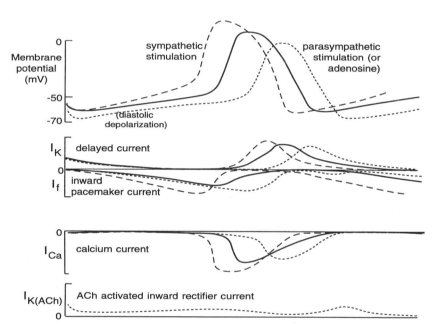

FIGURE 1-7. The currents responsible for pacemaker potentials in the sinoatrial (and also the atrioventricular) node. As indicated in the center panel, the DD is the result of the turn off of the delayed potassium current (I_K), as well as the hyperpolarization-induced activation of the inward pacemaker current (I_f).[136] The decline in the potassium conductance combined with the inward current result in the gradual depolarization until the action potential threshold is approached and calcium channels open. The I_{Ca} probably represents T-type channels initially activated during the latter half of the DD with the L-type channels providing the major current responsible for the AP upstroke.[66] In the presence of ß-adrenergic stimulation the I_f is enhanced as are the L-type calcium currents. The activated pacemaker current results in a rapid diastolic depolarization; the enhanced $I_{Ca,L}$ results in a larger and faster phase 0 of the AP. This later effect may result in faster conduction of the action potential through the nodal tissue. In the presence of muscarinic stimulation, the pacemaker potential is depressed while an inward rectifier current ($I_{K(ACh)}$) is stimulated by the acetylcholine (also by adenosine). This channel tends to maintain the cells in a hyperpolarized condition, decreasing the rate of diastolic depolarization by requiring a larger pacemaker current. Furthermore, the pacemaker current itself appears to be depressed by muscarinic stimulation.[134]

THE CHLORIDE CURRENTS

While not thought to play a prominent role at the present time, influx of Cl$^-$ through specific anion channels may contribute a repolarizing current to cardiac APs under various conditions.[139] These channels are usually blocked by a variety of negatively charged aromatic organic acids (see Table 1-2).

The Ca^{2+}-Activated Chloride Current

A transient outward current (assigned I_{to2} by some) has been defined during the plateau of the AP. Although this current contributes modestly compared to the 4-aminopyridine-sensitive K$^+$ current (I_{to1}), this outward plateau current may contribute to the initial repolarization of myocytes. The current appears to be activated by Ca^{2+}, which is released from sarcoplasmic reticulum stores and has been also designated as $I_{Cl(Ca)}$.[120,121,139]

The Cyclic AMP-Activated Chloride Channel

While there is normally no conductance to Cl$^-$ in normal cardiac tissue, a channel has clearly been defined which is activated by β-adrenergic stimulation[140–144] via phosphorylation by protein kinase A (cyclic AMP-controlled). The channel appears to be similar in many characteristics to the chloride channel, which has been identified as the defective gene in cystic fibrosis (cystic fibrosis transmembrane regulator).[139] This channel is insignificant in sinoatrial and atrial cells and is more prominent in epicardial than endocardial ventricular cells.[143]

The Stretch-Activated Chloride Channel

A Cl$^-$ channel activated by stretch[145] or swelling[146] has been documented in atrial myocytes; however, its role is unclear.

OTHER MODULATORS OF CARDIAC MEMBRANE POTENTIAL

The Na$^+$-K$^+$ ATPase (Sodium Pump)

The maintenance of the Na$^+$ gradient is obviously critical, since it is this gradient that drives rapid depolarization and rapid conduction, and it is the Na$^+$-K$^+$ ATPase sodium pump that is responsible. As noted, the Na$^+$ gradient is also of critical importance to provide the energy for the Na-Ca exchanger, and the myocardium consumes substantial energy as ATP to maintain this electromotive force. However, by pumping 3 Na$^+$ out for 2 K$^+$ in, the pump generates a negative current of its own. This current is probably present in the background at all times, providing a bias toward more negative potentials.

The pump current can assume a special prominence after very rapid heart rates, particularly when externally paced. The K$^+$ efflux and Na$^+$ influx per unit time are almost directly proportional to the heart rate, so that at very rapid heart rates there is greater accumulation of K$^+$ outside the cell membrane and accumulation of Na$^+$ inside. These ionic changes lead to enhanced pump function. If the stimulation is suddenly stopped, the pump keeps working and generates a substantial negative potential, a phenomenon termed "overdrive suppression" or "afterdrive hyperpolarization."[147] This hyperpolarization can have particularly marked effects in the pacemaker tissues, which because of the absence of G_{K1} are sensitive to small currents. The hyperpolarization may cause a profound but transient decrease the pacing rate. As the accumulated K$^+$ and Na$^+$ decrease, the pump current slows so that the intrinsic pacing rate will accelerate toward normal. This behavior may be observed clinically when suddenly stopping rapid atrial pacing can lead to transient bradycardia or sinus arrest.

The Ca-Na Exchange Current, I_{NaCa}

Since Ca$^+$ enters with each action potential, its removal from the cell is critical for proper function of cardiac tissue. Ultimately, if Ca^{2+} is not eliminated from the cell, Ca^{2+} is accumulated into the sarcoplasmic reticulum until it overflows (see "Delayed Afterdepolarizations," p. 40), and Ca^{2+} is finally taken up by mitochondria, poisoning them and leading to cell death. While the Ca-Na exchange mechanism was first described in squid axons, its potential relevance to cardiac tissue was noted by Mullins[148] and has subsequently been well documented. This membrane protein exchanges 3 Na$^+$ for 1 Ca^{2+} in either direction[149] and gives rise to a specific and distinct current, I_{NaCa}.[150] During diastole, when the [Ca]$_i$ is ≤0.1 μM, the ion pathway has a "reversal potential" (E_{NaCa}) near −40 mV situated between the resting potential and the plateau; I_{NaCa} is therefore inward and mediates removal of Ca^{2+} from the cell. During systole, once [Ca^{2+}]$_i$ has increased to 1 to 2 μM, E_{NaCa} becomes positive to the plateau so that I_{NaCa} continues to be inward and may even increase, contributing a significant depolarizing current (3 Na$^+$ enter and 1 Ca^{2+} exits on each exchange) during most of the AP plateau[151] (Fig. 1-6). Although during phases 0 to 1 of the AP, V$_m$ may transiently exceed

E_{NaCa},[152] the influx of Ca^{2+} via the exchanger is modest if present at all.[151] Because of the interdependent and electrogenic changes in intracellular $[Ca^{2+}]$ and $[Na^+]$, I_{NaCa} has an important role in modulating rate-dependent changes and initiating late afterdepolarizations.

ALTERATION IN CARDIAC CURRENTS BY NEUROTRANSMITTERS AND HORMONES

The primary modulation of the heart rate and contractility is achieved by the balance of influences from the autonomic nervous system: norepinephrine (NE) from the sympathetic nervous system binding to β- and $α_1$-adrenoceptors; acetylcholine (ACh) released by the parasympathetic nervous system binding to muscarinic receptors. ACh is also released on the sympathetic nerve terminals, decreasing the amount of NE release (presynaptic inhibition), so that part of the action of ACh release involves presynaptic modulation of NE.

Sympathetic, β-Adrenergic

NE binding to β-adrenoceptors has as its primary action the activation of protein kinase A mediated via G protein activation of cyclic AMP production by adenylyl cyclase (Fig. 1-8*A*). When intracellular cyclic AMP rises, two molecules bind to the regulatory subunits of cyclic AMP-dependent protein kinase A, causing the complex to dissociate into a regulatory dimer and two free and active catalytic subunits.[153] Protein kinase A can phosphorylate serine and threonine residues on numerous proteins, but prominent substrates are the cytoplasmic regions of a number of ion channels and ion transport pumps, as well as contractile proteins and other protein kinases.

The L-type calcium channel is prominently altered by sympathetic stimulation. While modulatory phosphorylation of calcium channels has been suspected for a long period,[154,155] the mechanism for increased I_{Ca} has only recently been defined. Increased Ca^{2+} entry occurs because the gating charge of the channels moves more rapidly and at more negative potentials,[53] so that faster opening occurs with slightly less depolarization. Also of considerable importance is the fact that the channels stay open for a more sustained period further increasing the total Ca^{2+} influx.[27] Because of the high resting G_{K1} in ventricular muscle, the density and rate of activation of calcium channels is normally not sufficient to perpetuate a propagated AP, especially in elevated $[K^+]_o$ such as cardioplegic solution. However, β-adrenergic stimulation generates sufficient active calcium channels so that propagated slow APs are possible.

An increase in heart rate is achieved by the increased activation of I_f,[135] which increases the rate of diastolic depolarization of pacemaker tissue, thereby increasing heart rate, while the enhanced $I_{Ca,L}$ in pacemaker tissue contributes to enhanced nodal slow AP conduction.[66] However, as part of the coordinated actions to increase heart rate and contractility, systole must be shortened. This effect occurs via enhancement of repolarizing currents, specifically I_K and $I_{Cl,cAMP}$. β-Adrenergic stimulation results in marked enhancement of the delayed potassium current I_K,[28,156] although unlike $I_{Ca,L}$ enhancement, this effect is temperature dependent.[98] Also contributing to the increased repolarizing current in ventricular cells is the opening of normally closed chloride channels, which do not appear to be voltage gated. During systole I_{Cl} provides a modest repolarizing current, so that in combination with I_K the AP duration is actually shortened with β-adrenergic stimulation, in spite of the enhanced I_{Ca}.[140,142] Since the reversal potential for Cl⁻ is near –50 mV, such channels may provide a modest depolarizing current in diastole during β-adrenergic stimulation,[142,157] which may be able to initiate arrhythmias.[158] With such increased conductance and slight depolarization, a decrease in conduction velocity might be anticipated as a result of sodium channel inactivation. However, like I_{Ca}, I_{Na} also shows enhancement by β-adrenergic stimulation, which is mediated by protein kinases, an effect that will tend to maintain conduction in the face of increased conductance which might otherwise shunt depolarizing currents.[159,160]

The obvious effects of β-adrenergic stimulation on contractility are mediated by additional means. Increased SR uptake of Ca^{2+} by activation of the SR Ca^{2+}-ATPase enhances relaxation and increases the SR Ca^{2+} store available for the subsequent contraction.[161] Phosphorylation of troponin results in *decreased* affinity of for Ca^{2+}, which permits faster relaxation.[162,163] The requirement for increased Ca^{2+} to activate myofibrillar interaction and tension development is more than satisfied by the increased Ca^{2+} entry and SR uptake and release. The increased cytoplasmic Ca^{2+} also may contribute to activation of the delayed current,[96,97] to further abbreviate the AP. Both the increased calcium channel activity (supporting slow AP conduction) and the increased SR store contribute to dysrhythmias (see below). By similar G-protein mediated stimulation of adenylyl cyclase, histamine and glucagon may have similar effects.

Sympathetic, α-Adrenergic

Although not nearly as profound, $α_1$-adrenergic activation also has effects on cardiac ion channel behavior, as well as having a positive inotropic action. Prolongation of the AP, most prominent in atrial tissue, appears to be mediated by blockade of the transient outward current.[164] The positive inotropic action of $α_1$-adrenoceptor activation appears related to increased myofibril Ca^{2+} sensitivity[165] (opposite to β-adrenergic effects), which is at least in part due to alkalization of the cell[166,167] mediated by increased activity of the Na^+/H^+ exchanger.[168] Normally, the β-adrenergic effects seems the dominant process.[165]

Parasympathetic Tone

The major ionic mechanism causing the negative chronotropic, dromotropic, and inotropic effects of the parasympathetic ACh release is the ACh-activated inwardly rectifying K^+ current ($I_{K(ACh)}$). This process represents a now classic example of specific receptors being directly coupled to a separate ion channel by guanine nucleotide binding proteins (G proteins).[169] ACh binds to a muscarinic receptor (M2), which is one of five muscarinic receptors in the extensive G protein-coupled receptor superfamily.[170,171] The ACh-bound receptor then catalyzes formation of the activated (GTP bound), a subunit of G_i (G_i where i stands for inhibition of adenylyl cyclase). The activated $G_i\alpha$ then associates and opens the specific inwardly rectifying K^+ current ($I_{K(ACh)}$).[172–174] In the same manner as for G_{K1}, this rectifying effect is mediated by internal Mg^{2+}, although the ion has slightly less blocking potency.[175,176] Activation of these channels may also have a modest action in ventricular tissue of decreasing action potential duration, decreasing the amount of Ca^{2+} current, and thereby modestly decreasing contractions.[177] In the absence of β-adrenergic stimulation, the inhibition of I_{Ca} appears modest.[177] The inhibitory action of ACh on I_{Ca} is far more profound on isoproterenol- or forskolin (a direct adenylyl cyclase stimulator)-enhanced I_{Ca}.[178] Furthermore, a similar inhibitory action of ACh has been reported on isoproterenol-enhanced I_{Na}.[179]

While the activation of $I_{K(ACh)}$ may contribute to the negative chronotropic action of ACh, an even greater effect may be that mediated on the pacemaker current (I_f), which can be strongly inhibited by lower concentrations of ACh than appear to be required to activate $I_{K(ACh)}$.[134]

ALTERATIONS INDUCED BY METABOLIC CHANGES

In addition to modulation by autonomic input, the metabolic state of the heart contributes considerably to behavior of the various ionic pathways. Ischemia and depletion of energy stores bring about distinct changes, which will typically down-regulate myocardial energy expenditure (rate and contractility). While profound acidosis (pH 6.5 to 6.8) may inhibit I_{Ca},[180,181] the major mechanisms appear to involve activation of potassium channels.

The Adenosine-Activated K⁺ Channel

With increased activity, adenosine (Ado) is increasingly produced by the heart, which provides graduated control of coronary muscle tone. In addition, Ado binds to A_1 receptors on myocytes and, acting via G_i protein, activates a inwardly rectifying K⁺ current identical to $I_{K(ACh)}$, and is termed $I_{K(Ado)}$.[182–184] Either endogenous production or exogenous administration of Ado is capable of producing marked bradycardic effects by increasing the outward K⁺ conductance, similar to a marked vagal stimulus. Like ACh, A_1 receptor activation may also cause modest depression of I_{Ca}; however, the effect of shortening AP duration is due primarily to $I_{K(Ado)}$ activation.[185] In addition to ACh and Ado, a similar inwardly rectifying K⁺ current appears to be activated by platelet-activating factor (1-O-alkyl-2-acetyl-*sn* glycero-3-phosphoryl-choline),[186] an agent known to cause bradycardia.

The ATP-Regulated K⁺ Channel, $I_{K(ATP)}$

A K⁺ channel that activates when ATP is reduced was first described in cardiac muscle by Noma[187] in 1983. Clearly distinct from G_{K1}, with higher conductance,[188] these channels have been reported in vascular smooth muscle, brain, skeletal muscle, and pancreatic β-cells[189]; however, the properties of the channels in different tissues appear variable.[190] Like I_{K1} and $I_{K(ACh)}$, $I_{K(ATP)}$ also has some inwardly rectifying characteristics. Of further interest, a K_{ATP} channel has been cloned and expressed. In contrast to the voltage-gated K⁺ channels, the K_{ATP} channel shows only two transmembrane α-helices,[13] and with the inward rectifier channel(s) constitutes the first members of another distinct ion channel superfamily.[12]

When activated in the presence of intracellular ATP depletion, such as that caused by ischemia, it is postulated that this current plays a protective role by abbreviating the action potential, decreasing Ca^{2+} entry, and reducing further ATP depletion (Fig. 1-8). In this setting, the sodium pump may be inhibited, and K⁺ efflux via K_{ATP} channels will accumulate outside the cell, causing some depolarization. Blockade of K_{ATP} channels by glibenclamide increases heart infarct size in a rabbit model in the absence of preconditioning, but blockade did not prevent the beneficial effect of preconditioning.[191] The effect of K_{ATP} channel openers is also equivocal.[192] The role that such channels may play in arrhythmias and the effects of their activation and blockade remain to be fully defined.

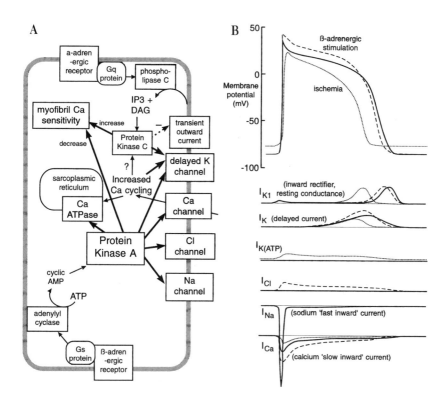

FIGURE 1-8.   Effects of adrenergic stimulation in the heart. *A,* cellular actions of sympathetic stimulation on electrophysiologic mechanisms. Most sympathetic effects result from β-adrenergic-stimulated protein phosphorylation by activation of protein kinase A activity (via increased cyclic AMP production). The phosphorylation of multiple ion channels results in increased ion fluxes, as well as increased Ca^{2+} cycling within the cell. A direct G-protein effect activating L-type calcium channels has been described,[171,173,220] although its physiologic relevance is debated.[221,222] α-Adrenergic stimulation has more modest effects, particularly upon the transient outward current. *B,* the effects of β-adrenergic activation of ion channels upon the cardiac action potential configuration. The increased I_{Ca} leads to a higher plateau. The activation of chloride current and enhancement of delayed rectifier currents result in a more abbreviated plateau. The inward rectifier appears to be little affected by adrenergic stimulation. The possible enhancement of I_{Na} is not shown.

Fatty Acid-Activated K⁺ Channels

In addition to those already noted, apparently two K⁺ channels with differing conductances can be observed in the presence of arachidonic acid and phosphatidylcholine. Since ischemia is known to increase levels of fatty acids, these channels may likewise serve a protective role.[193]

The Na⁺-Activated K⁺ Current

In addition to the numerous other currents, a high conductance K^+ channel has been demonstrated which is activated by increased cytosolic Na^+.[194,195] The physiologic relevance of this channel is unclear. It seems most likely to represent a protective, repolarizing current, which becomes activated when the cells become Na^+ overloaded as occurs with extremely high heart rates or sodium pump inhibition by cardiac glycosides and with ischemia.

DISTRIBUTION OF CURRENTS IN THE VARIOUS TISSUES OF THE HEART

Since each ion channel represents a distinct and particular membrane protein, each is clearly a candidate for distinct pharmacologic modulation. The variable expression of these different channels in different regions of the heart gives rise to the various and distinctive patterns of AP shown in Figure 1-3. The distribution of major channels in various cardiac tissues is listed in Table 1-3. Except for the bundle of His and proximal Purkinje fibers, most cardiac tissues do not express all of the various currents described. Furthermore, distributions may be subject to variation in species.

MECHANISMS OF ARRHYTHMIAS

Arrhythmias represent abnormal behavior of the ion channel fluxes, which result in abnormal impulse conduction (conduction block, re-entry, or reflection) or abnormal impulse initiation (abnormal automaticity or triggered rhythms), or possibly both (parasystole or phase 4 block).[196] The major categories, each with their own particular membrane micropathology and mechanism, are re-entry, triggered activity, and automaticity.[197]

Re-Entry

Since the myocardium, either atrial or ventricular, is an electrical syncytium, any mass of the tissue, microscopic or its whole, represents a potential circular path in which membrane potentials may be conducted. Normally, this tissue is depolarized rapidly and uniformly and then repolarizes with relative uniformity. A virtue of the heart as noted before is the fact that after its sustained depolarization, it is difficult to initiate another propagated depolarization. Re-entry is a phenomenon of recurring and self-perpetuating depolarization around a circuit (also termed a "circus" movement, as in Picadilly Circus, a traffic circle). Re-entry requires an area of one-way conduction or unidirectional block and a circumscribed conduction path around the block, so that the AP can be reintroduced and propagate retrograde through the blocked area. Although normal conduction velocity in a long circuit may support re-entry, an area of slowed conduction (caused by damaged tissue, K^+ accumulation, or ischemia) is typically present.[196]

Unidirectional block arises when a nonuniform injury exists in a conducting pathway, such as with graduated injury in one direction and an abrupt injury in the other. Conduction (from the forward direction) will fail when the depolarization enters an area of increasing injury or damage in which the depolarizing current dissipates before it reaches healthy tissue from which an AP can propagate. Conduction may succeed (from the "reverse" direction) when depolarization enters an abrupt region of injury, because the high density of depolarizing current does not have a region of partially injured tissue in which to dissipate before it depolarizes tissue from which an AP will propagate.[197]

Re-entry can exist with circuits of variable size. Clinical settings thought to involve re-entry are[196]: 1) atrial tachycardias, flutter, and fibrillation, with large circuits a number of centimeters in diameter; 2) sinus node and AV node re-entrant tachycardias, which may involve microscopic circuits within the nodal tissue itself; 3) AV re-entry tachycardia due to an accessory AV pathway; 4) bundle branch re-entrant tachycardia; and 5) ventricular tachycardia, which is thought to involve small (millimeter) circuits in damaged portions of the Purkinje system, endocardium, or ventricular wall.

Slowed Conduction

Because almost all the sodium channels inactivate with depolarization, a brief period of at least partial repolarization (negative to ~–50 mV) is necessary before they return to their activatible form. The period during which a second propagated AP cannot be elicited is the absolute (or effective) refractory period. Once repolarization is negative to –50 or –60 mV, a second AP may be elicited, but due to limited availability of working sodium channels, dV/dt and conduction velocity will be low. This period of time is the relative refractory period. Following repolarization, sufficient numbers of noninactivated (or activatible) sodium channels only gradually accumulate with phase 3 repolarization, so that 10 to 50 msec are required before the normal density of working sodium channels is available to support normal conduction. In the presence of ischemia or hypoxia and pericellular hyperkalemia, the increased conductance (due to K_{ATP} and Na^+-activated K channels) and partial depolarization will lead to most sodium channels being inactivated. However, if sufficient active calcium channels are present, they may contribute to depolarization at more positive membrane potentials. If depolarizing current arrives during repolarization when a few activatible sodium channels combine calcium channels, a slowly conducted AP can occur through the tissue, which may contribute to re-entry circuits and dysrhythmias. Even in the early stages of ventricular fibrillation, sodium currents still[198] appear to contribute to depolarizing current. In the presence of hypoxia, ischemia, and/or hyperkalemia, the effect of β-adrenergic stimulation may be significant in terms of its ability to restore APs by generating a larger number of activatible calcium channels.

Triggered Activity

Abnormal potentials, termed "afterdepolarizations," follow closely on a previous normal AP, occurring either with phase 3 repolarization (early afterdepolarizations, EADs) or after complete repolarization (late or delayed afterdepolarizations, DADs). In contractile tissue, these depolarizations may be accompanied by a small amount of tension development or after contractions. However, the mechanisms of these two forms of triggered activity differ markedly.

Early Afterdepolarizations

EADs arise during the late plateau or early phase 3 of AP depolarization.[196] They appear to occur under circumstances in which the resting K^+ conductance is blocked and in which calcium channels are not fully inactivated.[197] Circumstances in which this activity is observed include low or K^+-free medium, which inactivates G_{K1}, or blockade of G_{K1} by other means.[197,199,200] Slowly conducted APs arise which conduct to more fully repolarized regions and can then lead to a propagated ectopic beat. Such a decrease in K^+ conductance may explain the repolarization abnormality, which occurs in the long QT interval syndrome which is associated with torsade de pointes ventricular arrhythmia and sudden death.[199,201] Increasing K^+ conductance with potassium channel activators may prove to be beneficial in these circumstances.[202] While blockade of G_{K1} can be responsible, it seems likely that blockade of other K^+ currents can contribute to EADs. For example, the torsade de pointes rhythm observed with quinidine, sotalol, and bupivacaine overdose appear to be due to inhibition of I_K or I_{to}.[203,204] Furthermore, a window current of L-type calcium current has been recently identified that may contribute.

Delayed Afterdepolarizations

The best present explanation for DADs involves "overload" of the SR with Ca^{2+}.[196,197] Situations involving DADs usually include increased Ca^{2+} entry and/or increased SR Ca^{2+} uptake as occurs with very rapid beating rates or β-adrenergic stimulation. Ca^{2+} overload and DADs also occur with excessive cardiac glycosides poisoning the sodium pump, so that a decreased Na^+ gradient is no longer able to eliminate Ca^{2+} from the cell. The SR accumulates the Ca^{2+} until it can store no more. Once overloaded, the SR spontaneously releases Ca^{2+} into the myoplasm. The pattern of Ca^{2+} release in DADs is associated with a wavelike release of Ca in the cell, while in the case of EADs the second Ca^{2+} transient occurs uniformly throughout the cell.[205] The released Ca^{2+} is then eliminated by the Na-Ca exchanger, which in turn results in a net inward depolarizing current (3 Na^+ in for 1 Ca^{2+} out).[206–208] If the inward current causes sufficient depolarization to reach threshold, activation of an AP occurs. Such APs can result in a vicious cycle of continued rapid rhythms and continuing Ca^{2+} entry and overload. Such triggered tachyarrhythmias are thought to account for many ventricular tachycardias associated with acute myocardial infarctions, right ventricular outflow tract tachycardias, and certain atrial tachycardias.[196,209] Obviously, prevention of such tachycardias must be ultimately directed at prevention of Ca^{2+} overload.

Abnormal Automaticity

While the SA and AV nodes are typically automatic, any cardiac tissue has the potential to spontaneously depolarize to threshold at a regular rate. Automaticity may in many cases represent the appearance of pacemaker current (I_f), which had been dormant. Due to partial depolarization because of damage or drug/hormonal stimulation, other cardiac tissue may achieve a rate superseding that of the normal pacemaking tissue, particularly in settings where the normal pacemaker is depressed. Normally the depolarizing current is insufficient to bring the tissue to the AP threshold, at least in ventricular tissue. However, similar to early afterdepolarizations, in situations of reduced resting conductance, the inward I_f current may be sufficient to induce ectopic beats.

Considerations for Therapeutic Interventions

Normally the various K$^+$ currents ensure that depolarization and re-polarization occur in a coordinated and temporally unified fashion and provide considerable electrophysiologic stability so that it is difficult to induce sustained tachycardia in a normal heart. When APs in one region of the heart are shortened, leaving others unaltered, this creates greater temporal dispersion in AP duration. When such temporal dispersion is produced experimentally in ventricular tissue, tachycardia can be induced by an appropriate single stimulation.[210] When ischemia or hypoxia results in AP shortening, portions of myocardium may be repolarized at times when other areas may remain depolarized.[211] In hypertrophied myocardium, greater dispersion of refractory periods can be demonstrated and is associated with greater vulnerability for ventricular fibrillation.[212] If an adequate electrical pathway exists, then depolarizing current may flow to the more repolarized region and initiate an AP. When this occurs, a cyclical depolarization pattern may result.

To prevent such temporal dispersion of AP duration, the use of potassium channel blocking agents has been suggested,[213] and these agents have proven to have beneficial effects in certain models.[212,214] However, the specific potassium channel to be blocked is important. It is clear that under certain conditions, reduced K$^+$ conductance can lead to increased excitability and triggered dysrhythmias,[197,199,200] so that decreasing G_{K1}, which may lengthen the terminal aspect of the AP, may have proarrhythmic effects. While blockade of $I_{K(ATP)}$ or $I_{K,r}$ may extend the period of myocardial inexcitability and decrease temporal dispersion of AP duration, it must be remembered that it may also extend the period of Ca^{2+} entry and contractile activation, which may increase oxygen demand and worsen ischemia. Careful studies will be required to document which potassium channels may be blocked to provide a beneficial effect.

References

1. Giles WR, Imaizumi Y. Comparison of potassium currents in rabbit atrial and ventricular cells. J Physiol (Lond) 1988;405:123.
2. Catterall WA. Molecular properties of voltage-sensitive sodium channels. Annu Rev Biochem 1986;55:953.
3. Catterall WA, Seagar MJ, Takahashi M. Molecular properties of dihydropyridine-sensitive calcium channels in skeletal muscle. J Biol Chem 1988;263:3535.
4. Mikami A, Imoto K, Tanabe T, et al. Primary structure and functional expression of the cardiac dihydropyridine-sensitive calcium channel. Nature (Lond) 1989;340:230.

5. Folander K, Smith JS, Antanavage J, Bennett C, Stein RB. Cloning and expression of the delayed-rectifier I_{sK} channel from neonatal rat heart and diethylstilbestrol-primed rat uterus. Proc Natl Acad Sci USA 1990;87:2975.
6. Paulmichl MP, Nasmith P, Hellmiss R, et al. Cloning and expression of a rat cardiac delayed rectifier potassium channel. Proc Natl Acad Sci USA 1991;88:7892.
7. Tamkun MM, Knoth KM, Walbridge JA, Kroemer H, Roden DM, Glover DM. Molecular cloning and characterization of two voltage-gated K^+ channel cDNAs from human ventricle. FASEB J 1991;5:331.
8. Tsien RW, Ellinor PT, Horne WA. Molecular diversity of voltage-dependent Ca^{2+} channels. Trends Pharmacol Sci 1991;12:349.
9. Miller C. 1990: annus mirabilis of potassium channels. Science 1991; 252:1092.
10. Brehm P, Okamura Y, Mandel G. Ion channel evolution. Semin Neurosci 1991;3:355.
11. Barnard EA. Receptor classes and the transmitter-gated ion channels. Trends Biochem Sci 1992;17:368.
12. Kubo Y, Baldwin TJ, Jan YN, Jan LY. Primary structure and functional expression of a mouse inward rectifier potassium channel. Nature (Lond) 1993;362:127.
13. Ho K, Nichols CG, Lederer WJ, et al. Cloning and expression of an inwardly rectifying ATP-regulated potassium channel. Nature (Lond) 1993;362:31.
14. Tomaselli GF, Backx PH, Marban E. Molecular basis of permeation in voltage-gated ion channels. Circ Res 1993;72:491.
15. Katz AM. Cardiac ion channels. N Engl J Med 1993;328:1244.
16. Stühmer W. Structure-function studies of voltage-gated ion channels. Annu Rev Biophys Biophys Chem 1991;20:65-78.
17. Po S, Roberds S, Snyders DJ, Tamkum MM, Bennett PB. Heteromultimeric assembly of human potassium channels: molecular basis of a transient outward current? Circ Res 1993;72:1326.
18. Catterall WA. Molecular properties of voltage-gated ion channels in the heart. In: Fozzard HA, Haber E, Jennings RB, Katz AM, Morgan HE, eds. The heart and cardiovascular system—scientific foundations. 2nd ed. New York: Raven Press, 1992:945.
19. Heinemann SH, Terlau H, Stuhmer W, Imoto K, Numa S. Calcium channel characteristics conferred on the sodium channel by single mutations. Nature (Lond) 1992;356:441.
20. Bayer R, Hennkes R, Kaufmann R, Manhold R. Inotropic and electrophysiologic actions of verapamil and D600 in mammalian myocardium. III. Effects of optical isomers on the transmembrane action potentials. Naunyn-Schmiedebergs Arch Pharmacol 1975;290:49.
21. Yatani A, Brown AM. The calcium channel blocker nitrendipine blocks sodium channels in neonatal rat cardiac myocytes. Circ Res 1985;56:868.
22. Gilliam FR III, Rivas PA, Wendt DJ, Starmer CF, Grant AO. Extracellular pH modulates block of both sodium and calcium channels by nicardipine. Am J Physiol 1990;259:H1178.
23. Scheuer T, Kass RS. Phenytoin reduces calcium current in the cardiac Purkinje fiber. Circ Res 1983;53:16.
24. Scamps F, Undrovinas A, Vassort G. Inhibition of I_{Ca} in single frog cardiac cells by quinidine, flecainide, ethmozin, and ethacizin. Am J Physiol 1989;25:C549.

25. Lynch C III. Depression of myocardial contractility in vitro by bupivacaine, etidocaine, and lidocaine. Anesth Analg 1986;65:551.
26. Shenolikar S. Protein phosphorylation: hormones, drugs, and bioregulation. FASEB J 1988;2:2753.
27. Yue DT, Herzig S, Marban E. β-Adrenergic stimulation of calcium channels occurs by potentiation of high-activity gating modes. Proc Natl Acad Sci USA 1990;87:753.
28. Walsh KB, Kass RS. Regulation of a heart potassium channel by protein kinase A and C. Science 1988;242:67.
29. Armstrong CM, Bezanilla F. Currents related to the movement of gating particles of the sodium channel. Nature (Lond) 1973;242:459.
30. Bean BP, Rios E. Nonlinear charge movement in mammalian cardiac ventricular cells. J Gen Physiol 1989;94:65.
31. Field AC, Hill C, Lamb GD. Asymmetric charge movement and calcium currents in ventricular myocytes of neonatal rat. J Physiol (Lond) 1988; 406:277.
32. McLaughlin S, Szabo G, Eisenman G. Divalent ions and the surface potential of charged phospholipid membrane. J Gen Physiol 1971;58:667.
33. Ji S, Weiss JN, Langer GA. Modulation of voltage-dependent sodium and potassium currents by charged amphiphiles in cardiac ventricular myocytes. Effects via modification of surface potential. J Gen Physiol 1993;101:355.
34. Hamill OP, Marty A, Neher E, Sakmann B, Sigworth FJ. Improved patch-clamp techniques for high-resolution current recording from cells and cell-free membrane patches. Pfluegers Arch 1981;391:85.
35. Reuter H, Scholz H. A study of the ion selectivity and the kinetic properties of the calcium dependent slow inward current in mammalian cardiac muscle. J Physiol (Lond) 1977;264:17.
36. Lee KS, Tsien RW. Mechanism of calcium channel blockade by verapamil, D600, diltiazem and nitrendipine in single dialysed heart cells. Nature (Lond) 1983;302:790.
37. Hess P, Tsien RW. Mechanism of ion permeation through calcium channels. Nature (Lond) 1984;309:453.
38. Kohlhardt M, Bauer B, Krause H, Fleckenstein A. Selective inhibition of the transmembrane Ca conductivity of mammalian myocardial fibres. Pfluegers Arch 1973;338:115.
39. Lee KS, Tsien RW. Reversal of current through calcium channels in dialysed single heart cells. Nature (Lond) 1982;297:498.
40. Fozzard HA, Hanck DA. Sodium channels. In: Fozzard HA, Haber E, Jennings RB, Katz AM, Morgan HE, eds. The heart and cardiovascular system—scientific foundations. 2nd ed. New York: Raven Press, 1992:1091.
41. Kameyama M, Kiyosue T, Soejima M. Single channel analysis of the inward rectifier K current in the rabbit ventricular cells. Jpn J Physiol 1983;33:1039.
42. Veldkamp MW, Ginneken ACGV, Bouman LN. Single delayed rectifier channels in the membrane of rabbit ventricular myocytes. Circ Res 1993;72:865.
43. Bleeker WK, McKaay AJC, Masson-Pevot M, Bouman LN, Becker AE. The functional and morphological organization of the rabbit sinus node. Circ Res 1980;46:11.

44. Rogart RB, Cribbs LL, Muglia LK, Kephart DD, Kaiser MW. Molecular cloning of a putative tetrodotoxin-resistant rat heart Na+ channel isoform. Proc Natl Acad Sci USA 1989;86:8170.

45. Baer M, Best PM, Reuter H. Voltage-dependent action of tetrodotoxin in mammalian cardiac muscle. Nature (Lond) 1976;263:344.

46. Atwell D, Cohen I, Eisner D, Ohba M, Ojeda C. The steady state TTX sensitive ("window") sodium current in cardiac Purkinje fibers. Pfluegers Arch 1979;379:147.

47. Grant AO, Starmer CF. Mechanisms of closure of cardiac sodium channels in rabbit ventricular myocytes: single-channel analysis. Circ Res 1987;60:897.

48. Coraboeuf E, Deroubaix E, Colombe A. Effect of tetrodotoxin on the action potentials of the conducting system in the dog heart. Am J Physiol 1979;236:H561.

49. Nilius B, Hess P, Lansman JB, Tsien RW. A novel type of cardiac calcium channel in ventricular cells. Nature (Lond) 1985;316:443.

50. Hirano Y, Fozzard HA, January CT. Characteristics of L- and T-type Ca^{2+} currents in canine cardiac Purkinje fibers. Am J Physiol 1989; 256:H1478.

51. Rose WC, Balke CW, Wier WG, Marban E. Macroscopic and unitary properties of physiological ion flux through L-type Ca^{2+} channels in guinea-pig heart cells. J Physiol (Lond) 1992;456:267.

52. Balke CW, Rose WC, Marban E, Wier WG. Macroscopic and unitary properties of physiological ion flux through T-type Ca^{2+} channels in guinea-pig heart cells. J Physiol (Lond) 1992;456:247.

53. Josephson IR, Sperelakis N. Phosphorylation shifts the time-dependence of cardiac Ca^{++} channel gating currents. Biophys J 1991;60:491.

54. Yue DT, Backx PH, Imredy JP. Calcium-sensitive inactivation in the gating of single calcium channels. Science 1990;250:1735.

55. Bates SE, Gurney AM. Ca^{2+}-dependent block and potentiation of L-type calcium current in guinea-pig ventricular myocytes. J Physiol (Lond) 1993;466:345.

56. DeFelice LJ. Molecular and biophysical view of the Ca channel: a hypothesis regarding oligomeric structure, channel clustering, and macroscopic current. J Membr Biol 1993;133:191.

57. Kass RS, Tsien RW. Control of action potential duration by calcium ions in cardiac Purkinje fibers. J Gen Physiol 1976;67:599.

58. Sanguinetti MC, Kass RS. Voltage-dependent block of calcium channel current in the calf cardiac Purkinje fiber by dihydropyridine calcium channel antagonists. Circ Res 1984;55:336.

59. Hille B. Local anesthetics: hydrophilic and hydrophobic pathways for the drug-receptor interactions. J Gen Physiol 1977;69:497.

60. Hondeghem LM, Katzung BG. Time- and voltage-dependent interactions of antiarrhythmic drugs with cardiac sodium channels. Biochim Biophys Acta 1977;472:373.

61. Ehara T, Kaufmann R. The voltage and time-dependent effects of (-) verapamil on the slow inward current in isolated cat ventricular myocardium. J Pharmacol Exp Ther 1978;207:49.

62. McDonald TF, Pelzer D. Cat ventricular muscle treated with D600: characteristics of calcium channel block and unblock. J Physiol (Lond) 1984;352:217.

63. Siegelbaum SA, Tsien RW, Kass RS. Role of intracellular calcium in the transient outward current of calf Purkinje fibres. J Physiol (Lond) 1977; 269:611.
64. Bean BP. Two kinds of calcium channels in canine atrial cells: differences in kinetics, selectivity, and pharmacology. J Gen Physiol 1985;86:1.
65. Mitra R, Morad M. Two types of Ca2+ in guinea pig ventricular myocytes. Proc Natl Acad Sci USA 1986;83:5340.
66. Hagiwara N, Irisawa H, Kameyama M. Contribution of two types of calcium currents to the pacemaker potentials of rabbit sino-atrial node cells. J Physiol (Lond) 1988;395:233.
67. Wu J, Lipsius SL. Effects of extracellular Mg^{2+} on T- and L-type Ca^{2+} currents in single atrial myocytes. Am J Physiol 1990;259:H1842.
68. Doerr T, Denger R, Trautwein W. Calcium currents in single SA nodal cells of the rabbit heart studied with action potential clamp. Pfluegers Archiv 1989;413:599.
69. Lynch C III, Vogel S, Sperelakis N. Halothane depression of myocardial slow action potentials. Anesthesiology 1981;55:360.
70. Ikemoto Y, Yatani A, Arimura H, Yoshitake J. Reduction of the slow inward current of isolated rat ventricular cells by thiamylal and halothane. Acta Anaesthesiol Scand 1985;29:583.
71. Terrar DA, Victory JGG. Influence of halothane on membrane currents associated with contraction in single myocytes isolated from guinea-pig ventricle. Br J Pharmacol 1988;94:500.
72. Terrar DA, Victory JGG. Isoflurane depresses membrane currents associated with contractions in myocytes isolated from guinea-pig ventricle. Anesthesiology 1988;69:742.
73. Blanck TJJ, Runge S, Stevenson RL. Halothane decreases calcium channel antagonist binding to cardiac membranes. Anesth Analg 1988;67:1032.
74. Nakao S, Hirat H, Kagawa Y. Effects of volatile anesthetics on cardiac calcium channels. Acta Anaesthesiol Scand 1989;33:326.
75. Bosnjak ZJ, Supan FD, Rusch NJ. The effects of halothane, enflurane and isoflurane on calcium currents in isolated canine ventricular cells. Anesthesiology 1991;74:340.
76. Sakmann B, Trube G. Conductance properties of single inwardly rectifying potassium channels in ventricular cells from guinea-pig heart. J Physiol (Lond) 1984;347:641.
77. Hille B. Ionic channels of excitable membranes. 2nd ed. Sunderland, MA: Sinauer Associates, Inc., 1992.
78. Sheu S-S, Korth M, Lathrup DA, Fozzard HA. Intra- and extra-cellular K^+ and Na^+ activities and resting membrane potential in sheep cardiac Purkinje strands. Circ Res 1980;47:692.
79. Vandenberg CA. Inward rectification of a potassium channel in cardiac ventricular cells depends on internal magnesium ions. Proc Natl Acad Sci USA 1987;84:2560.
80. Matsuda H, Saigusa A, Irisawa H. Ohmic conductance through the inwardly rectifying K channel and blocking by internal Mg^{2+}. Nature (Lond) 1987;325:156.
81. Matsuda H. Magnesium gating of the inwardly rectifying K^+ channel. Annu Rev Physiol 1991;53:289.
82. Ishihara K, Mitsuiye T, Noma A, Takano M. The Mg^{2+} block and intrinsic gating underlying inward rectification of the K^+ current in guinea-pig cardiac myocytes. J Physiol (Lond) 1989;419:297.

83. Luo C, Rudy Y. A model of the ventricular cardiac action potential: depolarization, repolarization, and their interaction. Circ Res 1991;68:1501.
84. Pancrazio JJ, Frazer MJ, Lynch C III. Barbiturate anesthetics depress the resting K^+ conductance of myocardium. J Pharmacol Exp Ther 1993; 265:358.
85. Backx PH, Marban E. Background potassium current active during the plateau of the action potential in guinea pig ventricular myocytes. Circ Res 1993;72:890.
86. Yue DT, Marban E. A novel cardiac potassium channel that is active and conductive at depolarized potentials. Pfluegers Arch 1988;413:127.
87. McAllister RE, Noble D. The time and voltage dependence of the slow outward current in cardiac Purkinje fibres. J Physiol (Lond) 1966;186:632.
88. Noble D, Tsien RW. The kinetics and rectifier properties of the slow potassium current in cardiac Purkinje fibres. J Physiol (Lond) 1968;195:185.
89. Noble D, Tsien RW. Outward membrane currents activated in the plateau range of potentials in cardiac Purkinje fibres. J Physiol (Lond) 1969; 200:205.
90. McDonald TF, Trautwein W. The potassium current underlying delayed rectification in cat ventricular muscle. J Physiol (Lond) 1978;274:217.
91. Gintant GA, Datyner NB, Cohen IS. Gating of delayed rectification in acutely isolated canine cardiac Purkinje myocytes. Biophys J 1985; 48:1059.
92. Sanguinetti MC, Jurkiewicz NK. Two components of cardiac delayed rectifier K^+ current: Differential sensitivity to block by class III antiarrhythmic agents. J Gen Physiol 1990;96:195.
93. Sanguinetti MC, Jurkiewicz NK. Lanthanum blocks a specific component of I_K and screens membrane surface charge in cardiac cells. Am J Physiol 1990;28:H1881.
94. Jurkiewicz NK, Sanguinetti MC. Rate-dependent prolongation of cardiac action potentials by a methanesulfonanilide class III antiarrhythmic agent. Circ Res 1993;72:75.
95. Shibasaki T. Conductance and kinetics of delayed rectifier potassium channels in nodal cells of the rabbit heart. J Physiol (Lond) 1987;387:227.
96. Tohse N, Kameyama M, Irisawa H. Intracellular Ca^{2+} and protein kinase C modulate K^+ current in guinea pig heart cells. Am J Physiol 1987; 22:H1321.
97. Tohse N. Calcium-sensitive delayed rectifier potassium current in guinea pig ventricular cells. Am J Physiol 1990;27:H1200.
98. Walsh KB, Begenisich TB, Kass RS. β-Adrenergic modulation in the heart: independent regulation of K and Ca channels. Pfluegers Arch 1988;411:232.
99. Tohse N, Kameyama M, Sekiguchi K, Shearman MS, Kanno M. Protein kinase C activation enhances the delayed rectifier potassium current in guinea-pig heart cells. J Mol Cell Cardiol 1990;22:725.
100. Tohse N, Nakaya H, Kanno M. α_1-adrenoceptor stimulation enhances the delayed rectifier K^+ current of guinea pig ventricular cells through the activation of protein kinase C. Circ Res 1992;71:1441.
101. Boyle WA, Nerbonne JM. A novel type of depolarization-activated K^+ current in isolated adult rat atrial myocytes. Am J Physiol 1991;29:H1236.
102. Snyders DJ, Knoth KM, Robards SL, Tamkun MM. Time-, state- and voltage-dependent block by quinidine of a cloned human cardiac channel. Mol Pharmacol 1992;41:332-339.

103. Snyders DJ, Tamkun MM, Bennett PB. A rapidly activating and slowly inactivating potassium channel cloned from human heart: functional analysis after stable mammalian cell culture expression. J Gen Physiol 1993;101:513.

104. Fedida D, Wible B, Wang Z, et al. Identity of a novel delayed rectifier current from human heart with a cloned K+ channel current. Circ Res 1993;73:210.

105. Jeck CD, Boyden PA. Age-related appearance of outward currents may contribute to developmental differences in ventricular repolarization. Circ Res 1992;71:1390.

106. Siegelbaum SA, Tsien RW, Kass RS. Calcium-activated outward current in calf Purkinje fibres. J Physiol (Lond) 1980;299:485.

107. Coraboeuf E, Carmeliet E. Existence of two transient outward currents in sheep Purkinje fibres. Pfluegers Arch 1982;392:352.

108. Josephson IR, Sanchez-Chapula J, Brown AM. Early outward current in rat single ventricular cells. Circ Res 1983;54:157.

109. Kenyon JL, Gibbons TW. 4-Aminopyridine and the early outward current of sheep cardiac Purkinje fibres. J Gen Physiol 1979;73:139.

110. Kukushkin NI, Gainullin RZ, Sosunov EA. Transient outward current and rate dependence action potential duration in rabbit cardiac ventricular muscle. Pfluegers Arch 1983;399:87.

111. Giles WR, Van Ginneken ACG. A transient outward current in isolated cells from the crista terminalis of rabbit heart. J Physiol (Lond) 1985;368:243.

112. Litovsky SH, Antzelevitch C. Transient outward current prominent in canine ventricular epicardium but not endocardium. Circ Res 1988;62:116.

113. Dukes ID, Morad M. The transient K+ current in rat ventricular myocytes: evaluation of its Ca^{2+} and Na+ dependence. J Physiol (Lond) 1991;435:395.

114. Kass RS, Scheuer T, Malloy KJ. Block of outward current in cardiac Purkinje fibres by injection of quaternary ammonium ions. J Gen Physiol 1982;79:1041.

115. Isenberg G. Cardiac Purkinje fibers. [Ca++]$_i$ controls the potassium permeability via the conductance components g_{K1} and g_{KL}. Pfluegers Arch 1977;371:77.

116. Callewaert G, Vereecke J, Carmeliet E. Existence of calcium-dependent potassium channel in the membrane of cow cardiac Purkinje fibres. Pfluegers Arch 1986;406:424.

117. Wang Z, Fermini B, Nattel S. Repolarization differences between guinea pig atrial endocardium and epicardium: evidence for a role of I_{to}. Am J Physiol 1991;260:H1501.

118. Po S, Snyders DJ, Baker R, Tamkun MM, Bennett PB. Functional expression of an inactivating potassium channel cloned from human heart. Circ Res 1992;71:732.

119. Fozzard HA, Hiraoka M. The positive dynamic current and its inactivation properties in cardiac Purkinje fibres. J Physiol (Lond) 1973;234:569.

120. Zygmunt AC, Gibbons WR. Calcium-activated chloride current in rabbit ventricular myocytes. Circ Res 1991;68:424.

121. Zygmunt AC, Gibbons WR. Properties of the calcium-activated chloride current in heart. J Gen Physiol 1992;99:391.

122. Nakayama T, Irisawa H. Transient outward current carried by potassium and sodium in quiescent atrioventricular node cells of rabbit. Circ Res 1985;57:65.

123. Escande D, Coulombe A, Faivre J-F, Deroubaix E, Corabeouf E. Two types of transient outward currents in adult human atrial cells. Am J Physiol 1987;249:H142.
124. Clark RB, Giles WR, Imaizumi Y. Properties of the transient outward currents in rabbit atrial cells. J Physiol (Lond) 1988;405:147.
125. Kukushkin NI, Gainullin RZ, Sosunov EA. Transient outward current and rate dependence of action potential duration in rabbit cardiac ventricular muscle. Pfluegers Arch 1983;399:87.
126. Gilmour RF Jr, Zipes DP. Different electrophysiological responses of canine endocardium and epicardium to combined hyperkalemia, hypoxia, and acidosis. Circ Res 1980;46:814.
127. Kimura S, Bassett AL, Kohya T, Kozlovskis PL, Myerburg RJ. Simultaneous recording of action potentials from endocardium and epicardium during ischemia in the isolated cat ventricle: relation of temporal electrophysiologic heterogeneities to arrhythmias. Circulation 1986;74:401.
128. Franz MR, Bargheer K, Rafflenbeul W, Haverich A, Lichtlen PR. Monophasic action potential mapping in human subjects with normal electrocardiograms: direct evidence for the genesis of the T-wave. Circulation 1987;75:379.
129. Hauswirth O, Noble D, Tsien RW. Separation of the pace-maker and plateau components of delayed rectification in cardiac Purkinje fibres. J Physiol (Lond) 1972;225:211.
130. DiFrancesco D. A new interpretation of the pace-maker current in calf Purkinje fibres. J Physiol (Lond) 1981;314:359.
131. DiFrancesco D. A study of the ionic nature of the pace-maker current in calf Purkinje fibres. J Physiol (Lond) 1981;314:377.
132. Maylie J, Morad M, Weiss J. A study of pace-maker potential in rabbit sino-atrial node: measurement of potassium activity under voltage-clamp conditions. J Physiol (Lond) 1981;311:161.
133. Maylie J, Morad M. Ionic currents responsible for the generation of pace-maker current in the rabbit sino-atrial node. J Physiol (Lond) 1984;355:215.
134. DiFrancesco D, Ducouret P, Robinson RB. Muscarinic moducation of cardiac rate at low acetylcholine concentrations. Science 1989;243:669.
135. DiFrancesco D, Ferroni D, Mazzanti A, Tromba C. Properties of the hyperpolarizing-activated current (I_f) in cells isolated from the rabbit sino-atrial node. J Physiol (Lond) 1986;377:61.
136. Irisawa H, Brown HF, Giles W. Cardiac pacemaking in the sinoatrial node. Physiol Rev 1993;73:197.
137. Yu H, Chang F, Cohen IS. Pacemaker current exists in ventricular myocytes. Circ Res 1993;72:232.
138. DiFrancesco D. Pacemaker mechanisms in cardiac tissue. Annu Rev Physiol 1993;55:455.
139. Ackerman MJ, Clapham DE. Cardiac chloride channels. Trends Cardiovasc Med 1993;3:23.
140. Bahinski A, Nairn AC, Greengard P, Gadsby DC. Chloride conductance regulated by cyclic AMP-dependent protein kinase in cardiac myocytes. Nature (Lond) 1989;340:718.
141. Harvey RD, Hume JR. Autonomic regulation of a chloride current in heart. Science 1989;244:983.
142. Harvey RD, Clark CD, Hume JR. Chloride current in mammalian cardiac myocytes—novel mechanism for autonomic regulation of action potential duration and resting membrane potential. J Gen Physiol 1990;95:1077.

143. Takano M, Noma A. Distribution of the isoprenaline-induced chloride current in rabbit heart. Pfluegers Arch 1992;420:223.
144. Ehara T, Matsuura H. Single-channel study of the cyclic AMP-regulated chloride current in guinea-pig ventricular myocytes. J Physiol (Lond) 1993;464:307.
145. Hagiwara N, Masuda H, Shoda M, Irisawa H. Stretch-activated anion currents of rabbit cardiac myocytes. J Physiol (Lond) 1992;456:285.
146. Sorota S. Swelling-induced chloride-sensitive current in canine atrial cells revealed by whole-cell patch clamp. Circ Res 1992;70:679.
147. Vassalle M. The relationship among cardiac pacemakers: Overdrive suppression. Circ Res 1977;41:269.
148. Mullins LJ. The generation of electric currents in cardiac fibers by Na/Ca exchange. Am J Physiol 1979;236:C103.
149. Coraboeuf E, Gautier P, Guiradou P. Potential and tension changes induced by sodium removal in dog Purkinje fibres: role of an electrogenic sodium-calcium exchange. J Physiol (Lond) 1981;311:605.
150. Mechmann S, Pott L. Identification of Na-Ca exchange current in single cardiac myocyte. Nature (Lond) 1986;319:597.
151. Egan TM, Noble D, Noble SJ, Powell T, Spindler AJ, Twist VW. Sodium-calcium exchange during the action potential in guinea-pig ventricular cells. J Physiol (Lond) 1989;411:639.
152. Blaustein MP. Sodium/calcium exchange and the control of contractility in cardiac muscle and vascular smooth muscle. J Cardiovasc Pharmacol 1988;12(suppl 5):S56.
153. Knighton DR, Zheng J, TenEyck LF, et al. Crystal structure of the catalytic subunit of cyclic adenosine monophosphate-dependent protein kinase. Science 1991;253:407.
154. Sperelakis N, Schneider JA. A metabolic control mechanism for calcium ion influx that may protect the ventricular myocardial cell. Am J Cardiol 1976;37:1079.
155. Reuter H, Scholz H. The regulation of the calcium conductance of cardiac muscle by adrenaline. J Physiol (Lond) 1977;264:49.
156. Bennett P, McKinney L, Begenisich T, Kass RS. Adrenergic modulation of the delayed rectifier potassium channel in calf cardiac Purkinje fibers. Biophys J 1986;49:839.
157. Egan TM, Noble D, Powell T, Twist VW, Yamaoka K. On the mechanism of isoprenaline- and forskolin-induced depolarization of single guinea-pig ventricular myocytes. J Physiol (Lond) 1988;400:299.
158. Yamawake N, Hirano Y, Sawanobori T, Hiraoka M. Arrhythmogenic effects of isoproterenol-activated C1 current in guinea-pig ventricular myocytes. J Mol Cell Cardiol 1992;24:1047.
159. Matsuda JJ, Lee H, Shibata EF. Enhancement of rabbit cardiac sodium channels by β-adrenergic stimulation. Circ Res 1992;70:199.
160. Ono K, Fozzard HA, Hanck DA. Mechanism of cAMP-dependent modulation of cardiac sodium channel current kinetics. Circ Res 1993;72:807.
161. Tada M, Kirchberger MA, Repke DI, Katz AM. The stimulation of Ca^{++} transport in cardiac sarcoplasmic reticulum by adenosine 3':5'-monophosphate-dependent protein kinase. J Biol Chem 1974;249:6174.
162. Kurihara S, Konishi M. Effects of β-adrenoceptor stimulation on intracellular Ca transients and tension in rat ventricle muscle. Pfluegers Arch 1987;409:427.

163. Fleming JW, Wisler PL, Watanabe AM. Signal transduction by G proteins in cardiac tissues. Circulation 1992;85:420.
164. Fedida D, Shimoni Y, Giles WR. α-Adrenergic modulation of the transient outward current in rabbit atrial myocytes. J Physiol (Lond) 1990;423:257.
165. Endoh M, Blinks JM. Actions of sympathomimetic amines on the Ca^{2+} transients and contractions of rabbit myocardium: reciprocal changes in myofibrillar responsiveness to Ca^{2+} mediated through α- and β-receptors. Circ Res 1988;62:247.
166. Terzic A, Pucéat M, Clément O, Scamps F, Vassort G. α-Adrenergic effects on intracellular pH and calcium and on myofilaments in single rat cardiac cells. J Physiol (Lond) 1992;447:275.
167. Gambassi G, Spurgion hA, Lakatta EG, Blank PS, Capogrossi MC. Different effects of α- and β-adrenergic stimulation on cytosolic pH and myofilament responsiveness to Ca^{2+} in cardiac myocytes. Circ Res 1992;71:870.
168. Pucéat M, Terzic A, Clément O, Scamps F, Vogel SM, Vassort G. Cardiac $α_1$-adrenoreceptors mediate positive inotropy via myofibrillar sensitization. Trends Pharmacol Sci 1992;13:263.
169. Brown AM. Regulation of the heartbeat by G protein-coupled ion channels. Am J Physiol 1990;259:H1621.
170. Birnbaumer L, Abramowitz J, Brown AM. Receptor-effector coupling by G proteins. Biochim Biophys Acta 1990;1031:163.
171. Brown AM, Birnbaumer L. Ion channels and their regulation by G protein subunits. Annu Rev Physiol 1990;52:197.
172. Yatani A, Codina J, Brown AM, Birnbaumer L. Direct activation of mammalian atrial muscarinic potassium channels by GTP regulatory protein Gk. Science 1987;235:207.
173. Yatani A, Codina J, Imoto Y, Reeves JP, Birnbaumer L, Brown AM. A G protein directly regulates mammalian cardiac calcium channels. Science 1987;238:1288.
174. Cerbai E, Klöckner U, Isenberg G. The α subunit of the GTP binding protein activates muscarinic potassium channels of the atrium. Science 1988;240:1782.
175. Horie M, Irisawa H. Rectification of muscarinic K^+ current by magnesium ion in guinea pig atrial cells. Am J Physiol 1987;253:10.
176. Horie M, Irisawa H. Dual effects of intracellular magnesium on muscarinic potassium channel current in single guinea-pig atrial cells. J Physiol (Lond) 1989;408:313.
177. Boyett MR, Kirby MS, Orchard CH, Roberts A. The negative inotropic effect of acetylcholine on ferret ventricular myocardium. J Physiol (Lond) 1988;404:613.
178. Hartzell HC, Fischmeister R. Effect of forskolin and acetylcholine on calcium current in single isolated cardiac myocytes. Mol Pharmacol 1987;32:639.
179. Matsuda JJ, Lee H-C, Shibata EF. Acetylcholine reversal of isoproterenol-stimulated sodium currents in rabbit ventricular myocytes. Circ Res 1993;72:217.
180. Vogel S, Sperelakis N. Blockade of myocardial slow inward current at low pH. Am J Physiol 1977;233:C99.
181. Fry CH, Poole-Wilson PA. Effects of acid-base changes on excitation-contraction coupling in guinea-pig and rabbit cardiac ventricular muscle. J Physiol (Lond) 1981;313:141.

182. Belardinelli L, Isenberg G. Isolated atrial myocytes: adenosine and acetylcholine increase potassium conductance. Am J Physiol 1983; 224:H734.
183. Belardinelli L, Giles WR, West A. Ionic mechanisms of adenosine's actions in pacemaker cells from the rabbit heart. J Physiol (Lond) 1988;405:615.
184. Kurachi Y, Nakajima T, Sugimoto T. On the mechanism of of activation of muscarinic K+ channels by adenosine in isolated atrial cells: involvement of GTP-binding proteins. Pfluegers Arch 1986;407:264.
185. Visentin S, Wu S-N, Belardinelli L. Adenosine-induced changes in atrial action potential: contribution of Ca and K currents. Am J Physiol 1990; 258:H1070.
186. Ramos-Franco J, Lo CF, Breitwieser GE. Platelet-activating factor receptor-dependent activation of the muscarinic K+ current in bullfrog atrial myocytes. Circ Res 1993;72:786.
187. Noma A. ATP-regulated K+ channels in cardiac muscle. Nature (Lond) 1983;305:147.
188. Trube G, Hescheler J. Inward-rectifying channels in isolated patches of the heart cell membrane: ATP-dependence and comparison with cell-attached patches. Pfluegers Arch 1984;401:178.
189. Gross GJ, Auchampach JA. Role of ATP dependent potassium channels in myocardial ischaemia. Cardiovasc Res 1992;26:1011.
190. de Weille JR. Modulation of ATP sensitive potassium channels. Cardiovasc Res 1992;26:1017.
191. Thornton JD, Thornton CS, Sterling DL, Downey JM. Blockade of ATP-sensitive potassium channels increases infarct size but does not prevent preconditioning in rabbit hearts. Circ Res 1993;72:44.
192. Venkatesh N, Stuart JS, Lamp ST, Alexander LD, Weiss JN. Activation of ATP-sensitive K+ channels by chromakalim: effects on cellular K+ loss and cardiac function in ischemic and reperfused mammalian ventricle. Circ Res 1992;71:1324.
193. Kim D, Clapham DE. Potassium channels in cardiac cells activated by arachidonic acid and phospholipids. Science 1989;244:1174.
194. Kameyama M, Kakei M, Sato R, Shibasaki T, Matsuda H, Irisawa H. Intracellular Na+ activates a K+ channel in mammalian cardiac cells. Nature (Lond) 1984;309:147.
195. Luk H-N, Carmeliet E. Na+-activated K+ current in cardiac cells: rectification, open probability, block and role in digitalis toxicity. Pfluegers Arch 1990;416:766.
196. Waldo AL, Wit AL. Mechanisms of cardiac arrhythmias. Lancet 1993; 341:1189.
197. Janse M. The premature beat. Cardiovasc Res 1992;26:89.
198. Zhou X, Guse P, Wolf PD, Rollins DL, Smith WM, Ideker RE. Existence of both fast and slow channel activity during the early stages of ventricular fibrillation. Circ Res 1992;70:773.
199. Levine JH, Morganroth J, Kadish AH. Mechanisms and risk factors for proarrhythmia with type I_a compared with I_c antiarrhythmic drug therapy. Circulation 1989;80:1063.
200. Gilat E, Nordin CW, Aronson RS. The role of reduced potassium conductance in generating triggered activity in guinea-pig ventricular muscle. J Mol Cell Cardiol 1990;22:619.

201. Zipes DP. The long QT interval syndrome: a rosetta stone for sympathetic related ventricular tachyarrhythmias. Circulation 1991;84:1414.
202. Fish FA, Prakash C, Roden DM. Suppression of repolarization-related arrhythmias in vitro and in vivo by low-dose potassium channel activator. Circulation 1990;82:1362.
203. Imaizumi Y, Giles WR. Quinidine-induced inhibition of a transient outward current in rabbit heart. Am J Physiol 1987;253:H704.
204. Castle NA. Bupivacaine inhibits the transient outward K^+ current but not the inward rectifier in rat ventricular myocytes. J Pharmacol Exp Ther 1990;255:1038.
205. Miura M, Ishide N, Oda H, Sakarai M, Shinozaki T, Takishima T. Spatial features of calcium transients during early and delayed afterdepolarizations. Am J Physiol 1993;265:H439.
206. Lakatta EG, Capogrossi MC, Kort AA, Stern MD. Spontaneous myocardial calcium oscillations: overview with emphasis on ryanodine and caffeine. Fed Proc 1985;44:2977.
207. Lakatta EG. Functional implications of spontaneous sarcoplasmic reticulum Ca^{2+} release in the heart. Cardiovasc Res 1992;26:193.
208. Karagueuzian HS, Katzung BG. Voltage-clamp studies of transient inward current and mechanical oscillations induced by ouabain in ferret papillary muscle. J Physiol (Lond) 1982;327:255.
209. Campbell RWF. Ventricular ectopic beats and non-substained ventricular tachycardia. Lancet 1993;341:1454.
210. Kuo C-S, Munakata K, Reddy CP, Surawicz B. Characteristics and possible mechanism of ventricular arrhythmia dependent on the dispersion of action potential durations. Circulation 1983;67:1356.
211. Rosenbaum DS, Kaplan DT, Kanai A, et al. Repolarization inhomogeneities in ventricular myocardium change dynamically with abrupt cycle length shortening. Circulation 1991;84:1333.
212. Kowey PR, Friehling TD, Sewter J, et al. Electrophysiological effects of left ventricular hypertrophy: effect of calcium and potassium channel blockade. Circ Res 1991;83:2067.
213. Lynch Jr JJ, Sanguinetti MC, Kimura S, Bassett AL. Therapeutic potential of modulating potassium currents in the diseased myocardium. FASEB J 1992;6:2952.
214. Bacaner MB, Clay JR, Shrier A, Brochu RM. Potassium channel blockade: a mechanism for suppressing ventricular fibrillation. Proc Natl Acad Sci USA 1986;83:2223.
215. Miura DS, Hoffman BF, Rosen MR. The effect of extracellular potassium on the intracellular potassium ion activity and transmembrane potential of beating canine cardiac Purkinje fibers. J Gen Physiol 1977;69:463.
216. Lee CO, Fozzard HA. Membrane permeability during low potassium depolarization in sheep cardiac Purkinje fibers. Am J Physiol 1979; 237:C156.
217. Cohen CJ, Fozzard HA, Sheu SS. Increase in intracellular sodium activity during stimulation in mammalian cardiac muscle. Circ Res 1982; 50:651-662.
218. Baumgarten CM, Singer DH, Fozzard HA. Intra- and extracellular potassium activities, acetylcholine and resting potential in guinea pig atria. Circ Res 1984;54:65.

219. Surawicz B. Role of potassium channels in cycle length dependent regulation of action potential duration in mammalian cardiac Purkinje and ventricular muscle fibres. Cardiovasc Res 1992;26:1021.
220. Yatani A, Brown AM. Rapid β-adrenergic modulation of cardiac calcium channel currents by a fast G protein pathway. Science 1989;245:71.
221. Hartzell HC, Méry PF, Fischmeister R, Szabo G. Sympathetic regulation of cardiac calcium current is due exclusively to cAMP-dependent phosphorylation. Nature (Lond) 1991;351:573.
222. Hartzell HC, Fischmeister R. Direct regulation of cardiac Ca^{2+} channels by G proteins: neither proven nor necessary? Trends Pharmacol Sci 1992; 13:380.
223. Gadsby DC, Wit AL, Cranefield PF. The effects of acetylcholine on the electrical activity of canine cardiac Purkinje fibers. Circ Res 1978;43:29.
224. Campbell DL, Rasmusson RL, Strauss HC. Ionic current mechanisms generating vertebrate primary cardiac pacemaker activity at the single cell level: an integrative view. Annu Rev Physiol 1992;54:279.

Lawrence A. Turner
Zeljko J. Bosnjak

Autonomic and Anesthetic Modulation of Cardiac Conduction and Arrhythmias

2

Concern that the electrophysiologic actions of anesthetics may alter impulse initiation and conduction in the heart and thereby affect the cardiac rhythm was raised following electrocardiographic studies demonstrating a high incidence of arrhythmias during the perioperative period. Subsequently many studies have examined anesthetic actions on the electrophysiologic properties of cardiac tissues in an attempt to understand how these actions in combination with other proarrhythmic factors, including myocardial ischemia and endogenously released or exogenously administered catecholamines may precipitate arrhythmias during anesthesia. Classically, arrhythmias are considered to be caused by abnormal impulse initiation, conduction, or both.[1] However, demonstration of any potential single mechanism generating arrhythmias in vivo is extremely difficult, and the perioperative management of arrhythmias in individual patients is rarely based on firm knowledge of the responsible mechanism and influence of simultaneous drug therapy. Despite this difficulty, there is increasing evidence that volatile anesthetics have potent electrophysiologic influences on arrhythmias owing to both altered impulse initiation and conduction.[2] This chapter will focus largely on our current knowledge of the actions of anesthetics and catecholamines in the cardiac conduction system with the purpose of summarizing what is known and unknown about their com-

Clinical Cardiac Electrophysiology: Perioperative Considerations
Edited by Carl Lynch III. J.B. Lippincott Company, Philadelphia, PA ©1994

bined electrophysiologic influences on processes that may generate arrhythmias during anesthesia.

AUTONOMIC INNERVATION OF THE HEART

Supraventricular components of the cardiac conduction system include the sinoatrial (SA) node, atrial muscle fibers, with perhaps preferential conduction pathways, and the atrioventricular (AV) node. There is profuse innervation of the atria with sympathetic postganglionic fibers, coursing with mixed cardiac nerves from cervicothoracic ganglia that are innervated by upper thoracic preganglionic fibers and postganglionic cholinergic fibers originating from collections of parasympathetic ganglia in specific locations of the atrial epicardial plexus.[3] There is moderate "sidedness" to this innervation, the sympathetic ganglia exhibiting slight right to left dominance on SA nodal automaticity, but with less differential enhancement of AV nodal conduction. Stimulation of preganglionic parasympathetic fibers has marked effects on SA nodal rate and AV conduction and is capable of beat-to-beat modulation of impulse initiation and conduction to the ventricles.[4] The compact AV node penetrates into the ventricles as the bundle of His, located at the top of the interventricular septum, which divides into cordlike right and fanlike left branches.[5] The upper bundle branches are insulated by fibrous tissue and lack junctions with working muscle fibers in the upper half of the ventricles.[6] The right bundle branch courses to the base of the right anterior papillary muscle, first activating endocardial muscle at the septal apex, and continues as the free running moderator band to arborize over the anterior endocardium. The outer margins of the fan-shaped left bundle branch continue as the anterior and posterior false tendons to join and spread over the papillary muscles of the mitral valve, forming two functional divisions. Additional internal fibers of the left bundle branch spread apically over the septum such that the endocardium is first activated by an interconnected network of Purkinje fibers overlying the distal third of the left ventricular cavity. Conduction to muscle occurs with a few milliseconds delay at multiple discrete, endocardial sites (Purkinje fiber-muscle junctions).[7,8] Ventricular activation thereafter proceeds across muscle fibers from endocardium to epicardium and in an apical to basal direction.

There is relatively less parasympathetic than sympathetic innervation of the ventricles, the former coursing apically in the subepicardium across the AV groove, then passing deep to the subendocardium with some separate innervation of the interventricular septum and perhaps the specialized conduction system. Sympathetic fibers also traverse the AV groove subepicardially, because they are interrupted by phenol at that level, but continue apically in the subepicardium where they may be damaged by ischemia, resulting in sympathetic efferent denervation apical to an infarct.[9] Both vagal and sympathetic efferent activity produce highly localized refractory period responses in the ventricular myocardium that may significantly influence dysrhythmogenesis.[9,10–13] The cellular basis for the substantial differences between the intrinsic action potential and electrophysiologic properties of fibers at each level of the conduction system involves differences in the passive membrane properties and contributions of various ionic currents to the action potential in different cell types, as well as structural differences in electrical connections between fibers.

Activation of cardiac α_1, β_1 and muscarinic receptors by epinephrine, norepinephrine, and acetylcholine have marked effects on cardiac electrical activity mediated by different signal transduction systems,[14] as shown in Figure 2-1. Formation of the β_1-agonist-receptor complex is coupled to generation of a stimulatory G protein, which rapidly and directly modulates Ca^{2+} channel activity, increasing intracellular Ca^{2+}, and increases production of cyclic AMP by adenylyl cyclase, which further modulates cellular function by phosphorylation of proteins regulating contraction and L-type Ca^{2+}, K^+, and Na^+ channel activity.[15] The electrophysiologic effects mediated by β_1-adrenoceptors vary substantially in different parts of the conduction system, often including enhanced automaticity, shortening of refractory periods, and facilitation of conduction, and appear to be associated with changes in almost all transmembrane ionic currents known to contribute to the action potential.[16] Activation of α_1-receptors may either depress or enhance automaticity and usually prolongs repolarization. These actions may be due in part to coupling to G proteins, which stimulate hydrolysis of membrane lipids by phospholipase C. While not contributing prominently to normal cardiac excitation-contraction coupling, generation of the intracellular second messengers inositol 1,4,5-triphosphate (a regulator of calcium release from intracellular stores) and diacylglycerol (an activator of protein kinase C) may modulate of Ca^{2+} fluxes and K^+ conductances. Activation of cardiac muscarinic cholinergic receptors coupled to an inhibitory G protein slows automaticity due to the pacemaker current (I_f) and antagonizes most of the β-adrenergic responses related to generation of cyclic AMP by adenylyl cyclase. This intracellular modulation of the β-receptor-transduction system, in combination with indirect modulation of norepinephrine release by inhibitory presynaptic muscarinic receptors on cardiac sympathetic nerve endings, accounts for the phenomenon of "accentuated antagonism" by which the actions of vagal efferent activity opposite to those of sympathetic-mediated responses (e.g., modulation of myocardial repolarization and vulnerability to fibrillation) may be accentuated depending on the degree of ongoing β-receptor activation.[12,17–19]

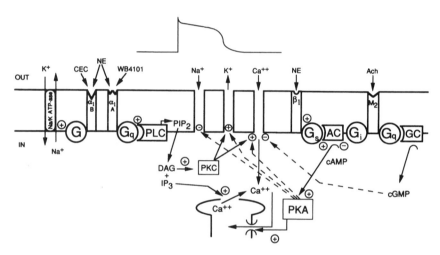

FIGURE 2-1. Simplified representation of G protein modulated voltage-dependent ionic currents in the heart. The regulation of adenylyl cyclase (AC) activity is under dual control of two G proteins: it can be stimulated by G_s protein and inhibited by G_i protein. Stimulation of β-adrenergic receptors generates activated $G_{s\alpha}$ to directly gate calcium, potassium, and sodium channels in the heart without involvement of cyclic AMP (not shown). Alternatively, $G_{s\alpha}$ may stimulate production of cyclic AMP via adenylyl cyclase. Following stimulation, adenylyl cyclase converts ATP into cyclic AMP, which in turn stimulates protein kinase A (PKA). The catalytic component of protein kinase A is liberated and able to phosphorylate and therefore modulate the functional activity of several channels. Cardiac α_1-adrenergic receptor stimulation activates the G_q protein that couples receptors to phospholipase C (PLC) and guanylyl cyclase (GC) (shown for acetylcholine (Ach) only). In addition, stimulation of the α_{1B}-receptor subtype activates the Na+-K+-ATPase. Stimulation of phospholipase C is responsible for production of diacylglycerol (DAG) and inositol 1,4,5-triphosphate (IP₃). The diacylglycerol activates protein kinase C (PKC), which phosphorylates ionic channels while inositol 1,4,5-triphosphate may act on the sarcoplasmic reticulum channel to release calcium. The α_1-receptor subtype blockers are chloroethylclonidine (CEC) and WB4101.

SUPRAVENTRICULAR IMPULSE INITIATION AND CONDUCTION

SA nodal automaticity is directly depressed by halothane, enflurane, and isoflurane in a manner antagonized by increasing extracellular Ca^{2+}.[20] These actions are consistent with anesthetic depression of T- and L-type Ca^{2+} currents in Purkinje and ventricular myocytes,[21,22] although no studies have directly examined anesthetic effects on the pacemaker current (I_f) underlying automatic phase 4 diastolic depolarization in SA nodal or Purkinje fibers. Similar slowing of the primary SA pacemaker is observed during anesthesia relative to conscious controls in the presence of pharmacologic autonomic blockade.[23] The direct actions of halothane on the action potentials of atrial muscle fibers[24,25] include depression of the plateau and slight prolongation of terminal repolarization while those of enflurane and isoflurane have not been reported. The effects of halothane, enflurane, and isoflurane anesthesia on specialized atrial and AV nodal refractory periods and AV nodal conduction time are influenced substantially by indirect changes in autonomic activity, are highly dependent on the specific model employed (paced or spontaneous), and suggest potential modest direct anesthetic effect increasing refractoriness and AV conduction delay that may be masked in vivo by other adrenergic- and reflex-mediated effects.[23,26–28] No study has directly examined anesthetic actions on the action potential characteristics of AV nodal cells.

Anesthetic depression of SA nodal automaticity in combination with changes in cardiac autonomic activity may be related to the emergence of subsidiary pacemaker function and atrial arrhythmias in patients without known heart disease. Normally the intrinsically more rapid rate of automatic phase 4 depolarization of the primary SA nodal pacemaker fibers maintains dominance over other potential sites of automaticity (atrial, AV junctional, and ventricular Purkinje fibers exhibiting spontaneous phase 4 diastolic depolarization) by a mechanism known as overdrive inhibition.[29] The consequence of more frequent depolarization of subsidiary pacemakers by the SA node is an increased Na^+ influx and K^+ efflux per unit of time. This produces relative activation of the electrogenic Na^+-K^+ pump and outward hyperpolarizing current, which increases the maximum diastolic potential (more negative) and reduces the slope of phase 4 depolarization of subsidiary automatic fibers. On cessation of overdrive, Na^+ influx decreases, intracellular Na^+ falls, pump current decreases, and the membrane potential during the pause gradually returns toward threshold with resumption of spontaneous activity at a low rate, which gradually increases to the rate present before overdrive. Thus, anesthetic depression of SA nodal automaticity, assuming less depression of automaticity or greater anesthetic inhibition of hyperpolarization by the Na^+-K^+ pump mechanism[30] in other portions of the conduction system, may permit escape of subsidiary pacemakers from SA nodal dominance.

Early studies of myocardial sensitization by anesthetics to the dysrhythmogenic effects of epinephrine[31,32] noted the occurrence of abnormal supraventricular foci in atrial and electrocardiographic recordings just prior to onset of ventricular arrhythmias. Atlee and Malkinson[33] characterized these shifts of activity (subsidiary atrial or wandering pacemaker, atrial ectopy, AV dissociation) as arrhythmias of "development" and demonstrated that they occur at lower doses of epinephrine than ventricular arrhythmias both with the more sensitizing agent halothane as well as with enflurane and isoflurane.[34] In superfused atrial preparations halothane, enflurane, and isoflurane produce a noncompetitive rightward shift in the positive chronotropic response of the SA node to epinephrine and isoproterenol.[35] In addition, in atrial preparations perfused through the SA nodal artery, both low concentrations of epinephrine and norepinephrine, either with or without halothane, frequently shift the site of earliest activation to subsidiary atrial pacemaker locations along the sulcus terminalis.[36] These shifts, suggesting that catecholamines may augment automaticity of subsidiary atrial pacemakers more so than the SA node, also occur in a chronically instrumented dog model.[37] The preliminary results of these studies indicate that halothane "sensitizes" the heart to atrial arrhythmias compared to awake control responses, increasing the degree or severity of shifts to subsidiary sites, including the His bundle, for a given epinephrine dose, and suggest that baroreceptor-mediated vagal inhibition of the SA node contributes to such shifts in the conscious state and at lower compared to higher halothane concentrations.

The role of atrial arrhythmia development as a harbinger of or contributor to generation of more severe arrhythmias (ventricular ectopy, bigeminy, and ventricular tachycardia) with anesthetics and catecholamines is unknown. Simultaneous β-adrenergic-mediated effects of catecholamines on AV conduction may include abbreviation of AV nodal conduction time at rapid atrial rates and shortening of the AV nodal functional refractory period, the minimum coupling interval or most premature response conducted to the His bundle.[38–40] These actions of epinephrine during anesthesia may result in conduction of irregularly coupled supraventricular activity through the AV node and the His-Purkinje system to the myocardium earlier during the relative refractory period than expected without adrenergic activation. Moe et al.[39] observed in pentobarbital-anesthetized dogs that actions of epinephrine on AV nodal refractoriness and conduction of premature atrial responses could produce aberrant ventricular conduction at rapid pacing rates in animals which usually only exhibit functional bundle branch block at low pacing rates. The importance of this "filtering" function of the AV node in preventing premature excitation of the ventricles during the vulnerable period is exemplified in certain animal models (neonatal goats, pigs) in which the lack of normal AV nodal conduction delays is associated with ventricular tachycardia or fibrillation induced by premature atrial stimuli.[41] However, no studies of anesthetic-catecholamine arrhythmias with sensitizing agents have examined the relationship between the prematurity of supraventricular impulses conducted to the His bundle and onset of ventricular ectopic beats with threshold doses of epinephrine.

VENTRICULAR IMPULSE INITIATION AND CONDUCTION

Ventricular Automaticity

The electrophysiologic actions of volatile anesthetics on ventricular tissues have been related to depression of transsarcolemmal ionic currents, particularly the inward Ca^{2+} current underlying the plateau phase following depolarization,[21,22,42–45] but also may involve inhibition of Na^+ current during depolarization[46,47] and various outward K^+ currents during repolarization.[48,49] Anesthetic inhibition of Ca^{2+} currents, which serve as the trigger for release of intracellular Ca^{2+} sequestered in the sarcoplasmic reticulum and reduction of Ca^{2+} stored in the SR,[50] is well established to be largely responsible for anesthetic-induced contractile depression[51] and may similarly underlie important antiarrhythmic actions on abnormal automatic activity associated with intracellular Ca^{2+} "overload." Cellular Ca^{2+} homeostasis[29] may be disturbed by excessive loading related to increased rate, producing increased Ca^{2+} influx per unit of time and increased inward Ca^{2+} current resulting from catecholamine exposure. In addition, Ca^{2+} overload may also be produced by diminished extrusion of Ca^{2+}, secondary to depression of the energy-dependent electrogenic Na^+-K^+ pump, accumulation of intracellular Na^+ ions, and reduced exchange of intracellular Ca^{2+} for extracellular Na^+, as occurs in digitalis toxicity and ischemia. Abnormal oscillatory Ca^{2+} release from the overloaded SR in early diastole produces aftercontractions, the transient inward current, and afterdepolarizations, which increase in amplitude with increased stimulation rate and may ultimately initiate bursts of triggered activity.

Early studies demonstrated anesthetic inhibition of isoproterenol-stimulated slow action potential responses, largely carried by Ca^{2+} ions, in K^+-depolarized papillary muscle preparations.[42,43] Subsequent studies utilizing voltage clamp techniques in single ventricular and Purkinje fibers have shown similar depression of Ca^{2+} channel currents by equianesthetic concentrations of halothane, enflurane, and isoflurane.[21,22] These actions are associated with dose-related reductions (except isoflurane) of the peak intracellular Ca^{2+} transient generated by Ca^{2+} influx and SR Ca^{2+} release,[52,53] which suggests that anesthetic actions would be expected to oppose or reduce Ca^{2+} overload.

The ability of volatile anesthetics to depress Ca^{2+} currents may be related to important antiarrhythmic actions on triggered dysrhythmias in several experimental models including digitalis intoxication, catecholamine stimulation, and ischemia. Reynolds and Horne,[54] Morrow,[55] and Gallagher and McClernan[56] have demonstrated antagonism by halothane of ventricular tachyarrhythmias because of digitalis toxicity in vivo, while enflurane and isoflurane may have similar antiarrhythmic actions.[57] Gallagher et al.[58] examined the actions of halothane on ouabain-intoxicated Purkinje fibers exhibiting typical afterdepolarization-related triggered arrhythmias potentiated by increasing extracellular Ca^{2+} and stimulation rate. Halothane reduced the amplitude of delayed afterdepolarizations in a manner antagonized by increasing Ca^{2+}. In enzymatically dispersed canine myocytes, Freeman and Li[59] similarly demonstrated antiarrhythmic actions of halothane on the amplitude of delayed afterdepolarizations, the accompanying abnormal intracellular Ca^{2+} transients, and triggered responses induced by isoproterenol. Monitoring cell length of single stimulated rat myocytes, Zuckerman and Wheeler[60] reported inhibition by halothane of spontaneous interbeat waves and late aftercontractions caused by isoproterenol and norepinephrine, as well as early aftercontractions induced by these agonists or phenylephrine. Halothane and isoflurane have also been shown to depress triggered afterdepolarizations induced by epinephrine in human atrial fibers obtained at the time of cardiac surgery.[61] The inhibition by halothane of these types of sympathomimetic-induced arrhythmogenic responses is probably a result of a reduction of the inward Ca^{2+} current and the cellular Ca^{2+} "load." Alteration and depletion of intracellular stores may also contribute. Subendocardial Purkinje fibers, which survive overlying the ischemic region of 24-hr-old infarcted canine hearts, are well known to exhibit triggered arrhythmias as a result of delayed afterdepolarizations potentiated by interventions which increase Ca^{2+} entry and loading of the sarcoplasmic reticulum (increased drive rate, elevated extracellular Ca^{2+}, and catecholamines).[62] Studies from this laboratory indicate that halothane reduces the rate of spontaneous arrhythmias originating in the ischemic region of infarcted hearts[63] and that halothane and enflurane, but not isoflurane, increase the number of drive stimuli required to induce triggered activity in ischemic fibers.[64] None of the anesthetics tested decreased the rate of slow (40 to 50 beats/min) sustained automaticity of Purkinje fibers derived from infarcts. Recent in vivo studies of spontaneous ventricular tachycardia in dogs 24 hr following experimental infarction demonstrate that halothane anesthesia tends to reduce the overall heart rate and markedly decreases the percentage of ventricular ectopic beats (97% to 35%) relative to the conscious control state.[56] This substantial antiarrhythmic action in ischemia could in part reflect direct anesthetic inhibition of abnormal automaticity as well as indirect anesthetic effects reducing cardiac sympathetic efferent activity[65,66] and plasma catecholamine concentrations.[67]

A number of in vitro studies have examined anesthetic actions on normal automaticity in ventricular tissues, although less is known about their actions on automaticity in vivo. Reynolds et al.[68] observed depression of spontaneous phase 4 diastolic depolarization by halothane in Purkinje fibers both with or without epinephrine, although in our laboratory, halothane, isoflurane, and particularly enflurane were found to modestly increase spontaneous rate (from about 30 to <60 beats/min) in Purkinje fibers previously exposed to high (2 or 15 μM) epinephrine concentrations.[69] The actions of enflurane, and less so halothane and isoflurane, were associated with shortened recovery time following overdrive, consistent with a report of Pratila et al.[30] that enflurane may directly increase Purkinje fiber automaticity by an action suppressing the normal mechanism of overdrive inhibition. Logic and Morrow[70] examined the effects of halothane on the emergence of ventricular pacemakers in vivo utilizing supramaximal vagal stimulation to produce complete heart block. Relative to basal pentobarbital anesthesia, halothane prolonged the escape time and slowed the rate of idioventricular pacemakers. However, assessment of anesthetic effects on the automaticity of ventricular escape pacemakers may be complicated by cholinergic depression of automaticity in the proximal bundle branches[71] and probably adrenergically mediated enhancement of automaticity due to baroreflex activation during AV block. No studies have examined direct anesthetic actions on idioventricular pacemaker function following experimental complete heart block or on phenomena such as "overdrive excitation." The latter occurs with rapid ventricular pacing during norepinephrine infusions and may represent induction of triggered activity, with acute AV block in pentobarbital-anesthetized animals.[72]

Ventricular Conduction and Refractoriness

The role of abnormal conduction during anesthesia in the nonischemic heart, leading presumably to facilitation of reentrant ventricular arrhythmias, is not clear. Reentry generally requires a site of unidirectional block,[73] which may be anatomical or functional owing to specific electrical properties of the tissues involved, and sufficiently slow conduction over a fixed or variable circuitous pathway of adequate length to permit reexcitation of tissues proximal to the site of block after their refractory period has passed. Propagation is dependent on local current flow from a source, usually an activated cell generating the fast inward Na^+ current in normally polarized Purkinje and ventricular myocytes, sufficient to produce a threshold voltage drop across the capacitance of connected fibers (the "sink"). The more depolarized the membrane potential, either as a consequence of disease, drug-induced phase 4 depolarization, or premature stimulation during early repolarization, the smaller the peak inward current and the conduction velocity. In addition, the degree of intercellular coupling through low resistance gap junctions varies with the specific geometry of cellular connections (divergence, convergence) and fiber orientation such that conduction velocity is normally two to three times faster in the direction of the long axis of myocardial fibers than in a direction perpendicular to the fibers.[74] The greater resistance to current flow due to cellular uncoupling, as may occur during ischemia, acidosis, and Ca^{2+} overload, the lower the conduction velocity. Conduction velocity in the normal ventricle varies from >2 m/sec in the false tendons, which have a high density of gap junctional proteins, to about 1.5 m/sec in the Purkinje layer overlying the papillary muscles, to <0.05 m/sec across the Purkinje fiber muscle junctions,[7,75] and then 0.5 to 0.8 m/sec in ventricular muscle; it is faster in longitudinal than transverse directions relative to fiber orientation. While propagation is continuous at a macroscopic level, conduction may actually be discontinuous at sites of increased intercellular resistance, such as the Purkinje fiber muscle junctions, at which the propagating wavefront nearly stops for the 3 to 5 msec required to excite endocardial muscle.

Anesthetic Effects

Hauswirth[24] investigated the direct actions of 1 and 2% halothane on Purkinje and ventricular fibers and reported larger decreases of action potential duration (APD) in the former than in the latter. There was also a reduction of the rate of phase 0 depolarization (V_{max}) at higher concentrations in both tissues and prolongation of conduction times in Purkinje fibers, suggesting that anesthetics may depress conduction and shorten refractory periods in ventricular tissues. Compared to conscious control dogs, halothane, enflurane, and isoflurane slightly (5 to 10%) and similarly prolong ventricular conduction intervals measured by His bundle recordings, but with little added depression at high compared to lower concentrations.[23,26,76–79] The prolongation of conduction intervals is not rate dependent.[78,80] In Purkinje fibers the depression of conduction velocity by high concentrations of halothane may be related to reduction of the peak inward Na^+ current,[46,47] because the anesthetic produces proportional reduction of V_{max} and (conduction velocity)^2 as predicted by cable theory.[81] Additional increases in longitudinal resistance,[82] in part due to depression of cell-to-cell coupling at cardiac gap junctions,[83–85] may contribute to more marked depression of conduction at Purkinje fiber-muscle junctions,[81] sites of intrinsically poor cell coupling.[7,75] In contrast to Purkinje fibers, a similar study of conduction in ventricular muscle fibers[86] indicated that halothane and enflurane depress conduction velocity with less influence on V_{max}, again suggesting an important anesthetic influence on cell-to-cell coupling that could potentially affect anisotropic propagation in cardiac muscle, the variation of conduction velocity with the direction of propagation. The effects of equianesthetic concentrations of halothane, enflurane, and isoflurane on conduction in Purkinje fibers, across the Purkinje-muscle junctions, and through the myocardium on peak inward Na^+ current during phase 0 and on intercellular resistance or cell-to-cell coupling have not been reported.

Under steady-state conditions (constant rate), APD and refractory periods are longer in certain regions of the His-Purkinje system (the false tendons) than in the endocardium, longer in the endocardium than in the epicardium, and longer in the epicardium at the apex than at the base of the heart.[39,87–90] The refractory periods of the bundle branches shorten more than those of the myocardium at a higher steady-state rate[88] and both Purkinje and myocardial fibers exhibit alternans of APD (alternating short and long durations which are damped over time) at rapid stimulation rates and after abrupt changes in rate before reaching steady-state values.[91,92] In addition, at high compared to low steady-state rate, there is decreased dispersion of myocardial refractory periods,[93,94] a major determinant of vulnerability to fibrillation during the relative refractory period.[95] Finally, the refractoriness of the myocardium can also be influenced importantly by mechanical factors, such as volume loading and dilation, which can heterogeneously shorten refractory periods and increase vulnerability to induction of arrhythmias.[96] Therefore, the significance of anesthetic shortening of refractory periods in ventricular tissues and their potential relationship to the occurrence of reentry may involve complex indirect as well as direct effects of anesthesia on heart rate, cardiac autonomic efferent activity, and hemodynamic changes that may not be readily assessed in vitro.

At a constant stimulation rate, volatile anesthetics decrease Purkinje fiber APD in the false tendons more so than in apical fibers exhibiting shorter APD, and at equianesthetic concentrations the actions of enflurane and isoflurane decrease false tendon fiber APD more so than halothane, as shown in Figure 2-2.[97] Although the actions of enflurane and isoflurane on refractoriness in the His-Purkinje system in vivo have not been determined, halothane was found to decrease refractoriness in the bundle branches at high compared to lower concentrations and more so at low (125 beats/min) than at high (175 beats/min) paced rate.[78] Myocardial fibers are less sensitive than Purkinje fibers to anesthetic effects on refractory characteristics,[24] and the comparative effects

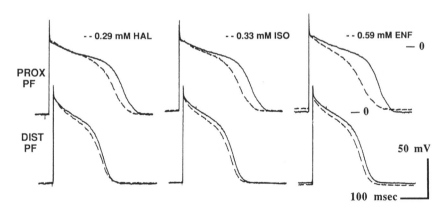

FIGURE 2-2. Simultaneously recorded single proximal (*PROX*) and distal (*DIST*) Purkinje fiber (*PF*) action potentials from the dog under control conditions (*solid tracings*) and after exposure (*dashed tracings*) to halothane (*HAL*), isoflurane (*ISO*), and enflurane (*ENF*). Each anesthetic produces a larger decrease of proximal than distal fiber action potential duration. (Reproduced with permission from Polic S, Bosnjak ZJ, Marijic J, Hoffmann RG, Kampine JP, Turner LA. Actions of halothane, isoflurane and enflurane on the regional action potential characteristics of canine Purkinje fibers. Anesth Analg 1991;73:603.)

of volatile anesthetics on the APD of myocardial fibers are not well established. Hauswirth[24] noted that 2% halothane decreased APD and the effective refractory period of sheep myocardial fibers and Lynch et al.[42] noted shortening of APD with 3% halothane in the guinea pig. Enflurane at high concentrations (3%) decreases APD[43] while isoflurane increases APD in guinea pig papillary muscles.[98] No studies have systematically examined actions of volatile anesthetics on myocardial repolarization in the normal heart in vivo compared to the awake control state. Han and Moe[95] originally observed in dogs following cardiac denervation that chloroform added to basal pentobarbital anesthesia prolonged the duration and increased the dispersion of myocardial refractory periods. Denniss et al.[99] reported that 2% halothane increased myocardial effective refractory periods relative to awake control values, without a change in QT intervals or vulnerability to induction of tachyarrhythmias by programmed stimulation techniques. Similar refractory period prolongation by halothane was observed in the nonischemic zone of animals with chronic (1- to 3-week) infarction and is a major mechanism responsible for reduction of the inducibility of ventricular tachycardia in this model.[56,99–101] Recent studies clearly indicate that halothane, isoflurane, and enflurane have important direct effects on myocardial repolarization as manifested by prolongation of the QT or QT_c intervals in animal models following pharmacologic autonomic nervous system blockade, and similar actions have been reported in human studies during anesthesia.[102,103] The mechanism underlying anesthetic actions on the uniformity of myocardial repolarization, reflected by the QT interval, is not known but may involve changes in activation time due to slowing of conduction in the Purkinje system and myocardium as well as potential differences of anesthetic actions on repolarization of myocardial fibers in different regions of the heart.

Autonomic Effects

The actions of catecholamines in the ventricles are different in the conduction system and myocardium and include effects mediated by activation of both α- and β-adrenergic receptors. Norepinephrine and epinephrine, acting at β-adrenergic receptors, have little effect on conduction at constant heart rate in Purkinje fibers either in vivo or in vitro, except perhaps to improve conduction in depressed fibers.[104,105] Conduction across the Purkinje-muscle junctions and in the myocardium is rapidly enhanced by β-adrenergic receptor activation because of phosphorylation of intracellular proteins by cyclic AMP-dependent protein kinases and increased gap junctional conductance.[106–109] β-Receptor activation and cyclic AMP-dependent phosphorylation may also modulate peak inward Na^+ current, and potentially conduction, in a voltage-dependent fashion.[15,110] Catecholamines, acting at α-adrenergic receptors, are not known to modulate Purkinje fiber conduction velocity, except in the presence of halothane.[111] One preliminary report suggests that high concentrations of phenylephrine may produce cell-to-cell uncoupling between paired myocytes,[112] although α-adrenergic effects on myocardial conduction velocity have not been reported. In Purkinje fibers at constant pacing rate, β-adrenergic receptor activation decreases while α-adrenergic receptor activation increases APD.[113] Figure 2-3 illustrates effects of the agonists isoproterenol and phenylephrine as well as those of norepinephrine on the APD of Purkinje fibers isolated from proximal and distal regions of the ventricular conduction system. In the myocardium, β-adrenergic receptors mediate a decrease of APD and effective refractory period, which may be antagonized by acetylcholine or vagal stimulation.[12,114] The β-adrenergic effects of catecholamines shortening APD in Purkinje and myocardial fibers are associated with increases of inward Ca^{2+} current, as well as marked increases in outward K^+ current (the delayed rectifier, I_K) and Cl^- currents[16,115] shortening APD. In contrast, α-adrenergic agonists (phenylephrine) prolong refractory periods in the conduction system but not in the myocardium.[116] α-Adrenergic prolongation of Purkinje fiber APD is thought to involve inhibition of outward K^+ currents.[117]

The actions of endogenously released and infused catecholamines have important influences on cardiac conduction, refractory periods, and vulnerability to fibrillation during premature stimulation of the

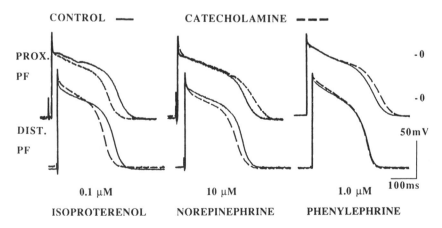

FIGURE 2-3. Simultaneously recorded single proximal (*PROX*) and distal (*DIST*) Purkinje fiber action potentials from the dog under control conditions (*solid tracings*) and after exposure (*dashed tracings*) to isoproterenol, norepinephrine, and phenylephrine.

ventricles. In Purkinje fibers, norepinephrine prolongs the conduction time of premature impulses[118] and sympathetic efferent activity, by simultaneously activating α- and β-adrenergic receptors, produces nonuniform changes in refractory periods of the His-Purkinje system in vivo as determined by premature stimulation of the His bundle.[119] Stimulation of the left stellate ganglion produces nonuniform shortening of myocardial refractory periods, QT interval prolongation, nonuniform conduction in the relative refractory period, and increased vulnerability to fibrillation by a second premature stimulus.[95,120,121] High dose infusions of epinephrine and norepinephrine (2 to 3 µg/kg/min) produce a biphasic response in the fibrillation threshold, initially (1 to 2 min) increasing vulnerability in association with increased dispersion and shortening of myocardial refractory periods. During continued infusion (5 to 10 min) the fibrillation threshold decreases with reduction of the dispersion of myocardial refractory periods and sustained refractory period shortening. Low dose infusions of norepinephrine (0.2 to 0.5 µg/kg/min) consistently increase vulnerability in a manner antagonized by bilateral vagal stimulation,[18] the latter probably being a result of antagonism of adrenergically mediated decreases of myocardial refractoriness.[12,19] The combination of acute coronary artery occlusion with left stellate stimulation produces a high incidence of fibrillation due to reentry and is widely utilized as a model for evaluation of the antiarrhythmic properties of pharmacologic agents.[122] Recent studies[123,124] clearly demonstrate important arrhythmogenic and antiarrhythmic influences of cardiac sympathetic and vagal efferent activity on the onset of fibrillation and sudden death in a conscious canine model of chronic infarction (4 weeks) with exercise and superimposed transient ischemia. The results in this model clearly implicate a role of adrenergic tone on induction of fibrillation during stress with ischemia as well as a role of baroreflex vagal modulation of rate and ventricular electrophysiologic properties. Other studies of the mechanism of altered fibrillation threshold by the narcotic agonist fentanyl also indicate that its reduction in vulnerability is related to a combination of withdrawal of cardiac sympathetic activity and increased vagal efferent activity during hemorrhagic stress and following acute coronary artery occlusion.[125,126]

Anesthetic-Catecholamine Interaction

The ability of some anesthetics to reduce the dose of epinephrine, usually measured as an infusion rate or product of rate and duration (range 0.5 to 10 µg/kg/min for 2 to 3 min) required to induce ventricular arrhythmias (typically ≥4 premature contractions/15 sec[127]), is a classic example of an adverse drug interaction. The electrophysiologic basis for this phenomenon, referred to as "sensitization," as well as the mechanism(s) generating the arrhythmias, remains unknown. Although the arrhythmogenic dose of epinephrine (ADE) in the conscious canine model is about 35 µg/kg, and about 5 µg/kg during 1.25 minimal alveolar concentration (MAC) halothane anesthesia,[128] the corresponding arrhythmogenic plasma concentrations (APCs) of epinephrine with threshold doses in the "awake" control state have not been reported. Representative doses and APCs of epinephrine (assumed as free base) in neurally intact ventilated dogs with different types of anesthesia are shown in Table 2-1. These studies indicate that the epinephrine required to produce this particular arrhythmia endpoint (≥4 premature ventricular contractions/15 sec) varies over about a 10- to 20-fold range with different anesthetics. Other reports indicate that the range of doses producing different types of arrhythmias, e.g., atrial arrhythmias versus premature ventricular contractions versus ventricular bigeminy, tachycardia, or fibrillation, is also reduced with more sensitizing agents,[33,34] although the APCs of epinephrine for less and more severe arrhythmias have not been reported.

Several recent studies indicate that the phenomenon of sensitization to the arrhythmogenic effects of catecholamines occurs at remarkably low anesthetic concentrations. Hayashi et al.[132] have shown that halothane, relative to a nonsensitizing etomidate basal anesthetic,[133] produces a dose-related 4-fold reduction of APC of epinephrine at halothane concentrations of 0 to 0.5 vol%, with little or no additional reduction at higher concentrations. The same group[131] reported about an 8-fold reduction of APC by thiopental, again relative to basal etomidate anesthesia, at plasma thiopental concentrations up to about 50 µg/ml, a

TABLE 2-1. ARRHYTHMOGENIC DOSES (ADE) AND PLASMA CONCENTRATIONS (APC) OF EPINEPHRINE DURING ANESTHESIA IN THE DOG

Anesthetic	ADE (µg/kg/min)	APC ng/ml	APC µM	Reference
Thiopental	0.8	11	0.058	131
Halothane (1.1%)	2.2	39	0.21	129
Halothane (1.1%)	2.7	48	0.26	132
Enflurane (2.75%)	5.2	100	0.55	130
Enflurane (2.75%)	11.4	206	1.13	129
Etomidate	10.7	222	1.21	131
Isoflurane (1.5%)	9.8	207	1.13	130
Pentobarbital	15.3	296	1.62	129

concentration estimated to be equivalent to about 1 MAC of inhalational anesthesia. These studies suggest that 1) the direct electrophysiologic actions of volatile anesthetics on cardiac tissues, including their actions on transmembrane potentials and ionic currents, automaticity, and conduction which generally occur at concentrations ≥0.5 MAC, may be less important in the phenomenon of sensitization than neural influences at low anesthetic concentrations and 2) that direct anesthetic effects may only influence the underlying mechanism(s) generating arrhythmias at higher concentrations. Recent reports indicate that the central α_2-agonist dexmedetomidine (intravenous) and the imidazoline α_2-antagonist atipamezole (intracisternally), agents that alter the balance of central vagosympathetic outflow, substantially (2- to 3-fold) increase and decrease, respectively, the APC of epinephrine during halothane anesthesia.[134,135] These studies strongly suggest that reflex-mediated changes in cardiac autonomic efferent activity have an important influence on the arrhythmic doses of epinephrine during anesthesia and probably also in the awake control state. However, the effect of thiopental may be related to the specific barbiturate effect on the resting potassium conductance.[135a]

The characteristics of catecholamine-anesthetic ventricular arrhythmias during anesthesia have been investigated in a variety of models. Early studies in the canine heart-lung preparation[136] indicated that induction of ventricular tachycardia requires the combined presence of halothane, epinephrine, and elevation of aortic pressure, without production of arrhythmias by any two of these elements alone, with sensitivity of the arrhythmias to β-adrenergic (pronethalol) but not to α-adrenergic (phenoxybenzamine) blockade. In isolated cat hearts,[137] halothane added to epinephrine produces fusion complexes between atrial P waves and a ventricular focus with development of bigeminal arrhythmias only on elevation of the ventricular pressure. There is also a positive correlation between the prematurity of bigeminal complexes and the left ventricular pressure of the preceding normal beat in halothane-anesthetized vagotomized dogs, suggesting either a pressure-dependent automatic focus or a reentrant circuit in which the returning impulse is conducted faster with shorter coupling intervals at higher intraventricular pressure.[31]

Initial studies using His bundle recordings[138,139] during cyclopropane-epinephrine arrhythmias in vagotomized dogs demonstrated that epinephrine shifts the supraventricular pacemaker to the His bundle and that the onset of bigeminy is characterized by early activation of the septal myocardium underneath the His bundle electrode during the premature complex. The role of the His bundle in maintenance of stable bigeminal rhythms was also explored by Sasyniuk and Dresel.[140] Destruction of the His bundle and proximal bundle branches by formalin injection abolished bigeminal rhythms, so that only multifocal ventricular tachycardia could be produced at higher epinephrine doses. Cauterizing the AV node, which maintained a normal ventricular activation sequence, did not prevent development of bigeminal rhythms and only moderately increased the epinephrine dose. A similar pattern of early septal activation during the first premature complex is observed in His bundle recordings with epinephrine and halothane[141] or isoflurane (M. Vicenzi, personal communication). Other characteristics of halothane-epinephrine arrhythmias are more controversial. Zink et al.[142] examined the role of pressure and heart rate in thiopental-halothane-anesthetized vagotomized dogs and found that either increasing atrial pacing rate or elevation of blood pressure by aortic occlusion consistently induced bigeminy with fixed coupling intervals during subthreshold infusions of epinephrine. In addition, increasing atrial pacing rate could both induce and "overdrive" the arrhythmia at higher rate, suggesting a reentrant mechanism rather than a triggered arrhythmia, and sustained bigeminy could be converted to sinus rhythm by stimulation of the right vagus nerve during atrial pacing. However, other studies in neurally intact halothane-anesthetized dogs (without thiopental) suggest no pressure dependence based on the observation that artificially reducing the blood pressure response with nitroprusside does not elevate the arrhythmogenic dose of epinephrine.[143] On the other hand, Hayashi et al.[144] found that elevating blood pressure with angiotensin II largely substitutes for the pressure elevation caused by phenylephrine in the production of arrhythmias by isoproterenol with halothane. In the latter study, rapid atrial pacing did not substitute for the presence of isoproterenol in production of arrhythmias by phenylephrine, suggesting that increased rate may not play an important role in the halothane-epinephrine interaction.

Studies utilizing relatively selective β- and α-adrenergic receptor antagonists have clarified the adrenergic receptor mechanisms by which epinephrine induces ventricular arrhythmias during halothane anesthesia. Maze and Smith[143] examined the relative contribution of cardiac β_1- and α_1-adrenergic receptor activation on the ADE during halothane anesthesia and reported that the threshold dose was elevated 13-fold by the α_1-antagonist prazosin but only an estimated 5-fold by β_1-adrenergic blockade with metoprolol. Subsequent studies by the same group[145] demonstrated a strong correlation between the individual ADE during halothane anesthesia and responsiveness to the pressor effects of phenylephrine, but no correlation between ADE and responsiveness to the positive chronotropic effects of isoproterenol. In addition, α_1-adrenergic blockade (droperidol or doxazosin) produced dose-related elevation of the ADE with halothane, strongly suggesting an important contribution of cardiac α_1-adrenergic receptor activation to induction of halothane-epinephrine arrhythmias.[146] Finally, Hayashi et al.[144] have elegantly demonstrated that full expression of catecholamine-induced arrhythmias during halothane anesthesia requires activation of both α_1- and β-adrenergic receptors in the heart and that the adrenergic mechanism involves a synergistic interaction between both receptors as assessed by simultaneous administration of the agonists phenylephrine and isoproterenol.

While it has long been appreciated that a combination of α- and β-adrenergic blockade is efficacious in management of patients with pheochromocytoma, the role of direct α_1-adrenoceptor-mediated electrophysiologic effects of catecholamines in modulation of ventricular arrhythmias is not clear. Older studies suggested that prazosin, but not propranolol, abolished ventricular fibrillation in a feline coronary artery occlusion/reperfusion model.[147] In isolated Purkinje fibers under conditions simulating ischemia (pH 6.7, PO_2 <25 mm Hg, 2.7 to 10.8 mM Ca^{2+}), the actions of phenylephrine (10 μM) inducing abnormal automaticity (maximum rate <30 beats/min) are related to α_1-mediated prolongation of the action potential and increased cellular Ca^{2+} loading.[148] In addition, feline Purkinje fibers exhibit typical triggered afterdepolarizations and tachyarrhythmias in the presence of markedly elevated extracellular Ca^{2+} (8.1 mM) with phenylephrine (10 μM) or norepinephrine (0.1 μM).[149] Activation of α_1-adrenergic receptors by norepinephrine can also modulate normal automaticity in Purkinje fibers, decreasing spontaneous rate at low concentrations (≤0.01 μM) and increasing rate at higher concentrations (maximum rate <30 beats/min with 10 μM norepinephrine). These chronotropic effects are related to activation of the two separate subtypes of α_1-receptors in Purkinje fibers[150]: a chloroethylclonidine-sensitive subtype mediating decreases of automaticity, and a WB4101-sensitive subtype mediating increases of automaticity and Purkinje fiber action potential duration.[117] However, the requirements for ischemia or high Ca^{2+} concentrations in addition to high catecholamine levels in these models suggest that induction of abnormal automaticity in normal hearts is unlikely to explain the role of α_1-receptor activation in anesthetic-catecholamine arrhythmias.

Recent efforts in our laboratory have been directed toward examining the possible role of abnormal conduction in the Purkinje system with halothane and catecholamines.[151] This role was suggested by earlier findings that halothane slows conduction and shortens refractory periods in the conduction system[78] and that α- and β-adrenergic effects of sympathetic efferent activity produce nonuniform refractory period changes in the conduction system in vivo.[119] In addition, one older report of Reynolds and Chiz[111] indicated that an α-adrenergic effect of epinephrine (4.5 μM) potentiated the modest slowing of conduction produced by halothane in Purkinje fibers. The interaction of catecholamines and anesthetics on Purkinje fiber conduction is readily examined in vitro by measuring the conduction time between action potentials recorded from false tendon fibers several millimeters apart in a fast flow-low volume tissue chamber. Figure 2-4 illustrates the changes in conduction velocity that occur at constant rate (150 beats/min) with 5 μM epinephrine alone and in combination with halothane (0.4 mM, about 1.6 MAC). Epinephrine alone does not significantly influence conduction velocity while halothane alone produces slight (<3%) slowing and a decrease of action potential duration relative to a drug-free control. However, epinephrine with halothane produces marked transient depression of conduction (–17%) without significant reduction of the rate of phase 0 depolarization (V_{max}), a major determinant of conduction velocity. Similar depression of conduction is observed with isoflurane, but to a smaller degree than with halothane. Also, with the norepinephrine in the presence of either halothane or isoflurane, the effect is smaller than with epinephrine.

Figure 2-5 shows the effect of changing the order of administration of high doses of halothane and epinephrine on conduction velocity. Epinephrine alone induces minimal biphasic changes of conduction velocity, while halothane added with continued epinephrine

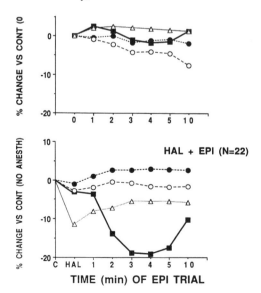

FIGURE 2-4. Simultaneous changes of canine Purkinje fiber conduction velocity (solid square) and action potential characteristics during trials of 5 μM epinephrine in the absence (*upper graph*) and presence (*lower graph*) of halothane: solid circle, amplitude; open circle, V_{max}; open triangle, action potential duration at 50% repolarization; *HAL*, halothane; *EPI*, epinephrine; *CONT*, control. (Reproduced with permission from Vodanovic S, Turner LA, Hoffmann RG, Kampine JP, Bosnjak ZJ. Transient negative dromotropic effects of catecholamines on canine Purkinje fibers exposed to halothane and isoflurane. Anesth Analg 1993;76:592.)

FIGURE 2-5. Changes of canine Purkinje fiber conduction velocity in one group of preparations after the addition of epinephrine (*E*) alone, followed by halothane (*H*) in the presence of epinephrine, compared to halothane alone and epinephrine in the presence of halothane. Standard deviations were omitted for clarity. Differences of mean conduction velocity between times of >0.08 m/sec were significant at P < 0.05.

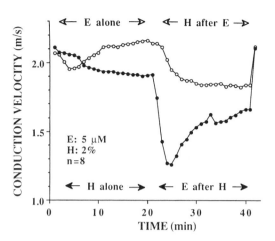

exposure decreased velocity to about 1.8 m/sec from a control of 2.1 m/sec. However, the usual clinical order of administration produced markedly different effects. Halothane alone decreased the velocity to about 1.9 m/sec, while epinephrine added in the presence of halothane transiently decreases conduction velocity to about 1.3 m/sec within 3 to 5 min, with a return to 1.7 m/sec by 20 min in the presence of both anesthetic and catecholamine. Thus, this negative dromotropic interaction is characterized by 1) depression of conduction velocity with the combination of epinephrine and halothane, regardless of order, to a steady-state value (about 1.75 m/sec at 40 min), which differs from the additive effects of either epinephrine (minimal) or halothane (moderate) alone and 2) substantially greater transient depression of conduction when the catecholamine is administered to fibers previously exposed to the anesthetic. The time course of this interaction suggests two competing processes: one rapid and transiently slowing conduction within a few minutes, similar to the time of onset of halothane-epinephrine arrhythmias in vivo; and the other gradually returning conduction velocity to more normal values over 10 to 15 min despite continued exposure to both agents. In addition, the marked decrease and gradual increase of velocity caused by epinephrine after halothane appears similar to the small negative and positive changes of conduction velocity produced by epinephrine alone. This observation suggests that the negative dromotropic interaction may represent an amplification of a "normal" effect of high epinephrine doses, rather than potentiation by epinephrine of the modest slowing of conduction produced by halothane.

The adrenergic receptor mechanisms and dose-response relationship for depression of conduction by epinephrine and halothane has been evaluated in additional groups of preparations. The results indicate that the depression of conduction is not attenuated by either β_1 (metoprolol)-adrenergic or β_1- and β_2 (propranolol)-adrenergic receptor blockade, that the responses are completely attenuated by the α_1-antagonist prazosin, and that similar slowing of conduction occurs with the α_1-agonist phenylephrine but not with the α_2-agonist clonidine.[152] Figure 2-6 illustrates the responses of the nadir of conduction velocity slowing (*filled circles*) with epinephrine at a high halothane concentration (0.7 µM, about 3 MAC). The depression of conduction velocity by epinephrine is dose related and produces about a 20% depression (to 1.7 m/sec), relative to halothane alone (1.9 m/sec), at a concentration of epinephrine (0.2 µM) comparable to arrhythmogenic plasma concentrations in vivo. This action is attenuated by the α_1-subtype antagonist WB4101 more so than by chloroethylclonidine,[153] indicating that the modulation of conduction by epinephrine with halothane involves activation of the same α_1-WB4101-sensitive adrenoceptor as that which has been reported to increase Purkinje fiber APD and automaticity. Under steady-state conditions (at 150 beats/min), the degree of conduction slowing in the false tendons is only modest (–5% relative to halothane) with 0.2 µM epinephrine and 1.7 MAC halothane, and conduction block does not occur even at unphysiologic concentrations of epinephrine (5 µM) and halothane (3 MAC). Conservatively, the results indicate that the α_1-mediated negative dromotropic effect of epinephrine would not contribute to arrhythmias in the absence of anesthetics and that it may produce only

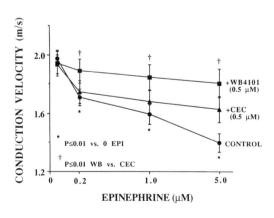

FIGURE 2-6. Average depression of Purkinje fiber conduction velocity at 2 to 5 min of epinephrine (*EPI*) administration in the presence of high halothane (0.7 µM or 3 MAC). The Krebs' solution contained 0.2 µM propranolol. The responses of 12 preparations are compared to control (*0 EPI*). Thereafter, two groups of six preparations each were treated with the α_1-subtype receptor blockers chloroethylclonidine (*CEC*) and WB4101 and the dose-response curves to epinephrine were repeated.

minimal depression of conduction in combination with subanesthetic (0.5%) "sensitizing" concentrations of halothane. On the other hand, this action at higher epinephrine and halothane concentrations, by delaying endocardial activation by impulses descending through different parts of the His-Purkinje system, may contribute to increased disparity of repolarization times in the conduction system and myocardium and thereby potentially facilitate abnormal conduction and generation of reentrant ventricular tachycardia or fibrillation by single premature ventricular impulses.

The mechanisms by which halothane and activation of the α_1-WB4101-sensitive adrenergic receptor depress conduction are not known but may involve a combination of actions on active (largely peak Na^+ current) and passive (membrane capacitance, intracellular, and gap junctional resistance) membrane properties, which determine conduction velocity. Since the depression of conduction occurs without large changes in V_{max},[151] modulation of cell-to-cell coupling may be involved.[81] Actions of halothane that could contribute to slowing of conduction include inhibition of inward Na^+ currents,[46,47] increased membrane capacitance and longitudinal resistance,[82] and reduction of gap junctional conductance.[83,85] In addition, it is possible that indirect anesthetic actions on second messengers levels, reducing cyclic AMP and increasing cyclic GMP,[154–156] in combination with adrenergic effects, may affect Na^+ channel activity[15] or modulate gap junctional conductance and cell-to-cell coupling.[107,108] At present, there is no evidence that the signal transduction pathway activated by the α_1-WB4101-sensitive subtype receptor, resulting in generation of inositol triphosphate and diacylglycerol, is involved in modulation of Na^+ channel activity or cell-to-cell coupling. However, one preliminary report indicates that high concentrations of phenylephrine (10 mM) can depress coupling between pairs of rat myocytes.[112] Therefore, it is possible that a synergistic interaction between the effects of WB4101-sensitive α_1-adrenergic receptor activation and volatile anesthetics on cell-to-cell coupling, both in the His-Purkinje system and myocardium, may be responsible for the negative dromotropic effects of catecholamines in combination with volatile anesthetics.

CONCLUSION

The volatile anesthetics have potent electrophysiologic actions at most levels of the cardiac conduction system that alone or in combination with endogenous neurotransmitters or epinephrine have important antiarrhythmic and dysrhythmogenic influences on the cardiac rhythm. Depression of inward Ca^{2+} currents by the volatile anesthetics probably contributes to largely antiarrhythmic actions on mechanisms generating abnormal impulses as a result of intracellular Ca^{2+} "overload" resulting from catecholamine exposure, digitalis intoxication, and ischemia. On the other hand, the actions of volatile anesthetics in combination with catecholamines depressing conduction and altering refractory characteristics in the ventricular conduction system and myocardium may potentially facilitate the induction of reentrant arrhythmias. The adverse interaction between volatile anesthetics and catecholamines remains a clinically relevant problem as exemplified by a recent case report[157] of a child with unsuspected congenital QT prolongation anesthetized with halothane who developed torsades de pointes and fibrillation following injection of a modest dose of epinephrine for hemostasis.

ACKNOWLEDGMENTS

This work was supported in part by Grants HL 39776, GM 25064, and HL 34708 from the National Institutes of Health and and Anesthesiology Research Training Grant GM 08377.

References

1. Hoffman BF, Crandfield PF. The physiological basis of cardiac arrhythmias. Am J Med 1964;37:670.
2. Atlee JL, Bosnjak ZJ. Mechanisms for cardiac arrhythmias during anesthesia. Anesthesiology 1990;72:347.
3. Randall WC, Ardell JL. Nervous control of the heart: anatomy and pathophysiology. In: Zipes DP, Jalife J, eds. Cardiac electrophysiology: from cell to bedside. Philadelphia: WB Saunders, 1990;33:291.
4. Salata JJ, Gill RM, Gilmour RF, Zipes DP. Effects of sympathetic tone on vagally induced phasic changes in heart rate and atrioventricular node conduction in anesthetized dog. Circ Res 1986;58:584.
5. Myerburg RJ, Gelband H, Castellanos A, et al. Electrophysiology of endocardial intraventricular conduction: the role and function of the specialized conducting system. In: Wellens HJJ, Lie KI, Janse MJ, eds. The conduction system of the heart. Lea & Febiger, Philadelphia, 1976;19:336.
6. Bailey JC, Lathrop DA, Pippenger DL. Differences between proximal left and right bundle branch block action potential durations and refractoriness in the dog heart. Circ Res 1977;40:464.
7. Overholt ED, Joyner RW, Veenstra RD, Rawling D, Wiedmann R. Unidirectional block between Purkinje and ventricular layers of papillary muscles. Am J Physiol 1984;247:H584.

8. Rawling DA, Joyner RW, Overholt ED. Variations in the functional electrical coupling between the subendocardial Purkinje and ventricular layers of the canine left ventricle. Circ Res 1985;57:252.

9. Zipes DP, Miyazaki T. The autonomic nervous system and the heart: basis for understanding interactions and effects on arrhythmia development. In: Zipes DP, Jalife J, eds. Cardiac electrophysiology: from cell to bedside. Philadelphia: WB Saunders, 1990;36:312.

10. Yanowitz F, Preston JB, Abildskov JA. Functional distribution of right and left stellate innervation to the ventricles: production of neurogenic electrocardiographic changes by unilateral alteration of sympathetic tone. Circ Res 1966;18:416.

11. Kralios FA, Martin L, Burgess MJ, Millar K. Local ventricular repolarization changes due to sympathetic nerve-branch stimulation. Am J Physiol 1975;228:1621.

12. Martins JB, Zipes DP. Effects of sympathetic and vagal nerves on recovery properties of the endocardium and epicardium of the canine left ventricle. Circ Res 1980;46:100.

13. Nattel S, Euler DE, Spear JF, Moore EN. Autonomic control of ventricular refractoriness. Am J Physiol 1981;241:H878.

14. Katz AM. Physiology of the heart. New York: Raven Press, 1992:274.

15. Schubert B, Vandongen AMJ, Kirsch GE, Brown AM. Inhibition of cardiac Na^+ currents by isoproterenol. Am J Physiol 1990;258:H977.

16. Gadsby DC. Effects of beta adrenergic catecholamines on membrane currents in cardiac cells. In: Rosen MR, Janse MJ, Wit AL, eds. Cardiac Electrophysiology: A Textbook. Mount Kisco, NY: Futura Publishing, 1990: 857.

17. Levy MN. Sympathetic-parasympathetic actions in the heart. Circ Res 1971;29:437.

18. Rabinowitz SH, Verrier RL, Lown B. Muscarinic effects of vagosympathetic trunk stimulation on the repetitive extrasystole (RE) threshold. Circulation 1976;53:622.

19. Kolman BS, Verrier RL, Lown B. Effect of vagus nerve stimulation upon excitability of the canine ventricle; role of sympathetic-parasympathetic interactions. Am J Cardiol 1976;37:1041.

20. Bosnjak ZJ, Kampine JP. Effects of halothane, enflurane and isoflurane on the SA node. Anesthesiology 1983;58:314.

21. Bosnjak ZJ, Supan FD, Rusch NJ. The effects of halothane, enflurane and isoflurane on calcium current in isolated canine ventricular cells. Anesthesiology 74:340-45, 1991.

22. Eskinder H, Rusch NJ, Supan FD, Kampine JP, Bosnjak ZJ. The effects of volatile anesthetics on L- and T-type calcium channel currents in canine cardiac Purkinje cells. Anesthesiology 1991;74:919.

23. Atlee JL, Brownlee SW, Burstrom RE. Conscious-state comparisons of the effects of inhalation anesthetics on specialized atrioventricular conduction times in dogs. Anesthesiology 1986;64:703.

24. Hauswirth O. Effects of halothane on single atrial, ventricular and Purkinje fibers. Circ Res 1969;24:745.

25. Hirota K, Momose Y, Takeda R, Nakanishi S, Ito Y. Prolongation of the action potential and reduction of the delayed outward K^+ current by halothane in single frog atrial cells. Eur J Pharmacol 1986;126:293.

26. Atlee JL, Rusy BF. Atrioventricular conduction times and atrioventricular nodal conductivity during enflurane anesthesia in dogs. Anesthesiology 1977;47:498.

27. Atlee JL, Rusy BF, Kreul JF, Eby T. Supraventricular excitability in dogs during anesthesia with halothane and enflurane. Anesthesiology 1978; 49:407.
28. Atlee JL, Yeager TS. Electrophysiologic assessment of the effects of enflurane, halothane and isoflurane on properties affecting supraventricular re-entry in chronically instrumented dogs. Anesthesiology 1989;71:941.
29. Vassalle M. Overdrive suppression and overdrive excitation. In: Rosen MR, Janse MJ, Wit AL, eds. Cardiac electrophysiology: a textbook. Mount Kisko, NY: Futura Publishing Company, 1990:175.
30. Pratila M, Vogel S, Sperelakis N. Inhibition by enflurane and methoxyflurane of postdrive hyperpolarization in canine Purkinje fibers. J Pharmacol Exp Ther 1984;229:603.
31. Reynolds AK, Chiz JF, Tanikella TK. On the mechanism of coupling in adrenalin-induced bigeminy in sensitized hearts. Can J Physiol Pharmacol 1975;53:1158.
32. Tucker WK, Rackstein AD, Munson ES. Comparison of arrhythmic doses of adrenaline, metaraminal, ephedrine and phenylephrine during isoflurane and halothane anaesthesia in dogs. Br J Anaesth 1974;46:392.
33. Atlee JL, Malkinson CE. Potentiation by thiopental of halothane-epinephrine induced arrhythmias in dogs. Anesthesiology 1982;57:285.
34. Atlee JL, Roberts FL. Thiopental and epinephrine-induced arrhythmias in dogs anesthetized with enflurane or isoflurane. Anesth Analg 1986;65:437.
35. Stowe DF, Dujic Z, Bosnjak ZJ, Kalbfleisch JH, Kampine JP. Volatile anesthetics attenuate sympathomimetic actions on guinea pig SA node. Anesthesiology 1988;68:887.
36. Polic S, Atlee JL, Laszlo A, Kampine JP, Bosnjak ZJ. Anesthetics and automaticity in latent pacemaker fibers. II. Effects of halothane and epinephrine or norepinephrine on automaticity of dominant and subsidiary atrial pacemakers in the canine heart. Anesthesiology 1991;75:298.
37. Woehlck HJ, Vicenzi MN, Bosnjak ZJ, Atlee JL. Halothane and epinephrine on stability of atrial pacemakers: atropine methylnitrate effects. Anesthesiology 1992;77:A595.
38. Mendez C, Han J, Moe GK. A comparison of the effects of epinephrine and vagal stimulation upon the refractory periods of the A-V node and the bundle of His. Naunyn-Schmiedebergs Arch Exp Pathol Pharmakol 1964;248:99.
39. Moe GK, Mendez C, Han J. Aberrant A-V impulse propagation in the dog heart: a study of functional bundle branch block. Circ Res 1965;16:261.
40. Ferrier GR, Dresel PE. Relationship of the functional refractory period to conduction in the atrioventricular node. Circ Res 1974;35:204.
41. Preston JB, McFadden S, Moe GK. Atrioventricular transmission in young mammals. Am J Physiol 1959;197:236.
42. Lynch C, Vogel S, Sperelakis N. Halothane depression of myocardial slow action potentials. Anesthesiology 1981;55:360.
43. Lynch C, Vogel S, Pratila MG, Sperelakis N. Enflurane depression of myocardial slow action potentials. J Pharmacol Exp Ther 1982;222:405.
44. Ikemoto Y, Yatani A, Arimura H, Yoshitake J. Reduction of the slow inward current of isolated rat ventricular cells by thiamylal and halothane. Acta Anaesthesiol Scand 1985;29:583.
45. Terrar DA, Victory JGG. Effects of halothane on membrane currents associated with contraction in single myocytes isolated from guinea-pig ventricle. Br J Pharmacol 1988;94:500.

46. Ikemoto Y, Yatani A, Imoto Y, Arimura H. Reduction in the myocardial sodium current by halothane and thiamylal. Jpn J Physiol 1986;36:107.
47. Buljubasic N, Berczi V, Supan DF, Marijic J, Turner LA, Kampine JP, Bosnjak ZJ. Depression of the Na$^+$ current by halothane and isoflurane in the rabbit Purkinje fiber. Anesthesiology 1993;79:A392.
48. Supan DF, Eskinder H, Buljubasic N, Kampine JP, Bosnjak ZJ. Effects of inhalational anesthetics on K$^+$ current in canine cardiac Purkinje cells. FASEB J 1991;5:A1742.
49. Supan DF, Buljubasic N, Marijic J, Eskinder H, Kampine JP, Bosnjak ZJ. Effects of volatile anesthetics on transient outward potassium current in Purkinje fibers. Anesthesiology 1991;75:A368.
50. Katsuoka M, Kobayashi K, Ohnishi T. Volatile anesthetics decrease calcium content of isolated myocytes. Anesthesiology 1989;70:954.
51. Rusy BF, Komai H. Anesthetic depression of myocardial contractility: a review of possible mechanisms. Anesthesiology 1987;67:745.
52. Bosnjak ZJ, Kampine JP. Effects of halothane on transmembrane potentials, Ca^{2+} transients, and papillary muscle tension in the cat. Am J Physiol 1986;251:H374.
53. Bosnjak ZJ, Aggarwal A, Turner LA, Kampine JM, Kampine JP. Differential effects of halothane, enflurane and isoflurane on Ca^{2+} transients and papillary muscle tension in guinea pigs. Anesthesiology 1992;76:123.
54. Reynolds AK, Horne ML. Studies on the cardiotoxicity of ouabain. Can J Physiol Pharmacol 1969:47:165.
55. Morrow DF. Anesthesia and digitalis toxicity. VI: effect of barbiturates and halothane on digoxin toxicity. Anesth Analg 1970;49:305.
56. Gallagher JD, McClernan CA. The effects of halothane on ventricular tachycardia in intact dogs. Anesthesiology 1991;75:866.
57. Ivankovitch AD, Miletch DJ, Grossman RK, et al. The effect of enflurane, isoflurane, fluroxene, methoxyflurane and diethyl-ether anesthesia on ouabain tolerance in the dog. Anesth Analg 1976;55:360.
58. Gallagher JD, Bianchi JJ, Gessman LJ. Halothane antagonizes ouabain toxicity in isolated canine Purkinje fibers. Anesthesiology 1989;71:695.
59. Freeman LC, Li Q. Effects of halothane on delayed afterdepolarizations and calcium transients in dog ventricular myocytes exposed to isoproterenol. Anesthesiology 1991;74:146.
60. Zuckerman RL, Wheeler DM. Effect of halothane on arrhythmogenic responses induced by sympathomimetic agents in single rat heart cells. Anesth Analg 1991;72:596.
61. Luk HN, Lin CI, Wei J, Chang CL. Depressant effects of isoflurane and halothane on isolated human atrial fibers. Anesthesiology 1988;69:667.
62. El-Sherif N, Gough WB, Zeiler RH, Mehra R. Triggered ventricular rhythms in 1-day-old myocardial infarction in the dog. Circ Res 1983; 52:566.
63. Turner LA, Bosnjak ZJ, Kampine JP. Actions of halothane on the electrical activity of Purkinje fibers derived from normal and infarcted canine hearts. Anesthesiology 1987;67:619.
64. Laszlo A, Polic S, Kampine JP, Turner LA, Atlee JL, Bosnjak ZJ. Halothane, enflurane and isoflurane on abnormal automaticity and triggered rhythmic activity of Purkinje fibers from 24-hour-old infarcted canine hearts. Anesthesiology 1991;75:847.
65. Schmeling WT, Bosnjak ZJ, Kampine JP. Anesthesia and the autonomic nervous system. Semin Anesth 1990;9:223.

66. Martins JB. Autonomic control of ventricular tachycardia: sympathetic neural influence on spontaneous tachycardia 24 hours after coronary occlusion. Circulation 1985;72:933.
67. Roizen MF, Moss J, Henry DP, Kopin IJ. Effects of halothane on plasma catecholamines. Anesthesiology 1974;41:432.
68. Reynolds AK, Chiz JF, Pasquet AF. Halothane and methoxyflurane—a comparison of their effects on cardiac pacemaker fibers. Anesthesiology 1970;33:602.
69. Laszlo A, Polic S, Atlee JL, Kampine JP, Bosnjak ZJ. Anesthetics and automaticity in latent pacemaker fibers: I. Effects of halothane, enflurane and isoflurane on automaticity and recovery of automaticity from overdrive suppression in Purkinje fibers derived from canine hearts. Anesthesiology 1991;75:98.
70. Logic JR, Morrow DH. The effect of halothane on ventricular automaticity. Anesthesiology 1972;36:107.
71. Bailey JC, Greenspan K, Elizari MV, Anderson GJ, Fisch C. Effects of acetylcholine on automaticity and conduction in the proximal portion of the His-Purkinje specialized conduction system of the dog. Circ Res 1972;30:210.
72. Vassalle M. Overdrive excitation: the onset of spontaneous activity following a fast drive. In: Zipes DP, Jalife J, eds. Cardiac electrophysiology and arrhythmias. Orlando, FL: Grune & Stratton, 1985:97.
73. Quan W, Rudy Y. Unidirectional block and reentry of cardiac excitation: a model study. Circ Res 1990;66:367.
74. Wit AL, Dillon SM. Anisotropic Reentry. In: Zipes DP, Jalife J, eds. Cardiac electrophysiology: from cell to bedside. Philadelphia, WB Saunders, 1990;39:353.
75. Joyner RW, Overholt ED. Effects of octanol on canine subendocardial Purkinje-to-ventricular transmission. Am J Physiol 1985;249:1228.
76. Atlee JL, Rusy BF. Halothane depression of A-V conduction studied by electrograms of the bundle of His in dogs. Anesthesiology 1972;36:112.
77. Atlee JL, Alexander SC. Halothane effects on conductivity of the AV node and His-Purkinje system in the dog. Anesth Analg 1977;56:378.
78. Turner LA, Zuperku EJ, Purtock RV, Kampine JP. In vivo changes in canine ventricular cardiac conduction during halothane anesthesia. Anesth Analg 1980;59:327.
79. Atlee JL, Hamann SR, Brownlee SW, Kreigh C. Conscious state comparisons of the effects of the inhalation anesthetics and diltiazem, nifedipine or verapamil on specialized atrioventricular conduction time in spontaneously beating dog hearts. Anesthesiology 1988;68:519.
80. Atlee JL, Homer LD, Tobey RE. Diphenylhydantoin and lidocaine modification of A-V conduction in halothane-anesthetized dogs. Anesthesiology 1975;43:49.
81. Freeman LC, Muir WW. Effects of halothane on impulse propagation in Purkinje fibers and at Purkinje-muscle junctions: relationship of V_{max} to conduction velocity. Anesth Analg 1991;72:5.
82. Hauswirth O. The influence of halothane on the electrical properties of cardiac Purkinje fibers. J Physiol (Lond) 1968;201:42P.
83. Terrar DA, Victory JGG. Influence of halothane on electrical coupling in cell pairs isolated from guinea-pig ventricle. Br J Pharmacol 1988;94:509.
84. Niggli E, Rudisuli A, Maurer PK, Weingart R. Effects of general anesthetics on current flow across membranes in guinea pig myocytes. Am J Physiol 1989;256:C273.

85. Burt JM, Spray DC. Volatile anesthetics block intercellular communication between neonatal rat myocardial cells. Circ Res 1989;65:829.
86. Ozsaki S, Nakaya H, Gotoh Y, Azuma M, Kemmotsu O, Kanno M. Effects of halothane and enflurane on conduction velocity and maximum rate of rise of action potential upstroke in guinea pig papillary muscles. Anesth Analg 1989;68:219.
87. Mendez C, Mueller WJ, Merideth J, Mow GK. Interaction of transmembrane potentials in canine Purkinje fibers and at Purkinje fiber-muscle junctions. Circ Res 1969;24:361.
88. Han J, Moe GK. Cumulative effects of cycle length on refractory periods of cardiac tissues. Am J Physiol 1969;217:106.
89. Burgess MJ, Green LS, Millar K, et al. The sequence of normal ventricular recovery. Am Heart J 1972;84:660.
90. Autenrieth G, Surawicz B, Kuo CS. Sequence of repolarization on the ventricular surface in the dog. Am Heart J 1975;89:463.
91. Saitoh H, Bailey JC, Surawicz B. Alternans of action potential duration after abrupt shortening of cycle length: differences between dog Purkinje and ventricular fibers. Circ Res 1988;62:1027.
92. Tchou PJ, Lehmann MH, Dongas J, et al. Effect of sudden rate acceleration on the human His-Purkinje system: adaptation of refractoriness in a dampened oscillatory pattern. Circulation 1986;73:920.
93. Han J, Millet D, Chizzonitti B, Moe GK. Temporal dispersion of recovery of excitability in atrium and ventricle as a function of heart rate. Am Heart J 1966;71:481.
94. Rosenbaum DS, Kaplan DT, Kanai A, et al. Repolarization inhomogeneities in ventricular myocardium change dynamically with abrupt cycle length shortening. Circulation 1991;84:1333.
95. Han J, Moe GK. Nonuniform recovery of excitability in ventricular muscle. Circ Res 1964;14:44.
96. Reiter MJ, Synhorst DP, Mann DE. Electrophysiological effects of acute ventricular dilatation in the isolated rabbit heart. Circ Res 1988;62:554.
97. Polic S, Bosnjak ZJ, Marijic J, Hoffmann RG, Kampine JP, Turner LA. Actions of halothane, isoflurane and enflurane on the regional action potential characteristics of canine Purkinje fibers. Anesth Analg 1991;73:603.
98. Lynch C. Differential depression of myocardial contractility by halothane and isoflurane in vitro. Anesthesiology 1986;64:620.
99. Denniss AR, Richards DA, Taylor AT, Uther JB. Halothane anesthesia reduces inducibility of ventricular tachyarrhythmias in chronic canine myocardial infarction. Basic Res Cardiol 1989;84:5.
100. Hunt GB, Ross DL. Comparison of effects of three anesthetic agents on induction of ventricular tachycardia in a canine model of myocardial infarction. Circulation 1988;78:221.
101. Deutsch N, Hantler CB, Tait AR, Uprichard A, Schork MA, Knight PR. Suppression of ventricular arrhythmias by volatile anesthetics in a canine model of chronic myocardial infarction. Anesthesiology 1990;72:1012.
102. Riley DC, Schmeling WT, Al-Wathiqui MH, Kampine JP, Warltier DC. Prolongation of the QT interval by volatile anesthetics in chronically instrumented dogs. Anesth Analg 1988;67:741.
103. Schmeling WT, Warltier DC, McDonald DJ, Madsen KE, Atlee JL, Kampine JP. Prolongation of the QT interval by enflurane, isoflurane and halothane in humans. Anesth Analg 1991;72:137.

104. Wallace AG, Sarnoff SJ. Effects of cardiac sympathetic nerve stimulation on conduction in the heart. Circ Res 1964;14:86.
105. Hoffman BF, Singer DH. Appraisal of the effects of catecholamines on cardiac electrical activity. Ann NY Acad Sci 1967;139:914.
106. DeMello WC Modulation of junctional permeability. Fed Proc 1984; 43:L2692.
107. Burt JM, Spray DC. Inotropic agents modulate gap junctional conductance between cardiac myocytes. Am J Physiol 1988;254:H1206.
108. DeMello WC. Effect of isoproterenol and 3-isobutyl-1-methylxanthine on junctional conductance in heart cell pairs. Biochim Biophys Acta 1989; 1012:291.
109. Veenstra RD. Physiological modulation of cardiac gap junction channels. J Cardiovasc Electrophysiol 1991;2:168.
110. Matsuda JJ, Lee H, Shibata EF. Enhancement of rabbit cardiac sodium channels by β-adrenergic stimulation. Circ Res 1992;70:199.
111. Reynolds AK, Chiz JF. Epinephrine-potentiated slowing of conduction in Purkinje fibers. Res Commun Chem Pathol Pharmacol 1974;9:633.
112. Burt JM, Spray DC. Adrenergic control of gap junction conductance in cardiac myocytes. Circulation 1988;78(suppl II):258.
113. Giotti A, Ledda F, Mannaioni PF. Effects of noradrenalin and isoprenaline, in combination with α- and β-receptor blocking substances, on the action potential of cardiac Purkinje fibers. J Physiol (Lond) 1973;229:99.
114. Bailey JC, Watanabe AM, Besch HR, Lathrop DA. Acetylcholine antagonism of the electrophysiological effects of isoproterenol on canine cardiac Purkinje fibers. Circ Res 1979;44:378.
115. Bahinski A, Nairn AC, Greengard P, Gadsby DC. Chloride conductance regulated by cyclic AMP-dependent protein kinase in cardiac myocytes. Nature (Lond) 1989;340:718.
116. Martins JB, Wendt DJ. α-Adrenergic effects on relative refractory period in Purkinje system of intact canine left ventricle. Am J Physiol 1989; 257:H1156.
117. Lee JH, Steinberg SF, Rosen MR. A WB4101-sensitive alpha-1 adrenergic receptor subtype modulates repolarization in canine Purkinje fibers. J Pharmacol Exp Ther 1991;258:681.
118. Harrison LA, Wittig J, Wallace AG. Adrenergic influences on the distal Purkinje system of the canine heart. Circ Res 1973;32:329.
119. Turner LA, Zuperku EJ, Bosnjak ZJ, Kampine JP. Autonomic modulation of refractoriness in canine specialized His-Purkinje system. Am J Physiol 1985;248:R515.
120. Han J, de Jalon G, Moe GK. Adrenergic effects on ventricular vulnerability. Circ Res 1964;14:516.
121. Surawicz B. Dispersion of refractoriness in ventricular arrhythmias. In: Zipes DP, Jalife J, eds. Cardiac electrophysiology: from cell to bedside. Philadelphia: WB Saunders, 1990:377.
122. Schwartz PJ, Vanoli E, Zaza A, Zuanetti G. The effect of antiarrhythmic drugs on life-threatening arrhythmias induced by the interaction between acute myocardial ischemia and sympathetic hyperactivity. Am Heart J 1985;109:937.
123. Verrier RL. Autonomic substrates for arrhythmias. In: Zipes DP, Rowlands DJ, eds. Progress in cardiology. Philadelphia: Lea & Febiger, 1988;65.

124. Schwartz PJ, LaRovere MT, Vanoli E. Autonomic nervous system and sudden cardiac death: experimental basis and clinical observation for post-myocardial infarction risk stratification. Circulation 85(suppl I):I-77.

125. Saini V, Carr DB, Hagestad EL, Lown B, Verrier RL. Antifibrillatory action of the narcotic agonist fentanyl. Am Heart J 1988;115:598.

126. Saini V, Carr DB, Verrier RL. Comparative effects of the opioids fentanyl and buprenorphine on ventricular vulnerability during acute coronary artery occlusion. Cardiovasc Res 1989;23:1001.

127. Pace NL, Ohmura A, Wong KC. Epinephrine-induced arrhythmias: effects of exogenous prostaglandins and prostaglandin synthesis inhibition during halothane-O_2 anesthesia in the dog. Anesth Analg 1979; 58:401.

128. Joas TA, Stevens WC. Comparison of the arrhythmic doses of epinephrine during forane, halothane and fluroxene anesthesia in dogs. Anesthesiology 1971;35:48.

129. Sumikawa K, Ishizaka N, Suzaki M. Arrhythmogenic plasma levels of epinephrine during halothane, enflurane and pentobarbital anesthesia in the dog. Anesthesiology 1983;58:322.

130. Hayashi Y, Sumikawa K, Tashiro C, Yamatodani A, Yoshiya I. Arrhythmogenic threshold of epinephrine during sevoflurane, enflurane and isoflurane anesthesia in dogs. Anesthesiology 1988;69:145.

131. Hayashi Y, Sumikawa K, Yamatodani A, Tashiro C, Wada H, Yoshiya I. Myocardial sensitization by thiopental to arrhythmogenic action of epinephrine in dogs. Anesthesiology 1989;71:929.

132. Hayashi Y, Sumikawa K, Yamatodani A, Kamibayashi T, Kuro M, Yoshiya I. Myocardial epinephrine sensitization with subanesthetic concentrations of halothane in dogs. Anesthesiology 1991;74:134.

133. Metz S, Maze M. Halothane concentration does not alter the threshold for epinephrine-induced arrhythmias in dogs. Anesthesiology 1985; 62:470.

134. Hayashi Y, Sumikawa K, Maze M, et al. Dexmetomidine prevents epinephrine-induced arrhythmias through stimulation of central α_2 adrenoceptors in halothane-anesthetized dogs. Anesthesiology 1991;75:113.

135. Hayashi Y, Kambayashi T, Maze M, et al. Role of imidazoline-prefering receptors in the genesis of epinephrine-induced arrhythmias in halothane-anesthetized dogs. Anesthesiology 1993;78:524.

135a. Pancrazio JJ, Frazer MJ, Lynch C III. Barbiturate anesthetics depress the resting K^+ conductance of myocardium. J Pharmacol Exp Ther 1993; 265:358.

136. Somani P, Lum BKB. Blockade of epinephrine- and ouabain-induced cardiac arrhythmias in the dog heart-lung preparation. J Pharmacol Exp Ther 1966;152:235.

137. Reynolds AK. Cardiac arrhythmias in sensitized hearts—primary mechanisms involved. Res Commun Chem Pathol Pharmacol 1983;40:3.

138. Moore EN, Morse HT, Price HL. Cardiac arrhythmias produced by catecholamines in anesthetized dogs. Circ Res 1964;15:77.

139. Sasyniuk BI, Dresel PE. Mechanism and site of origin of bigeminal rhythms in cyclopropane-sensitized dogs. Am J Physiol 1971;220:1857.

140. Sasyniuk BI, Dresel PE. The effect of destruction of the bundle of His or the atrioventricular node on cyclopropane-adrenalin cardiac arrhythmias. Can J Physiol Pharmacol 1970;48:207.

141. Smith ER, Dresel PE. Site of origin of halothane-epinephrine arrhythmia determined by direct and echocardiographic recordings. Anesthesiology 1982;57:98.
142. Zink J, Sasyniuk BI, Dresel PE. Halothane-epinephrine-induced cardiac arrhythmias and the role of heart rate. Anesthesiology 1975;43:548.
143. Maze M, Smith CM. Identification of receptor mechanism mediating epinephrine-induced arrhythmias during halothane anesthesia in the dog. Anesthesiology 1963;59:322.
144. Hayashi Y, Sumikawa K, Tashiro C, Yoshiya I. Synergistic interaction of α_1- and β-adrenoceptor agonists on induction arrhythmias during halothane anesthesia in dogs. Anesthesiology 1988;68:902.
145. Spiss CK, Maze M, Smith CM. α-Adrenergic responsiveness correlates with epinephrine dose for arrhythmias during halothane anesthesia in dogs. Anesth Analg 1084;63:297.
146. Maze M, Hayward E, Gaba DM. Alpha$_1$-adrenergic blockade raises epinephrine-arrhythmia threshold in halothane-anesthetized dogs in a dose-dependent fashion. Anesthesiology 1985;63:611.
147. Sheridan DJ, Penkoske PA, Sobel BE. Alpha adrenergic contributions to arrhythmia during myocardial ischemia and reperfusion in cats. J Clin Invest 1980;65:161.
148. Anyukhovsky EP, Rosen MR. Abnormal automatic rhythms in ischemic Purkinje fibers are modulated by a specific α_1-adrenergic receptor subtype. Circulation 1991;83:2076.
149. Kimura S, Cameron JS, Kozlovskis PL, Bassett AL, Myerburg RJ. Delayed afterdepolarizations and triggered activity induced in feline Purkinje fibers by α-adrenergic stimulation in the presence of elevated calcium levels. Circulation 1984;70:1074.
150. Del Balzo U, Rosen MR, Malfatto G, Kaplan LM, Steinberg SF. Specific α_1-adrenergic subtypes modulate catecholamine-induced increases and decreases in ventricular automaticity. Circ Res 1990;67:1535.
151. Vodanovic S, Turner LA, Hoffmann RG, Kampine JP, Bosnjak ZJ. Transient negative dromotropic effects of catecholamines on canine Purkinje fibers exposed to halothane and isoflurane. Anesth Analg 1993;76:592.
152. Vodanovic S, Turner LA, Kampine JP, Bosnjak ZJ. α_1-Mediated transient negative dromotropic effect in canine Purkinje fibers exposed to halothane. FASEB J 1993;7:A97.
153. Turner LA, Vodanovic S, Kampine JP, Bosnjak ZJ. WB4101 sensitive α_1-adrenergic depression of conduction in canine Purkinje fibers exposed to halothane. FASEB J 1993;7:A96.
154. Vulliemoz Y, Verosky M, Triner L. Effect of halothane on myocardial cyclic AMP and cyclic GMP content of mice. J Pharmacol Exp Ther 1986;236:181.
155. Vulliemoz Y. Volatile anesthetics and second messengers in cardiac tissues. In: Blanck TJJ, Wheeler DM, eds. Mechanisms of anesthetic action in skeletal, cardiac and smooth muscle. New York: Plenum Press, 1991:169.
156. Hirota K, Ito Y, Kuze S, Momose Y. Effects of halothane on electrophysiologic properties and cyclic adenosine 3',5'-monophosphate content in isolated guinea pig hearts. Anesth Analg 1992;74:564.
157. Richardson MG, Roark GL, Helfaer MA. Intraoperative epinephrine-induced torsades de pointes in a child with long QT syndrome. Anesthesiology 1992;76:647.

Michael J. Barber

3 | Class I Antiarrhythmic Agents

In cardiac tissue, the propagation of an impulse is critically dependent upon the ability of individual myocardial cells to transiently increase their membrane conductance to ions. While several ions play a key role in the cardiac action potential, sodium ions and obviously the channels involved in promoting their transmembrane travel are focal in normal and aberrant impulse conduction and formation in cardiac muscle.

Local anesthetics block the propagation of action potentials in nerve tissue by blocking sodium channels. Antiarrhythmic drugs with local anesthetic action block sodium channels in cardiac tissues. It is this effect which is the presumed mechanism of action of the local anesthetic class of antiarrhythmic drugs.[1,2] Several reviews[3-6] have described, in detail, the chemistry, structure-activity relationships, and neural mechanisms of action of local anesthetics. While it is beyond the scope of this chapter to cover all topics referable to local anesthetic drugs, I shall attempt to review general concepts of the structure of the sodium channel, the mechanism of action of local anesthetic agents on the sodium channel, the scheme currently used for classification of sodium channel blocking agents, the modulation of action of local anesthetic agents, and some of the more common side effects and toxicities of local anesthetics.

Clinical Cardiac Electrophysiology: Perioperative Considerations
Edited by Carl Lynch III. J.B. Lippincott Company, Philadelphia, PA ©1994

SODIUM CHANNEL STRUCTURE AND FUNCTION

Blockade of membrane sodium ion conductance is the presumed mechanism of action of antiarrhythmic drugs in the local anesthetic class.[1,2] As reviewed in Chapter 1, the cardiac action potential requires the presence of several types of functioning ion channels, the most important of which may be the voltage-sensitive sodium channel.

In order to understand voltage-sensitive sodium channel function, I first shall consider the structure of the channel.[7,8] Simplistically, the channel is a transmembrane protein sitting in a lipid bilayer of the membrane (Figure 3-1). With the use of isolation and quantitative binding techniques, this structure has been found to be composed of several thousand amino acids forming a 295-kDa molecule, which is arranged as subunits into four repeating homologous segments. It is felt that these "homology repeats" of the molecule fold into domains that subsequently assume a tetrameric arrangement forming the channel. It appears that about 60% of the polypeptide mass of this molecule may be organized in this membrane-spanning tetrameric arrangement while a majority of the remaining portion of the molecule is exposed to cytoplasm.[7] Because only the homologous repeats seem to span the cell membrane, they are felt most likely to contain the domains forming the ion pathway, i.e., the pore of the channel. Additionally, the sites of interaction of the ion channel (binding sites) with the variety of neurotoxins, local anesthetics, and antiarrhythmic agents known to effect the channel are felt to be located within these homologous segments.[6-8]

Internally, the pore of the sodium channel is complex (Figure 3-2). As a result of interactions of ions with the wall of the pore and limited access of ions to the channel orifice, sodium channels do not obey Ohm's Law ($E = IR$), i.e., sodium currents do not increase proportionally as the concentration of external sodium ions increases.[7,9] Ultimately sodium ion flow, and hence the amount of current carried into the cell, plateaus, most likely due to saturation of specific binding sites within the ion channel. These specific binding sites move the ion through the channel in a "hopscotch" fashion from one site to the next with ions on both sides of the membrane competing for channel binding sites.

The rate at which an ion passes through a channel is determined by energy differences at the binding site and adjacent energy barriers.[10] Permanent ions will have deep energy wells (Figure 3-3) at the binding site and relatively small energy barriers to overcome to leave the site such that they will move quickly and easily through the channel. Less permeant ions or compounds have shallow energy wells because of low affinity for the binding site and proportionally higher energy barriers to

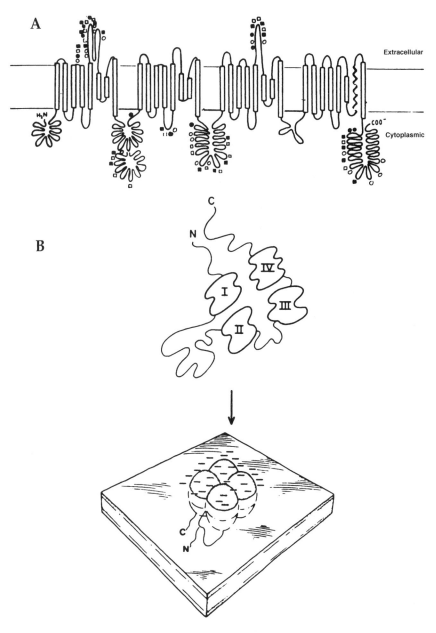

FIGURE 3-1. Schematic representation of the structure of the sodium channel. A is a linear schematic diagram of the amino acid array showing the four homologous repeats each with seven membrane spanning segments. These segments fold into domains (I, II, III, and IV) and assume the tetrameric arrangement of the sodium channel (B) as viewed from the top (B) and in three dimensions (C).

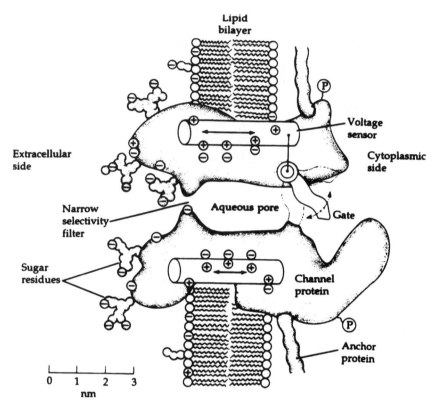

FIGURE 3-2. Schematic depiction of the components of a functioning sodium channel. The extracellular surface of the molecule contains multiple sugar residues and a narrow opening which aid in the "selectivity" of the channel. A portion of the channel's amino acid structure functions as a voltage sensor within the channel and contributes to opening and inactivation of the channel by operating the channel gate. The pore itself is an aqueous connection between the extracellular and cytoplasmic portion of the cell. (Reproduced with permission from Hille B. Ionic channels in excitable membranes. Sunderland, MA: Sinauer Associates, 1992:261.)

overcome to leave the binding site. This results in "selectivity" of the channel with the exclusion of many ions or drugs from the pore. Blocking agents such as local anesthetics have deep energy wells at the binding sites and almost insurmountable energy barriers to overcome to leave the wells. Hence, once bound they tend to stay bound and block sodium ions from their binding sites within the channel. This decreases sodium ion flux. The decrease in ion flux results in a decrease in sodium current and a subsequent decreased rate of cellular depolarization and ultimately a decrease in impulse conduction through the myocardium.

FIGURE 3-3. Schematic depiction of the energy profile of an ion in a channel. Channels are believed to have one or more binding sites within the pore (A). Simplistically, an ion (labeled ++) moves from binding site to binding site (*cross-hatched areas*) in a hopscotch fashion with an associated energy profile (B). Permeant ions (*solid energy profile*) have deep, negative energy wells and low energy barriers to overcome to move from site to site. Less permeant ions (*broken energy profile*) have relatively shallow binding energies depicting less affinity for the ion by the binding site and higher (often positive) energy barriers to overcome to move from site to site. This makes the likelihood that a less permeant ion will bind within the channel and pass through successfully much lower than that of the permeant ion. (Reproduced with permission from Hille B. Ionic channels in excitable membranes. Sunderland, MA: Sinauer Associates, 1992:261.)

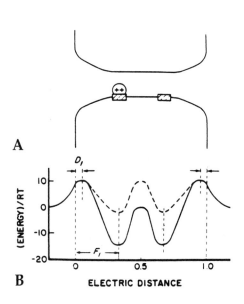

FIGURE 3-4. Schematic diagram of the Modulated Receptor Hypothesis. This model proposes that the channel may exist in any of its three states—open, inactivated, or resting—under normal (no drug-bound) or modified (drug-bound) conditions. It is assumed that the modified or drug-bound channel displays different abilities to conduct sodium ions than the channel in the normal or unbound state. (Reproduced with permission from Hondeghem LM. Antiarrhythmic agents: modulated receptor applications. Circulation 1987;75:514.)

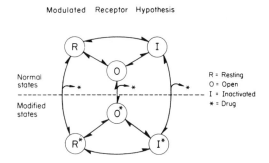

LOCAL ANESTHETIC INTERACTION WITH THE SODIUM CHANNEL

The concepts of channel gating, inactivation, and binding as related to the protein structure of the channel are important when the mechanism of action of the local anesthetic class of antiarrhythmic medications is considered. While the exact mechanism(s) of this interaction remains unknown, several theories exist as to how antiarrhythmic drugs interact at the level of the sodium channel. The model which at the present time best explains antiarrhythmic drug effects on the sodium channel and seems most feasible based on existing data is the "Modulated Receptor Model" of Hille[11] and Hondeghem and Katzung.[2,12] The Modulated Receptor model (Figure 3-4) focuses on the interaction of drugs with binding sites of the sodium channel in each of these three known states of the channel: open (O), resting (R), and inactivated (I).

Sodium channels, when activated, are considered open and therefore allow the flow of ion species. During depolarization of the cell (Figure 3-5), there is the rapid opening of large numbers of sodium channels, allowing the flow of ions. As sodium ions stream into the cell, the transmembrane potential rises, causing the upstroke of the action potential (phase 0). With sufficient depolarization, the open (activated) state of the sodium channel changes to the closed (inactivated) state, which corresponds to the plateau (phase 2) of the action potential. As the cell repolarizes (phase 3) due to activation of the ion pumps, the sodium channel again changes conformation from the inactivated state to the resting state. While channels are in the inactivated state they cannot easily reopen to allow more sodium ions to pass, but once they have returned to the resting state the cell again may readily depolarize. In terms of cardiac tissue, the resting state of the channel is prevalent during diastole while the open or activated state predominates during depolarization. The inactivated state is present when the tissue is depolarized either by the process of a normal action potential or during a pathophysiologic event such as ischemia.

In the Modulated Receptor model (Figure 3-4), antiarrhythmic drugs interact with the ion binding site or receptor within the channel pore. The central premise of the model is that channel binding sites occupied by antiarrhythmic (local anesthetic) agents (bound state) do not conduct ions, i.e., they are blocked, and will remain in this state until the antiarrhythmic drug leaves the receptor site (unbound state). Hondeghem and Katzung[2] schematically depicted the states of the sodium channel as either drug-free or drug-associated. The transitions between the states of drug-associated and drug-free are reversible.

There are multiple ways in which the blocking agents can gain access to the receptor site (Figure 3-6). Most local anesthetics are weak

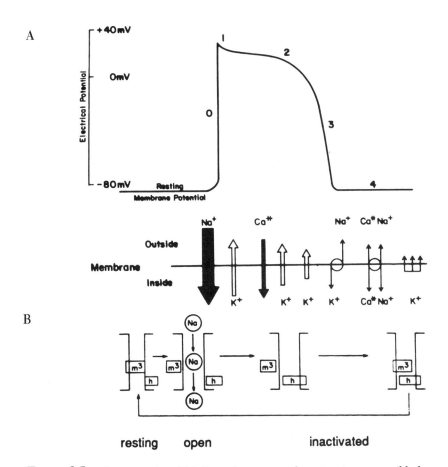

FIGURE 3-5. The *top portion* of this figure demonstrates the various ions responsible for a single cardiac cellular action potential. The *arrows* point in the direction of ion flow and their size qualitatively depicts the relative amount of ions which cross the membrane during an individual action potential. Local anesthetic (class I) drugs alter the influx of sodium ions by impeding the flow of ions and decreasing phase 0 of the action potential (A). The *bottom portion* of this figure demonstrates the state of the sodium channel associated with the phase of the cardiac action potential (B). During depolarization of the cell (phase 0), the resting gate (m^3) opens and allows the free passage of sodium ions. As the cell depolarizes, the voltage sensors within the channel structure close the inactivation gate (h) and block further ion movement. As the cell repolarizes (phase 3), the resting and inactivation gates reset to their original position as seen during phase 4.

bases that exist either in the cationic (positively charged) or neutral (uncharged) form at physiologic pH. In open channels, cationic compounds, which tend to be hydrophilic, may interact with the binding site directly by entering the pore.[13,14] Neutral compounds, because of the absence of charge and their more lipophilic nature, most likely access the binding site by diffusing across the membrane protein and

(A) CHARGED DRUG   **(B) NEUTRAL DRUG**

 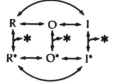

(C) LOCAL ANESTHETIC RECEPTOR

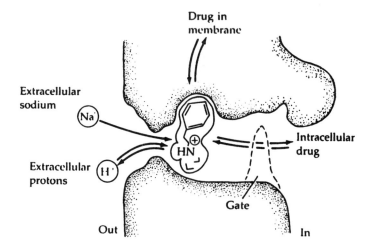

FIGURE 3-6.   Modification of a local anesthetic's ability to access the channel occurs if the drug exists in a charged (A) or neutral (B) form. The charged form of the drug is hydrophilic and therefore can interact only with the channel binding site when the site is accessible via the pore (i.e., an open channel). Neutral forms of the drug are more lipophilic and therefore may access the binding site either directly through the pore or indirectly by diffusion through the membrane (C). (Reproduced with permission from Hille B. Ionic channels in excitable membranes. Sunderland, MA: Sinauer Associates, 1992:261.)

may interact with either open or inactivated channels.[13,14] Although at high enough concentration, sodium channel blockers may bind to the receptor site in any state, clinically useful antiarrhythmic agents usually block either activated (quinidine), inactivated (amiodarone), or both (lidocaine) types of channels.

Most clinically important antiarrhythmic agents have a relatively low affinity for resting or repolarized sodium channels present in normal cardiac tissue.[1,15] The concept of binding to resting sodium channels is termed "tonic" or "resting" block.[13] Each time the channel opens, drug, if present, may enter the pore and associate with the ion binding site; therefore, sodium ion flux is decreased and block develops (Figure 3-7). This is termed "phasic" or "frequency (use)-dependent" block.[16,17] The amount of block that develops depends upon the permeability of the agent, the amount of drug present, the binding affinity of the drug, and the duration and frequency of channel opening. With repolarization of the cell, the drug-associated channels slowly "reprime" due to the diffusion of the antiarrhythmic drug from the receptor site (unblocking). In clinical applications, if the diastolic interval is too short for full repriming of the channel, then block will accumulate until a steady-state level of block is achieved. If the diastolic interval is long, drug will dissociate almost completely from all channels and relatively little block results. Keeping in mind that most useful antiarrhythmic agents block open or inactivated channels, then a drug that preferentially blocked resting or normal channels would in all likelihood be a poor antiarrhythmic agent.[1,12,15] as it would reduce

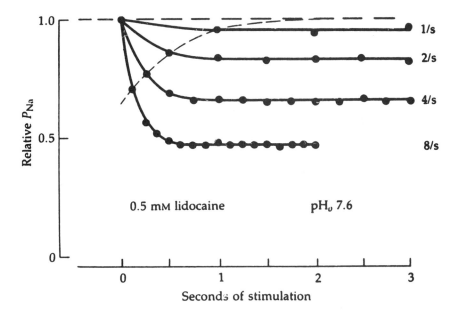

FIGURE 3-7. Classic demonstration of use-dependent block. In the presence of 0.5 mM lidocaine, the relative amount of sodium current which flows with each successive depolarization decreases until the amount of block occurring with each depolarization equals the amount of recovery from block occurring during repolarization. This is termed the steady state. As the rate of stimulation increases, there is less time for the drug to diffuse from the binding site (unbinding kinetics depicted as the *broken line*) and accumulation of more block occurs until the new steady state is achieved. The more the channel is "used" the greater the degree of blockade. (Reproduced with permission from Hille B. Ionic channels in excitable membranes. Sunderland, MA: Sinauer Associates, 1992:261.)

conduction in normal tissue (increased resting block) while interfering little with conduction in abnormal (depolarized or ischemic) tissue containing inactivated channels.

CLASSIFICATION OF LOCAL ANESTHETIC (ANTIARRHYTHMIC) AGENTS

In 1970, Vaughan Williams[18] proposed perhaps the most useful scheme to categorize antiarrhythmic agents. This scheme was based on the presumed mechanism of action of the drugs in suppressing ventricular ectopy and arrhythmias (Table 3-1). Through the years, this system has been modified[19-21] to include subclassification of the type I antiarrhythmic (local anesthetic) drugs into subclasses a, b, and c (Table 3-2). This subclassification is based upon the effects of various types of local anesthetic-like agents on conduction velocity and repolarization properties of the action potential of canine Purkinje fibers. While this system provides a useful clinical tool, such classification, especially as related to local anesthetic type blocking agents, suffers from several inadequacies.[22,23] These notwithstanding, the Vaughan Williams schema of antiarrhythmic drug classification can aid in understanding the mechanisms of action of the local anesthetic type of antiarrhythmic agents. The function of the sodium channel with applications of the Vaughan Williams scheme and the Modulated Receptor model can be used to examine the varying effects of the class I antiarrhythmic agents. Specifically, looking at each subclass of these agents can enable us to predict the mechanisms by which they exert their effect.

The drugs categorized as class Ia agents (Tables 3-2 and 3-3) interact with the sodium channel slowly, having a binding constant in the

TABLE 3-1. VAUGHAN WILLIAM'S CLASSIFICATION OF ANTIARRHYTHMIC DRUGS

Class	Action(s)	Cardiac Effects
I	Local anesthetic (Na⁺ channel blockade)	Alter conduction and APD
II	β-Adrenergic blockade	Antagonize action of catecholamines
III	Variable/mixed	Prolongs APD
IV	Ca²⁺ channel blockade	Slow AV node conduction; inhibits triggered activity

APD, action potential duration; AV, artrioventricular.

TABLE 3-2. SUBCLASSIFICATION OF CLASS I ANTIARRHYTHMIC DRUGS

a: Prolong APD; Moderate ↓ in Conduction	b: Little Effect or Shortens APD; Mild ↓ in Conduction	c: No Effect on APD; Marked ↓ in Conduction
Quinidine	Lidocaine	Encainide
Procainamide	Mexiletine	Flecainide
Disopyramide	Phenytoin	Propafenone
Cibenzoline	Tocainide	Lorcainide
Pirmenol	Moricizine (?)	

APD, action potential duration.

range of 1 to 3 sec and a recovery time (unbinding) constant of 1 to 10 sec. This means that the accumulation of block at the level of the channel occurs gradually with a steady-state level of block being achieved in approximately 10 to 20 action potentials. Because block accumulates and is greater at faster than at slower heart rates, class Ia agents would seem to be particularly effective against tachycardias.

Class Ib agents (Tables 3-2 and 3-3) interact very rapidly with the sodium channel and have binding constants in the range of 0.1 to 1.0 seconds. These agents also unbind rapidly from the channel and have short recovery time constants (0.1 to 0.8 sec). The steady-state level of block is reached rapidly—usually in only a few beats—and recovery from block is rapid. As a result, these drugs move on and off the receptor site quickly in normal tissue. In the presence of prolonged channel inactivation or cellular depolarization, the time constant of recovery (unbinding) is slowed markedly. These agents bind preferentially to inactivated channels and exert their greatest effect at very rapid heart rates or in depolarized tissue as seen during ischemia.

Class Ic antiarrhythmic drugs (Tables 3-2 and 3-3) interact very slowly with the sodium channel receptor site, demonstrating a binding time constant of 1 to 8 sec. These agents require greater than 20 cellular depolarizations to reach their steady-state level of block. The amount of block per action potential is small, and the recovery from block (unbinding time constant) is also very long, exceeding 10 sec. These drugs would be expected to move on and off the receptor site in a very limited fashion and therefore affect channels in their resting state to a greater degree than the other two subclasses. This would result in the slow accumulation and relief of block with poor differentiation between normal and abnormal (depolarized) tissue.

TABLE 3-3. CHARACTERISTICS OF CLASS I ANTIARRHYTHMIC DRUGS

Drug	Class	pK$_a$	Amount Protein Bound (%)	Mode of Elimination	Estimated Half-Life (hr)	Therapeutic Serum Concentration (µg/ml)	Active Metabolites	Recovery Time Constant (sec)
Quinidine	Ia	8.0	90	H/R	4–8	1.5–4	Yes	4–5
Procainamide	Ia	8.9	15	R	2–4	4–16	Yes	2–3
Disopyramide	Ia		~40	R	6–8	2–4	Yes	2–3
Lidocaine	Ib	7.9	50	H	1–3	1–5	Yes	0.3–0.5
Mexiletine	Ib		~40	H	10–24	0.5–2	No	0.4–0.6
Phenytoin	Ib	10.1	90	H	18–30	10–20	No	0.6–0.9
Tocainide	Ib		50	H/R	10–15	6–12	No	0.2–0.3
Encainide	Ic		70	H	<2		Yes	10–15
Flecainide	Ic		40	H/R	14–20	0.4–0.8	No	20
Propafenone	Ic		95	H	6–13		Yes	15–16
Moricizine	Ib(?)		95	H/R	6–14		Yes	13–14
Bupivacaine		8.1	95	H	2–4		No	1.5
Mepivacaine		7.6	75	H	1–3		No	
Tetracaine		8.5					No	
Benzocaine		3.5					No	
Etidocaine		7.7	95	H	2–4		N	
Pirmenol	Ia				4–15		?	
Cibenzoline	Ia/III				4		?	

H, hepatic; R, renal.

The balance between the development of block during the action potential and the recovery from block during diastole determines the degree of channel blockade (Figure 3-8). When the recovery from block is very fast (as with lidocaine) channels may be almost 100% blocked at the start of diastole yet almost totally recovered by the beginning of the next depolarization at normal heart rates (≤120 beats/min). Conversely, when recovery from block is slow (as with flecainide), significant block will still exist when the next conducted impulse begins. As noted earlier, the time constant of recovery in most cases is strongly voltage and pH dependent[13,14,24] and under conditions when arrhythmias are most likely to occur, i.e., depolarized myocardium and decreased pH of ischemia, the recovery from block may be dramatically altered.[1,3–5,11,13,14,24] Acidosis, as occurs in ischemic tissue, promotes the cationic form of the drug. This form dissociates more slowly from the receptor site resulting in a greater amount of sodium channel block at a given heart

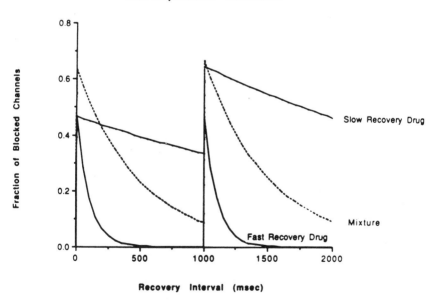

FIGURE 3-8. Demonstration of the concept of fast recovery and slow recovery drugs. The relative number of sodium channels blocked with each depolarization is shown on the ordinate. With the fast recovery drug, by the time the next depolarization occurs drug has diffused off all channels and little residual block remains. The slow recovery drug diffuses from the binding site much more slowly, and as a result there is still a significant amount of channel blockade present when the next depolarization occurs. Mixtures of drugs result in varying kinetics (see text).

rate. Most local anesthetic class drugs therefore can selectively depress conduction in acidotic tissue. Conversely, alkalinization promotes the neutral form of most drugs[13,14] and hence accelerates recovery and reduces sodium channel block. The difference between the sodium channel blocking capability of an agent under "arrhythmogenic" conditions compared to that under normal conditions determines the therapeutic safety of the drug. Agents that have short time constants of recovery such as lidocaine discriminate well between normal and abnormal myocardium. Agents with longer time constants of recovery do not discriminate as well between normal and abnormal tissues so they would be expected to have lower therapeutic indices with more potential for arrhythmogenic side effects (proarrhythmia).

In some cases, two agents with sodium channel blocking properties may be administered simultaneously and display effects at the level of the sodium channel. In the presence of two channel blocking agents that have different kinetic properties, one would expect modified amounts of diastolic block. This interaction of drugs could act in one of two ways—produce a synergistic or additive effect in the amount of total sodium channel blockade or an antagonistic effect with an overall decrease in the level of block.

A synergistic effect for a drug combination would produce more diastolic block than could be achieved with either drug alone. A good clinical example would be the combination of mexiletine (an oral lidocaine congener) with either quinidine or disopyramide. As discussed earlier (Table 3-2), the class Ia agents prolong action potential duration and therefore would be expected to increase the amount of time the sodium channel spends in the depolarized or inactivated state. In that the class Ib agents preferentially block inactivated sodium channels, one would expect a synergistic response in therapy[25] and an increase in overall channel blockade.

Conversely, an antagonistic effect may result when the choice of a drug combination results in a substantial "mismatch" in the binding and unbinding properties of the agents. Addition of a second antiarrhythmic agent may result in competition between the two agents for the channel receptor sites, resulting in the displacement or exclusion of one agent from binding and a subsequent overall reduction in drug effect. In this case, the drug with the shorter recovery time constant will competitively displace the agent with the slower kinetics from the receptor. The channels will recover more rapidly from block such that at the end of the diastolic interval, the total amount of block will be reduced. A clinical example of such competition is seen with lidocaine and one of its metabolites, glycine xylidide (GX).[26] In this report, patients with ventricular tachycardia that was originally controlled with intravenous lidocaine developed recurrence of tachycardia in spite of continued therapeutic lidocaine levels. These patients were found to have high levels of the GX metabolite. Subsequent cellular experiments demonstrated that the GX metabolite had significantly faster kinetics at the level of the sodium channel and was capable of displacing lidocaine from the binding site. As GX has no demonstrable antiarrhythmic action, such action could account for loss of efficacy of lidocaine despite therapeutic blood levels. Similar demonstrations of antagonism between two sodium channel blocking agents at both the clinical and experimental level have been reported for propoxyphene/lidocaine,[27] amitriptyline/phenytoin,[28] and aprindine/lidocaine.[29]

PROPERTIES OF INDIVIDUAL CLASS I AGENTS

In this and the following section, the electrophysiologic and toxicologic properties of selected class I antiarrhythmic agents will be discussed (Table 3-3). For a more complete discussion of individual drugs, readers are referred to any one of several reviews.[3,15,30]

Quinidine

Quinidine is a type Ia antiarrhythmic agent available in oral and injectable (intramuscular, intravenous) forms as a sulfate, polygalactoronate, or gluconate salt. Quinidine is well absorbed orally (70 to 90% bioavailability) and reaches peak serum concentrations within 1 to 3 hr after administration. It is metabolized via hydroxylation in the liver and has a serum half-life of 4 to 7 hr.

Experimentally, quinidine reduces automaticity in Purkinje fiber preparations by decreasing the rate of rise of phase 4 depolarization. Quinidine also depresses phase 0 of the action potential as do all sodium channel blockers. Quinidine prolongs both atrial and ventricular effective refractory period (ERP) and action potential duration (APD) measurements.

Clinically, quinidine has a well described "vagolytic effect," which can increase baseline heart rate, shorten atrioventricular (AV) nodal refractoriness, and enhance AV nodal conduction. The vagolytic effect coupled with the basic electrophysiologic properties tend to cancel each other and result in no change in the PR interval on the electrocardiogram (ECG). By using invasive electrophysiologic evaluation, it has been found that the ERPs of the atria, ventricles, and accessory pathways are prolonged from the predrug state.

Procainamide

Procainamide is also a type Ia antiarrhythmic agent available in oral and injectable (intravenous) forms. Procainamide is well absorbed orally (80 to 90% bioavailability) and reaches peak serum concentrations within 15 min to 2 hr after oral administration. It is metabolized via acetylation in the liver and has a serum half-life of 2 to 4 hr. Its metabolite (N-acetyl-procainamide) also has antiarrhythmic activity with an efficacy approaching 50 to 70% of the parent compound. N-Acetylprocainamide has a half-life almost twice as long as procainamide.

Experimentally, procainamide has basic electrophysiologic properties similar to quinidine. Of interest clinically, procainamide is much less vagolytic (though it still has some effect). ERPs of the atria, ventricles, His-Purkinje system, and accessory pathways are prolonged while AV nodal ERP remains unchanged or shortened. Procainamide slows conduction in ventricular muscle, which may be reflected by prolongation of the PR interval and widening of the QRS complex.

Disopyramide

This agent also is a type Ia drug with experimental and clinical electrophysiologic effects similar to quinidine and procainamide. Clinically, disopyramide prolongs refractoriness in the atria, ventricles, His-Purkinje system, and accessory bypass tracts. QRS and QT intervals on the ECG are lengthened while the PR interval may remain unchanged because of the vagolytic nature of the drug.

Disopyramide is available in an oral form only, is well absorbed (70 to 80% bioavailability) and reaches peak serum concentrations within 2 hr after administration. It is excreted unchanged in the urine (50 to 60%) but also undergoes limited metabolism in the liver, which produces a somewhat active metabolite (N-monodealkylated disopyramide). The serum half-life of the parent compound ranges from 4 to 10 hr. In patients with significant renal impairment, the half-life may double.

Clinically, disopyramide has a well-documented vagolytic effect that results in a number of side effects (see below). The drug may increase baseline heart rate, shorten AV nodal refractoriness, and enhance

AV nodal conduction indirectly while slowing conduction via a direct effect. The combined result is no change in the PR interval on the ECG.

Lidocaine

Lidocaine is a type Ib antiarrhythmic agent available in an injectable intravenous form. Oral absorption is poor (20% bioavailability) and over 90% of a dose is metabolized by the liver into two major metabolites: monoethyl glycine xylidide (70 to 80% as potent at lidocaine) and glycine xylidide (5 to 10% as potent as the parent). Its serum half-life is 1 to 2 hr.

Experimentally, lidocaine has such rapid blocking and unblocking characteristics that its effects on normal cardiac tissues is minimal. Small decreases in conduction are seen in His-Purkinje tissue and ventricular muscle. Lidocaine reduces automaticity in Purkinje fiber preparations and depresses phase 0 of the action potential in a use-dependent fashion. As discussed above, lidocaine has an affinity for inactivated sodium channels as found in ischemic tissue and under these conditions may significantly alter ERP, APD, and conduction velocity measurements.

Clinically, lidocaine has little or no effect on atrial or ventricular ERP. Effects on AV nodal refractoriness are slight. In a small percentage of patients (<5%), conduction over an accessory bypass tract may be enhanced with lidocaine administration. This may result in increased rates of ventricular response over the bypass tract during atrial fibrillation.

Mexiletine

Mexiletine is an oral congener of lidocaine and another type Ib antiarrhythmic agent. It is available in an oral form only. Oral absorption is excellent (90% bioavailability), and the drug is strongly protein bound. Hepatic metabolism in the major route of elimination (over 85% of the dose is metabolized by the liver). Its onset of action peaks within 2 hr and the serum half-life is 10 to 12 hr. Experimentally, mexiletine has electrophysiologic properties similar to lidocaine.

Clinically, mexiletine has little or no effect on atrial or ventricular ERP. Effects on AV nodal refractoriness are slight. Interestingly, as described above, the effects of mexiletine appear to be enhanced in the presence of other agents that prolong the repolarization phase of the action potential. This prolongation results in the sodium channel remaining in the inactivated state for a greater percentage of the APD. As type Ib drugs have their greatest affinity for inactivated sodium channels, enhanced efficacy of these drugs may be seen in combination therapy with type Ia agents.

Diphenylhydantoin (phenytoin)

Phenytoin is a type Ib antiarrhythmic agent available in an oral and injectable form. Oral absorption of the drug is slow, but bioavailability is good (90%). The half-life of the drug varies considerably between 24 and 36 hr. More than 85% of a dose of the drug is metabolized by the liver via hydroxylation. This metabolite undergoes glucuronidation with the product being excreted primarily by the kidneys.

Experimentally, diphenylhydantoin decreases resting membrane potential in a number of tissues and depresses conduction. APD and ERP are both shortened to varying degrees, and phase 4 depolarization is slowed.

Clinically, phenytoin has minor effects on atrial and ventricular ERP. Effects on AV nodal refractoriness are slight, but AV nodal conduction may be enhanced. QRS duration is not affected usually with phenytoin therapy, but the QT interval may shorten as with lidocaine administration. Along these lines, phenytoin has been used as a "competitive" sodium channel blocker in cases of digitalis and tricyclic antidepressant toxicity, displacing these drugs from the channel and decreasing the toxic effects. It also may be of benefit in treatment of the long QT syndrome though, from a clinical standpoint, it has fallen out of use (because of the development of other antiarrhythmic agents with better efficacy) except in the pediatric population.

Tocainide

Tocainide is a type Ib antiarrhythmic agent available in oral form. Oral absorption of the drug is excellent, and bioavailability is good (95%). The half-life of the drug ranges between 13 and 15 hr. Elimination is via both hepatic and renal mechanisms, and no major active metabolites are known. With oral administration, the onset of action occurs within 2 hr. In the presence of renal insufficiency or failure, the duration of action may be significantly prolonged.

Experimentally, tocainide is similar to lidocaine (see above). APD and ERP are both shortened, but there is little effect on intraatrial or intraventricular conduction or sinus node function.

Clinically, tocainide has minor effects on atrial and ventricular ERP. Effects on AV nodal refractoriness are slight. QRS duration is not affected, but the QT interval may shorten as with lidocaine administration. Combination therapy with type Ia agents may increase efficacy of arrhythmia suppression as described with mexiletine.

Encainide

Encainide is a type Ic agent used for therapy of life-threatening ventricular arrhythmias. Available in oral form, the drug is well absorbed orally (85% bioavailability) and reaches peak serum concentrations within 1 to 2 hr after administration. It is metabolized extensively in the liver (80%) such that the half-life of the parent compound is only 3 to 4 hr. In spite of the short half-life of the parent, one of the metabolites (*O*-desmethyl-encainide) has very active antiarrhythmic action and a half-life of 8 to 12 hr. In rapid metabolizers, the active drug predominantly is *O*-desmethyl-encainide, while in slow or poor drug metabolizers the active drug is encainide, but the half-life of the parent compound is extended to 6 to 12 hr. These convenient pharmacokinetics made the dosing regimens the same (three times per day) in extensive or poor metabolizers.

Experimentally, encainide reduces phase 4 automaticity and depresses phase 0 of the action potential. APD is increased as are the ERPs of all cardiac tissues. Sinus node automaticity is affected minimally, but conduction in atrial, AV nodal, His-Purkinje, and ventricular tissue preparations is slowed markedly.

Clinically, encainide lengthens PR, QRS, and QT duration on the ECG while the ERPs of the atria, AV node, and ventricles are prolonged. Accessory pathway tissue ERPs also are prolonged and make this drug an excellent choice in the medical therapy of Wolff-Parkinson-White syndrome.

Flecainide

The first of the type Ic agents approved by the Food and Drug Administration for therapy of life-threatening ventricular arrhythmias, flecainide also has lost popularity due to the Cardiac Arrhythmia Suppression Trial study. Available in oral form, the drug is well absorbed orally (95% bioavailability) and reaches peak serum concentrations within 2 to 4 hr after administration. It is metabolized in the liver (70%) and excreted unchanged in the urine (30%). The serum half-life is 12 to 24 hr.

Experimentally, flecainide reduces phase 4 automaticity while increasing ERP in ventricular and His-Purkinje tissues. Sinus node automaticity is affected minimally. There is marked depression of phase 0 of the action potential and overall APD is increased. Studies examining the effect on conduction velocity shows pronounced slowing in all cardiac tissues.

Clinically, flecainide significantly lengthens PR, QRS, and QT duration. By using invasive electrophysiologic evaluation, it has been found that the ERPs of the atria, ventricles, His-Purkinje, and accessory pathway tissue are prolonged from the predrug state.

Propafenone

Propafenone is a new type Ic agent which, in addition to sodium channel blocking activity, has β-blocking properties and at very high doses some calcium channel blocking effects. Available in oral form, the drug is well absorbed orally (95%) but undergoes extensive "first-pass" metabolism and elimination by the liver. Similar to a number of other antiarrhythmic agents, propafenone metabolism is genetically determined and varies extensively with more than 10% of patients being poor metabolizers. The parent compound therefore has a wide half-life range (1 to 12 hr; mean 6 hr). The major metabolites (5-hydroxy-propafenone and *N*-debutylpropafenone) have some sodium channel blocking activity. The drug reaches peak serum concentrations within 1 to 3 hr after administration.

Experimentally, propafenone has effects similar to those described for encainide and flecainide (see above). Its structure is similar to that of propranolol and its strong membrane-stabilizing and weak β-blocker properties are probably due to this resemblance. Calcium channel blockade occurs at high doses only, and the mechanism is unclear.

Clinically, propafenone acts similarly to the other type Ic agents. It significantly lengthens PR, QRS, and QT duration on the ECG and increases the ERP of the atria, ventricles, His-Purkinje, and accessory pathway tissue on electrophysiologic evaluation.

Moricizine

Moricizine is a phenothiazine derivative with local anesthetic activity, sodium channel blocking activity, and membrane-stabilizing activity similar to the type Ia agents but with effects on conduction and repolarization similar to the type Ib or Ic agents.

Available in oral form, the drug is well absorbed (95%) but undergoes "first-pass" metabolism and elimination by the liver. However, none of its 26 metabolites seems to have significant antiarrhythmic activity. Bioavailability of the parent drug is only 35 to 40%. The drug reaches peak serum concentrations within ½ to 2 hr after administration and has a half-life of approximately 6 hr.

Basic electrophysiologic studies show that moricizine shortens repolarization while decreasing V_{max} in Purkinje fiber preparations. In animal models, no effect on sinus or AV nodal function has been described and ventricular ERP does not appear to change.

In man, moricizine also does not appear to affect sinus node function, but conduction through the AV node is decreased and the AH interval lengthens. On the ECG, the PR and QRS duration are prolonged, but the QT interval does not change appreciably. Ventricular and atrial ERPs are not altered, but AV nodal and accessory pathway refractoriness has been described as being prolonged.

TOXICITIES AND SIDE EFFECTS OF INDIVIDUAL CLASS I AGENTS

From a basic toxicologic point of view, the local anesthetic class antiarrhythmic agents are quite toxic. Fortunately, in clinical usage the dosage required to achieve antiarrhythmic effects (or local anesthesia) generally falls well below the dosage at which side effects occur.

The primary effect of local anesthetic agents is the inhibition of propagated action potentials in peripheral nerve but, as discussed above, the effects of these agents are not limited to these tissues. Heart, central nervous system, and neuromuscular junction are all potential targets of local anesthetic action and as a result, may manifest toxic side effects.

Quinidine

Subjective toxicity from quinidine administration is quite common. More than 30% of all patients receiving short-term or long-term quinidine therapy will develop side effects. The predominant complaint with quinidine administration is gastrointestinal upset (nausea, vomiting, and diarrhea), some of which may be dose related. Other significant side effects include the production of drug fever and/or rash, tinnitus, and drug-induced thrombocytopenia (rare). An occasional patient will experience exacerbation of congestive heart failure symptoms, but in general the cardiodepressant side effects of quinidine are mild.

Quinidine has important interactions with at least two drugs—digoxin and amiodarone. In general, digoxin dosage should be halved when quinidine is added to the medical regimen, and quinidine dosages need to be adjusted (usually halved) when amiodarone is added.

The vagolytic actions of quinidine also may precipitate problems. Quinidine may increase the ventricular response to atrial fibrillation by enhancing AV nodal conduction in the absence of an accompanying AV nodal blocking agent. In the case of atrial flutter, quinidine may result in a slowing of the flutter rate and enhancement of AV nodal conduction such that 2:1 atrial:ventricular conduction will be converted to a 1:1 response. By slowing conduction in ventricular muscle, quinidine may stabilize an arrhythmia circuit and actually precipitate worsening of preexisting ventricular tachycardia (proarrhythmia). More frequently (up to 5% of cases), this slowed conduction may result in the prolongation of the QT interval and lead to torsade de pointes.

Procainamide

Toxicity from procainamide administration is usually minimal although significant side effects may be seen. Gastrointestinal upset (nausea, vomiting, and diarrhea) occurs in 5 to 15% of patients. Other potential side effects include the production of drug fever and/or rash and rare drug-induced thrombocytopenia. The side effect of most concern during procainamide administration is the well-documented production of a systemic lupus-like reaction in 10 to 20% of patients. There is some evidence that this is more likely to occur in patients who are slow acetylators of the drug, but this has not been confirmed. A majority (65 to 75%) of patients will develop a positive antinuclear antibody titer if procainamide treatment is continued for 1 year or more, but this in and of itself does not dictate discontinuation of the drug if no systemic side effects are present. The cardiodepressant side effects of procainamide are mild and occur predominantly with intravenous administration. In a manner similar to quinidine, procainamide may stabilize an arrhythmia circuit and precipitate worsening of a preexisting ventricular tachycardia. Prolongation of the QT interval and production of torsade de pointes also are risks.

Disopyramide

Disopyramide's subjective toxicity stems almost entirely from its profound anticholinergic activity (almost 1% that of atropine). Complaints of dry mouth, blurred vision, urinary retention (especially in older men), constipation, nausea, vomiting, and worsening of glaucoma occur in 10 to 30% of patients. The most significant adverse reaction to therapy with disopyramide, though, is worsening of congestive heart failure in almost half of all patients with a previous history of failure. This is a result of the drug's marked depression of myocardial contractility. Disopyramide also slows conduction, which can result in the production of complete AV block, bundle branch block, and prolonged QRS duration. Proarrhythmia and torsade de pointes have been seen with disopyramide therapy.

Lidocaine

Toxicity from lidocaine administration is a well-described, dose-related effect. The most common toxic response is in the form of central nervous system side effects. These include slurred speech, numbness, paresthesia, dulled sensorium or drowsiness, confusion, tremors, and seizures. Other side effects include nausea, vomiting, hypotension (rare), AV block (rare), and idiopathic drug reactions in the form of drug rash or fever.

Mexiletine

Central nervous system toxicity from mexiletine administration is a dose-related effect. The most common toxic central nervous system side effects are slurred speech, numbness, paresthesia, dulled sensorium or drowsiness. Other side effects seen with type Ib agents (nausea, vomiting, hypotension, AV block, and idiopathic drug reactions) are rare. Drug-induced hepatitis has been seen, and rare instances of proarrhythmic responses have been reported. There are no reported cases of torsade de pointes.

Diphenylhydantoin

Phenytoin has a number of side effects. When administered orally, the drug may cause dizziness, vertigo, diplopia, nystagmus, drowsiness, and ataxia. Idiosyncratic skin reactions and the Stevens-Johnson syndrome are reported also. Gingival hyperplasia is not uncommon. Intravenously, the major acute side effect is profound hypotension caused in part by myocardial depression. In most cases this may be avoided by infusing the drug slowly and carefully, monitoring hemodynamics.

Tocainide

Similar to lidocaine and mexiletine, the side effect spectrum of tocainide focuses basically on gastrointestinal and neurologic disturbances. Anorexia, nausea, vomiting, constipation, and abdominal discomfort occur in 10 to 20% of patients and tremors, dizziness, paresthesia, confusion, etc., in another 10 to 15%. Drug-induced rash and exacerbation of preexisting symptoms of congestive heart failure are rare. Of greatest significance are the rare (<1%) but devastating side effects of drug-induced agranulocytosis and pulmonary fibrosis. With other, similar, medications available, this makes tocainide an unappealing choice for chronic antiarrhythmic therapy.

Encainide

Encainide therapy usually is well-tolerated with less than 2 to 3% of patients stopping the drug because of side effects. Noncardiac side effects that result in termination of therapy include headache, dizziness, tinnitus, diplopia, vertigo, leg cramps, and a metallic or "tinny" taste in the mouth. While these generally are dose-related and resolve with dose reduction, some patients find them intolerable.

Otherwise, the most significant side effect of encainide therapy (and a major problem) is the potential to cause a proarrhythmic response. In patients with benign ventricular arrhythmias, proarrhythmia occurs in fewer than 4% of cases. In patients with more worrisome ventricular ectopy (nonsustained ventricular tachycardia with evidence of hemodynamic compromise; polymorphic nonsustained ventricular tachycardia; sustained ventricular tachycardia; ventricular fibrillation), development of proarrhythmia is more of a problem. The "classic" form of proarrhythmia induced by encainide therapy is a slow (140 to 170 beats/min) incessant ventricular tachycardia. This may occur in up to 10% of patients and often cannot be terminated even with cardioversion.

Encainide has little effect on the sinus node and usually will not exacerbate preexisting sinus node dysfunction. It can, however, aggravate or precipitate high-degree AV block at the level of either the AV node or the His bundle because of profound conduction depression in these areas. As with flecainide, an increase in acute and chronic pacing thresholds may be seen.

Flecainide

Toxicity from flecainide is usually minimal and in general the drug is very well tolerated with treatment being terminated in less than 10% of patients. Noncardiac side effects are dose related and are predominantly a result of central nervous system complaints (dizziness, visual disturbances, and headache). The cardiovascular side effects of flecainide are of greater concern. Flecainide depresses left ventricular function and, especially in patients with a history of previous congestive heart failure, presents a significant risk to the patient with underlying decreased left ventricular function.

Perhaps the greatest concern with flecainide therapy, though, is the occurrence of ventricular proarrhythmic responses with the drug. In patients with benign ventricular arrhythmias, proarrhythmia may occur in up to 5% of cases. In patients with a history of sustained ventricular tachycardia, development of proarrhythmia, including incessant ventricular tachycardia, may occur up to 20% of the time.

Flecainide may worsen preexisting sinus node dysfunction, aggravate or precipitate high-degree AV block, increase acute and chronic pacing thresholds, and increase (slightly) digoxin levels.

Propafenone

As with the other type Ic drugs, therapy with propafenone usually is well tolerated. Subjective complaints (2 to 3% of patients) include dizziness, tinnitus, a bitter metallic taste, nausea, vomiting, and constipation. Incidents of a drug-induced rash (2%) have been reported. Because of its β-blocker actions, some patients also complain of fatigue, tiredness, and general sluggishness. Exacerbation of AV (His-Purkinje) block, worsening of congestive heart failure resulting from the negative inotropic effects of the drug, and proarrhythmia (5–10%) are the major side effects.

Moricizine

The most common adverse effects of this medication include dry mouth, nausea, vomiting, diarrhea, dyspepsia, headache, dizziness, vertigo, paresthesia, and fatigue. No significant end-organ toxicities have, as yet, been described and the drug appears to have only minimal significant cardiodepressor effects. Proarrhythmia, while described less often with this agent than the other type Ic drugs (3 to 4%), does occur and can be life-threatening.

Due to its longer sodium channel blockade and its ability to elicit ventricular tachyarrhythmias,[31] bupivacaine has characteristics similar to the other type Ic drugs. Consequently, it should be administered with particular caution to patients already receiving a drug of this subclass, since the additive effects could increase the risk of arrhythmias.

References

1. Grant AO. On the mechanism of action of antiarrhythmic agents. Am Heart J 1992;123:1130.
2. Hondeghem LM, Katzung BG. Time- and voltage-dependent interactions of antiarrhythmic drugs with cardiac sodium channels. Biochim Biophys Acta 1977;472:373.
3. Carpenter RL, Mackey DC. Local anesthetics. In: Barash PG, Cullen BF, Stoelting RK, ed. Clinical anesthesia. Philadelphia, JB Lippincott, 1989:509.
4. Courtney KR, Strichartz GR. Structural elements which determine local anesthetic activity. In: Strichartz GR, ed. Local anesthetics: handbook of experimental pharmacology. New York: Springer-Verlag, 1987:53.
5. Butterworth JF, Strichartz GR. Molecular mechanisms of local anesthesia: a review. Anesthesiology 1990;72:711.
6. Catterall WA. Common modes of drug action on sodium channels: local anesthetics, antiarrhythmics and anticonvulsants. Trends Pharmacol Sci 1987;8:57.
7. Catterall WA. Structure and function of voltage-sensitive ion channels. Science 1988;242:50.
8. Numa S. A molecular view of neurotransmitter receptors and ionic channels. Harvey Lect 1989;83:121.
9. Hille B. Ionic channels in excitable membranes. Sunderland, MA: Sinauer Associates, 1992:261.
10. Hille B. Ionic selectivity, saturation and block in sodium channels. A four barrier model. J Gen Physiol 1975;66:535.
11. Hille B. Local anesthetics: hydrophilic and hydrophobic pathways for the drug-receptor reaction. J Gen Physiol 1977;69:497.
12. Hondeghem LM. Antiarrhythmic agents: modulated receptor applications. Circulation 1987;75:514.
13. Schwarz W, Palade PT, Hille B. Local anesthetics: effects of pH on use-dependent block of sodium channels in frog. Biophys J 1977;20:343.
14. Strauss HC, Broughton A, Starmer CF, Grant AO. pH potentiation of local anesthetic action in heart muscle. In: Zipes DP, Jalife J, eds. Cardiac electrophysiology and arrhythmias. New York: Grune and Stratton, 1985;217.
15. DiMarco JP. Antiarrhythmics. In: Chernow B, ed. Essentials of critical care pharmacology. Philadelphia: Williams & Wilkins, 1989;168.

16. Courtney KR. Frequency-dependent inhibition of sodium currents in frog myelinated nerve by GEA 968, a new lidocaine derivative. PhD dissertation. University of Washington, 1974. Ann Arbor, MI: University Microfilms International, No. 74-29, 393.
17. Courtney KR. Mechanism of frequency-dependent inhibition of sodium currents in frog myelinated nerve by lidocaine derivative GEA 968. J Pharmacol Exp Ther 1975;195:225.
18. Vaughan Williams EM. Classification of antiarrhythmic drugs. In: Sandoe E, Glensted-Jensen E, Olsen EH, eds. Symposium of cardiac arrhythmias. Elsinore, Denmark: AB Astra, 1970;449.
19. Vaughan Williams EM. A classification of antiarrhythmic actions reassessed after a decade of new drugs. J Clin Pharmacol 1984;24:129.
20. Vaughan Williams EM. Subdivision of class I antiarrhythmic drugs. In: Reiser HJ, Horowitz LN, eds. Mechanisms and treatment of cardiac arrhythmias. Relevance of basic studies to clinical management. Baltimore: Urban and Schwarzenberg, 1985;165.
21. Harrison DC. Is there a rational basis for the modified classification of antiarrhythmic drugs? In: Morganroth J, Moore EN, eds. Cardiac arrhythmias. New therapeutic drugs and devices. Boston: Martinus Nijhoff, 1985;36.
22. Campbell TJ. Subclassification of class I antiarrhythmic drugs. In: Vaughan Williams EM, ed. Antiarrhythmic drugs, handbook of experimental pharmacology. New York: Springer-Verlag, 1989;135.
23. Cobbe SM. Clinical usefulness of the Vaughan Williams classification system. Eur Heart J 1987;8(suppl A):65.
24. Courtney KR. pH and voltage dependence of INa recovery kinetics in atrial cells exposed to lidocaine. Am J Physiol 1988;24:H1554.
25. Duff HJ, Roden D, Primm RK, Oates JA, Woosley RA. Mexiletine in the treatment of resistant ventricular arrhythmias: enhancement of efficacy and reduction of dose-related side-effects by combination with quinidine. Circulation 1983;67:1124.
26. Bennett PB, Woosley RL, Hondeghem LM. Competition between lidocaine and one of its metabolites, glycylxylide, for cardiac sodium channels. Circulation 1988;78:692.
27. Whitcomb DC, Gilliam FR III, Starmer CF, Grant AO. Marked QRS complex abnormalities and sodium channel blockade by propoxyphene reversed with lidocaine. J Clin Invest 1989;84:1629.
28. Barber MJ, Starmer CF, Grant AO. Blockade of cardiac sodium channels by amitriptyline and diphenylhydantoin. Evidence for two use-dependent binding sites. Circ Res 1991;69:677.
29. Kodama I, Toyama J, Yamada K. Competitive inhibition of cardiac sodium channels by aprindine and lidocaine studied using maximum upstroke velocity of action potential in guinea-pig ventricular muscle. J Pharmacol Exp Ther 1987;241:1065.
30. Zipes DP. Management of cardiac arrhythmias: pharmacological, electrical and surgical techniques. In: Braunwald E, ed. Heart disease: a textbook of cardiovascular medicine. Philadelphia: WB Saunders, 1992;628.
31. Clarkson CW, Hondeghem LM. Mechanism for bupivacaine depression of cardiac conduction: fast block of sodium channels during the action potential with slow recovery from block during diastole. Anesthesiology 1985;62:396.

John D. Gallagher

4 | Class III Antiarrhythmic Agents: Bretylium, Sotalol, Amiodarone

In 1980, I anesthetized a 54-year-old man who traveled from the Netherlands to Philadelphia to undergo mapping guided endocardial resection and left ventricular aneurysmectomy for medically refractory ventricular tachycardia. His antiarrhythmic regimen included an experimental agent, amiodarone. Although we reviewed the available clinical literature, the cardiologists caring for the patient assured us that the only relevant effect of amiodarone would be mild β-adrenergic blockade. As surgery proceeded, we dealt with atropine- and isoproterenol-resistant heart block; severe bradycardia requiring pacing; peripheral vasodilation unresponsive to epinephrine, norepinephrine, and metaraminol; and myocardial depression despite catecholamines, calcium, and intra-aortic counterpulsation. Over the next week, the hemodynamic abnormalities gradually resolved, and the patient left the hospital 2 weeks later, cured of his ventricular tachycardia.[1] This experience provided me not only with a publication, but stimulated continued interest in electrophysiologic interactions between anesthetics and antiarrhythmics.[2]

Potentially, many antiarrhythmic agents can interact with anesthetics. Class III antiarrhythmic agents, which prolong action potential duration and QT interval as their primary mode of action,[3] coupled with general anesthesia-induced prolongation of QT interval,[4] could incite early afterdepolarizations and torsade de pointe-type ventricular tachycardia.[5] Other actions of the class III agents, such as adrener-

Clinical Cardiac Electrophysiology: Perioperative Considerations
Edited by Carl Lynch III. J.B. Lippincott Company, Philadelphia, PA ©1994

gic antagonism by amiodarone[6] and sotalol,[7] may also interact with anesthetics. The goal of this chapter is to review the actions and use of the class III antiarrhythmic agents amiodarone, bretylium, and sotalol, emphasizing the possible interactions with anesthetic agents and the perioperative period. Where possible, I have cited primary sources. Reference to review articles occurs when excessive numbers of primary citations exist or to credit particular insights. Much of the pharmacokinetic and toxicity data derive from product literature for Betapace (sotolol), Cordarone (amiodarone), and Bretylol (bretylium).

EFFECTS OF AMIODARONE, BRETYLIUM, AND SOTALOL ON IONIC CURRENTS

Class III agents increase action potential duration in isolated cardiac tissues, reflected in the patient by QT interval prolongation. Complex interplay between repolarizing and depolarizing currents governs the duration of the cardiac action potential. Table 4-1 summarizes the effects of amiodarone, bretylium, and sotalol on the ionic currents of the plateau and repolarization phases of the action potential. Persistence of the inward sodium current (I_{Na}) during the action potential plateau forms the window current.[8] Patch clamp studies attribute this persistent current to the occasional entry of cardiac sodium channels into a gating mode in which the channels inactivate very slowly, often over several hundred milliseconds.[9] These long bursts appear more frequently in inside-out patches, suggesting that cytoplasmic components not present in the inside-out configuration inhibit this gating mode.[10]

Amiodarone reduces I_{Na} in both Purkinje cells and ventricular myocytes (Fig. 4-1).[11] Block of I_{Na} is strongly use dependent and involves rested and inactivated, but not open sodium channels.[11,12] Amiodarone decreases single channel opening frequency, without changing conductance or channel open time.[11] The window current is also blocked.[13] Neither bretylium nor sotalol binds to sodium channels or antagonizes I_{Na}.[14,15]

Both the L- type and T-type components of the slow inward current (I_{Ca}) have roles in cardiac cells.[16] The L-type calcium channels activate at depolarizations positive to −30 mV, are regulated by β-adrenergic agents, and are blocked by class IV antiarrhythmics such as nifedipine.[17,18] The T-type channels activate at more negative potentials, show a "tiny" conductance relative to the L-type channels, and are not affected by traditional calcium blockers or β-adrenergic agents.[16] Current carried by T-type channels contributes to the late phases of pace-

TABLE 4-1. PLATEAU AND REPOLARIZING CURRENTS

Current	Direction	Effect	Amiodarone	Bretylium	Sotalol
I_{Na} (window)	Inward Na	Prolongs AP, raises plateau	↓	0	0
Na$^+$-K$^+$-ATPase (pump current)	Outward Na	Shortens AP	↓	↓	0
Na/Ca exchange	Inward Na	Prolongs AP	↓	0	0
$I_{Ca(L)}$	Inward Ca^{2+}	Prolongs AP, raises plateau	↓	0	0
$I_{Ca(T)}$					↓0
I_{to} (transient outward)	Outward K$^+$	Causes phase 1	↓ ($I_{K,s}$)		
I_K (delayed rectifier)	Outward K$^+$	Shortens AP	↓	↓	↓ ($I_{K,r}$)
I_{K1} (inward rectifier)	Outward K$^+$	Shortens AP	↓	↓0	↓0
$I_{K(Ca)}$	Outward K$^+$	Shortens AP			
$I_{K(ATP)}$	Outward K$^+$	Shortens AP during ischemia			0

AP, action potential. ↓ inhibits current; 0 no effect on current; ↓0 species or tissue differences in current inhibition.

A

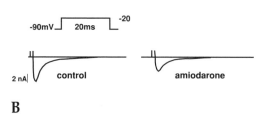

B

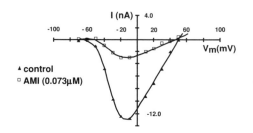

FIGURE 4-1. Block of I_{Na} by amiodarone. A, the effects of amiodarone (1.4 mM) on I_{Na} from a Purkinje fiber cell. Peak inward sodium current, the downward deflection in the current traces, is reduced by amiodarone during the step depolarization shown. B, repeating the experiment at several voltage steps allows construction of a current-voltage (I-V) curve. Amiodarone blocked I_{Na} at all test potentials without affecting the current voltage relationship. (Reproduced with permission from Follmer CH, Aomine M, Yeh JZ, Singer DH. Amiodarone-induced block of sodium current in isolated cardiac cells. J Pharmacol Exp Ther 1987;243:187.)

maker depolarization.[19] Although activation of I_{Ca} could increase action potential duration, I_{Ca} normally peaks within 3 msec and largely inactivates within 30 msec, suggesting a minimal contribution to the later portion of the plateau.[20,21]

Amiodarone blocks I_{Ca} in a use-dependent manner, binding to inactivated calcium channels (Fig. 4-2).[22] Because differing inactivation time courses allow separation of I_{Ca} into its L and T components, Cohen and co-workers[23] differentiated effects of amiodarone on each component of I_{Ca}. When binding equilibrates at normal diastolic potentials ($\leq$–70 mV), amiodarone more potently antagonizes T than L-type calcium channels (Fig. 4-3).[23]

In rabbit and guinea pig ventricular myocytes, sotalol does not reduce I_{Ca}.[15] Although norepinephrine release triggered by bretylium could augment I_{Ca},[18] the drug has no direct effects on the slow inward current.[24]

Blockade of potassium currents responsible for repolarization provides the primary mode of action of class III drugs. The inward rectifier (I_{K1}) maintains resting membrane potential by passing inward current at hyperpolarized membrane potentials. During depolarization, intracellular Mg^{2+} blocks I_{K1}, reducing outward current.[25] Inward rectification reduces potassium conductance during the action potential plateau, minimizing the magnitude of inward currents required to maintain or terminate the plateau phase. The repolarization of the cardiac action potential (phase 3) results initially from activation of the delayed rectifier current (I_K) terminating the action potential plateau and then the inward rectifier (I_{K1}), which provides the outward current

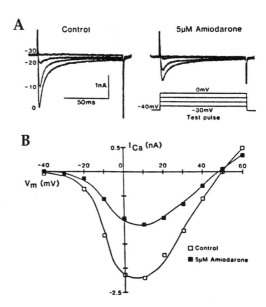

FIGURE 4-2. Amiodarone blocks I_{Ca}. A, Similar to Figure 4-1, amiodarone reduces the amplitude of the slow inward current during a series of depolarizations as shown. B, I-V relationship shows that amiodarone produces a 50% reduction in peak I_{Ca}. (Reproduced with permission from Nishimura M, Follmer CH, Singer DH. Amiodarone blocks calcium current in single guinea pig ventricular myocytes. J Pharmacol Exp Ther 1989;251:650.)

during the later half of the repolarization.[26] Estimation of respective current densities,[27] pharmacologic blockade of the delayed rectifier current(I_K) versus the inward rectifier current(I_{K1}),[28,29] and computational reconstructions of the action potential[26] support the combined importance of I_{K1} and I_K for repolarization.

Amiodarone in guinea pig ventricle[30] and sotalol in ventricular muscle and Purkinje fibers of the rabbit[31] as well as sheep Purkinje fibers[32] inhibit I_{K1}. In contrast, studies in guinea pig ventricular myocytes[33] and in rabbit and guinea pig ventricular myocytes[15] found that sotalol did not alter I_{K1}. Although all three drugs suppress hyperpolarization-induced inward currents (I_f) in rabbit sinus node cells,[34] inhibition by bretylium of I_{K1} in ventricular cells has not been demonstrated. In fact, Argentieri and co-authors[35] exclude bretylium from discussion as a class III agent because of its modest prolongation of action potential duration. Perhaps failure to block I_{K1} contributes to this limited effect on action potential duration.

Early descriptions of two components of the delayed rectifier—a rapidly activating component, and a kinetically slower, larger component[36]—were dismissed in favor of a one-component model. Class III antiarrhythmic agents, however, differentiate different components of I_K.[29] Table 4-2 summarizes the two components of the delayed rectifier. Amiodarone blocks the slowly activating component.[37] Differing effects of quinidine and amiodarone on I_K suggested that the drugs act by different mechanisms (Fig. 4-4).[37] Both bretylium and sotalol block I_K.[15,29,31,33,38] Sotalol blocks a component of I_K with a short time constant

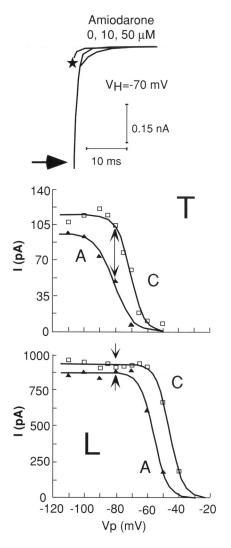

FIGURE 4-3. Amiodarone blocks both the T and L components of I_{Ca}. The *top trace* shows slow inward Ca^{2+} tail current traces from guinea pig atrial cells recorded in the presence of 0, 10 (peak shown by *arrow*), and 50 mM (peak shown by *star*) amiodarone. The current is reduced in a dose-dependent manner. The *lower panels* show the effects of amiodarone on T and L components of I_{Ca}, determined by analyzing the time course of the tail currents. Although both components are blocked, the *arrows*, at a normal membrane potential of –80 mV in each graph, show that block of the T-type current is more profound. (Reproduced with permission from Cohen CJ, Spires S, Van Skiver D. Block of T-type Ca channels in guinea pig atrial cells by antiarrhythmic agents and Ca channel antagonists. J Gen Physiol 1992;100:703.)

of 179 msec ($I_{K,r}$ for rapidly activating) (Fig. 4-5).[29] Surprisingly, while sotalol decreased I_{K1} in sheep Purkinje fibers, it only diminished I_K at the highest concentration used.[32]

Perhaps the clinically relevant actions of these antiarrhythmic agents will only be known when human tissues from various myocardial regions have been studied and their channels defined. Electrophysiologic characteristics of a delayed rectifier potassium channel cloned from human heart do not resemble either the rapid or slow components of the guinea pig delayed rectifier described above.[39]

TABLE 4-2. CHARACTERISTICS OF THE TWO COMPONENTS OF THE DELAYED RECTIFIER

Component	Rapidly Activating ($I_{K,r}$)	Slowly Activating ($I_{K,s}$)
Amplitude	Small	Large
Rectification	Inward	Linear
Run down	No	Yes
Quinidine block	Yes	Yes
Amiodarone block	No	Yes
Sotalol block	Yes	No
External divalent cations	Increases	No effect
Isoproterenol	No effect	Increases
Species[61]	Guinea pig ventricle	Guinea pig ventricle
	Sheep Purkinje fiber	Sheep Purkinje fiber
	Chick atrium	Chick atrium
	Rabbit sinoatrial node, atrioventricular node, Purkinje fiber	
	Cat ventricle	
		Frog atrium

Rather, the channel is similar to delayed rectifiers from adult rat atria and neonatal epicardial myocytes.[39]

Both voltage-dependent and Ca^{2+}-dependent transient outward potassium currents (I_{to}) exist.[40] In Purkinje fibers and epicardial ventricle, I_{to} governs phase 1 of the action potential and greatly influences action potential duration, especially in atrium. By reducing plateau voltage, I_{to} prevents both voltage-dependent activation of I_K and inactivation of $I_{Ca(L)}$.[41] In rabbit and guinea pig ventricular myocytes, sotalol did not affect I_{to}.[15] However, in sheep Purkinje fibers, sotalol decreased I_{to}.[32]

Ligand-gated potassium channels such as $I_{K(ATP)}$ and $I_{K(Ca)}$ are important determinants of action potential duration in a variety of circumstances. Increases in intracellular $[Ca^{2+}]$ activate $I_{K(Ca)}$[42] and modulate I_K,[43] thereby abbreviating the plateau and limiting further calcium influx. The ATP-gated channel[44] is closed in the presence of normal intracellular levels of ATP. However, during ischemia, ATP levels fall and the channel opens, causing efflux of potassium, hyperpolarization, and reduction in action potential duration. These effects protect the ischemic myocardium from damage by reducing Ca^{2+} entry, but may be arrhythmogenic.[45] During ischemia, elevation of extracellular $[K^+]$ depolarizes resting membrane potential, depresses I_{Na}, slows conduction, and along with activation of $I_{K(ATP)}$, shortens action potential duration.[46] Ventricular refractoriness initially decreases during

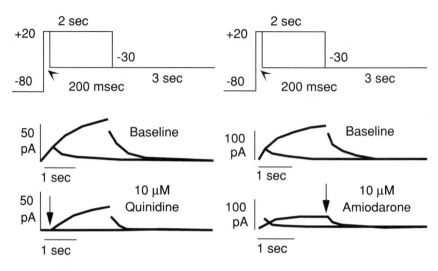

FIGURE 4-4. Effects of amiodarone and quinidine on rapidly and slowly activating components of the delayed rectifier current. For the short (200 msec) and long (2 sec) depolarizations shown in the *top traces*, outward potassium (I_K) recordings are shown superimposed in each current trace. Quinidine decreases the amplitude of both components; note especially the *arrow* showing block of the small rapidly activating component in the *lower left panel*. Amiodarone has little effect on the rapid component, but blocks the larger slowly activating component, shown as the *arrow* in the *lower right panel*. (Reproduced with permission from Balser JR, Bennett PB, Hondeghem LM, Roden DM. Suppression of time-dependent outward current in guinea pig ventricular myocytes. Actions of quinidine and amiodarone. Circ Res 1991;69:519.)

ischemia,[47] but later lengthens,[48] which can produce arrhythmogenic disparities in refractory period in the ischemic region.

Sotalol does not affect $I_{K(ATP)}$ channels.[49] However, this should not be misinterpreted. Glibenclamide, a K(ATP) channel antagonist when given before ischemia, prevents the shortening of action potential duration, but worsens postischemic dysfunction. Sotalol prolongs action potential duration, but neither slows the rate of action potential shortening during ischemia nor affects the recovery of systolic function.[49] In contrast, amiodarone strongly inhibits ATP-sensitive potassium channels.[50]

Action potential duration is also increased by Na^+ entry through Na^+-Ca^{2+} exchange mechanism.[51] The 3 Na^+:1 Ca^{2+} exchange ratio extrudes Ca^{2+} during the plateau of the action potential, producing a net inward Na^+ current.[51,52] The Na^+-K^+-ATPase pump generates a net outward sodium current. Normally membrane potential changes alter Na^+-K^+-pump activity minimally. However, if intracellular Na^+ concentration doubles, as it may in ischemia or cardiac glycoside toxicity, action potential duration may be shortened through increased activity of

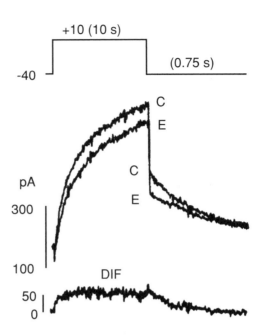

FIGURE 4-5. Sotalol blocks the rapidly activating component of the delayed rectifier. Current records before (control, C) and after drug administration are shown in the *middle traces*. In this example, the sotalol analog E-4031 was used (hence *E*), but results with sotalol were identical. The *lower trace* shows the difference between traces C and E, the component block by sotalol. Notice the small amplitude and the fast rise to a plateau, characteristic of $I_{K,r}$. (Reproduced with permission from Sanguinetti MC, Jurkiewicz NK. Two components of cardiac delayed rectifier K+ current: differential sensitivity to block by class III antiarrhythmic agents. J Gen Physiol 1990;96:195.)

the Na+-K+-pump.[53,54] Both amiodarone and bretylium inhibit Na+-K+-ATPase pump activity in guinea pig heart.[55-57] Bretylium competitively inhibits the actions of ouabain, suggesting similar binding sites.[57]

PROARRHYTHMIA AND CLASS III ACTIONS

The increased mortality produced by flecainide and encainide in the Cardiac Arrhythmia Suppression Trial (CAST)[58] suggested that excessive slowing of conduction could be proarrhythmic.[59] Hondeghem and Snyders[60] suggested that the usefulness of class III agents might be limited by their tendency to produce excessive QT prolongation, resulting in proarrhythmia. However, differentiation must be made between the homogeneous prolongation of action potentials produced by class III agents and the heterogeneous spatial variations in refractoriness due primarily to alterations in sodium conductance caused by class I agents.[61] Simulating the effects of class I and III drugs on the probability of unidirectional block, a prerequisite for reentry, Colatsky and colleagues[61] showed a marked difference between the drug classes, which increased as unbinding rate constant decreased (i.e., slower dissociation from the channel or a longer time constant) (Fig. 4-6). Among class III agents, those with more selective effects on a specific channel may be less arrhythmogenic than those with more diverse effects.[61]

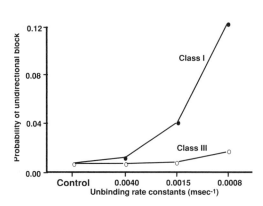

FIGURE 4-6. Simulation of the effect of class I and class III drugs on the probability of unidirectional block. Class I drugs are 16-fold more likely to cause one-way block required for reentry, especially when kinetically slow drugs are examined. (Reproduced with permission from Colatsky TJ, Follmer CH, Starmer CF. Channel specificity in antiarrhythmic drug action. Mechanism of potassium channel block and its role in suppressing and aggravating cardiac arrhythmias. Circulation 1990;82:2235.)

REVERSE USE DEPENDENCE

An interesting property of class III antiarrhythmics, especially sotalol, is reverse use-dependence, shown in Figure 4-7.[62] Lathrop[63] reported reverse use-dependence in Purkinje fibers, most notably observing the appearance of early afterdepolarizations when fibers were returned to slower paced rates.[63] Increasing prolongation of the QT interval by sotalol at slower paced rates (longer RR intervals) suggested to Hondeghem and Snyders[60] that class III agents would be increasingly proarrhythmic at slow heart rates, which is likely given the β-blocking properties of sotalol. In addition, reverse use dependence renders the drugs least effective at fast heart rates characteristic of life-threatening arrhythmias.[60] Amiodarone is an exception. It lengthens action potential duration to a similar extent at both slow and fast rates.[64]

Colatsky and colleagues[61] dispute this speculation regarding reverse use dependence. The bulk of clinical evidence suggests that agents that prolong repolarization without slowing conduction are extremely effective in suppressing programmed stimulation-induced tachyarrhythmias and that their blunted ability to prolong QT interval at short cycle lengths does not diminish their antiarrhythmic efficacy.[61] Although a variety of explanations are offered, most intriguing is the possibility that action potential duration is not directly proportional to I_K block because the contribution of other channels to repolarization may increase at short cycle lengths.[61]

HETEROGENEOUS ANISOTROPY

Conduction in cardiac tissues that differs whether conduction parallels or crosses myocardial fiber direction is called anisotropic.[65] In normal canine epicardium, longitudinal conduction velocity was 0.55 m/sec

FIGURE 4-7. Reverse dependence of sotalol. QT interval is plotted at increasing sotalol concentrations for each of several paced rates. Normally (sotalol concentration 0), QT interval increases as RR interval increases. Sotalol, in a concentration-dependent manner, accentuates this difference at slower paced rates. The effects of sotalol are largely abolished, and perhaps reversed, at faster paced rates. In contrast, class I agent effects typically increase as RR interval decreases. (Reproduced with permission from Funck-Brentano C, Kibleur Y, LeCoz, F, Poitier J-M, Mallet A, Jaillon P. Rate dependence of sotalol-induced prolongation of ventricular repolarization during exercise in humans. Circulation 1990;83:536.)

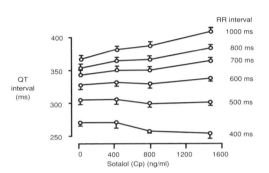

while transverse velocity was 0.23 m/sec.[66] Nonuniform anisotropy occurs when dyssynchronous propagation between adjacent groups of cells causes irregular activation and can lead to reentrant arrhythmias.[67] Anisotropy is relevant to a discussion of class III antiarrhythmic agents since action potential duration shows anisotropy. Action potentials are longest after adjacent site stimulation (centrifugal propagation), shortest after transverse propagation, and intermediate during longitudinal propagation.[68] In canine myocardial strips, amiodarone depressed longitudinal conduction at paced rates faster than 60 beats/min, but had no effect on transverse propagation, thus making conduction more uniform.[69] Anderson and co-workers,[64] however, found that 3 weeks of amiodarone therapy depressed both longitudinal and transverse conduction velocity.

OTHER EFFECTS

A variety of other effects of these three class III agents have been described. Amiodarone, but not standard sodium channel inhibitors, appears to decrease free radical production by neutrophils, considered to play an important role in arrhythmogenesis.[70] With a neutrophil chemiluminescence model to detect free radical production, Siminiak and colleagues[71] obtained somewhat similar results, finding that both amiodarone and bretylium inhibited free radical production, along with a variety of other drugs.

Class III antiarrhythmic agents prolong refractoriness by modulating ion channels, which may be sensitive to calcium regulatory proteins or enzymes.[72] Amiodarone potently inhibited calmodulin-regulated activity, but sotalol was inactive.[72] Both amiodarone and desethylamiodarone also inhibited protein kinase C activity,[72] and since protein kinase C enhances delayed rectifier currents,[73] this represents an additional mechanism of action.

Sudden Death and the Antifibrillatory Effects of Class III Antiarrhythmic Agents

Sudden cardiac death is a major cause of mortality, and ventricular fibrillation represents the most common terminal mechanism. Despite identification of ventricular ectopy after myocardial infarction as a significant risk factor for sudden death,[74] suppression of ectopy with standard antiarrhythmic agents fails to improve survival.[75] In contrast, β-adrenergic blockade reduces reinfarction and sudden death,[76] emphasizing the role of alterations in autonomic tone in precipitating arrhythmias.[77]

In 1982, Lucchesi and co-workers[78] began a series of studies in a conscious canine model that was susceptible to the initiation of stimulus-induced arrhythmias in the subacute phase of myocardial infarction. Table 4-3 summarizes the results of several of these studies. Immediately apparent is the discrepant behavior of certain agents when tested against the arrhythmias induced by programmed stimulation and ischemic ventricular fibrillation. Quinidine, a class Ia drug, for example, suppresses inducible arrhythmias but not ventricular fibrillation. Amiodarone, in contrast, though ineffective during programmed stimulation, improved survival. In this model, the three class III antiarrhythmic agents described were most protective.[79]

AGENT-SPECIFIC ACTIONS

Sotalol

l-Sotalol is a noncardioselective β-adrenoceptor blocker devoid of intrinsic sympathomimetic activity. The dextrorotatory isomer, *d*-sotalol, prolongs the action potential duration of isolated cardiac tissues, providing class III antiarrhythmic activity. The clinically available racemic mixture, sotalol, displays both properties.

TABLE 4-3. EFFECTS OF SELECTED ANTIARRHYTHMIC AGENTS IN A CANINE SUDDEN DEATH MODEL[79]

Drug	Class	Prevention of Programmed Stimulation Induced VT	24-Hr Survival (Prevention of VF)
None		No	6%
Quinidine	Ia	Yes	9%
Sotalol	II, III	Yes	65%
Bretylium	III	Yes	60%
Amiodarone	I, III	No	60% after acute treatment 80% after 10 days of therapy
Diltiazem	IV	No	10%

VT, ventricular tachycardia; VF, ventricular fibrillation.

Electrophysiologic Effects

Table 4-4 details the electrophysiologic effects of sotalol. Sotalol prolongs the action potential duration and refractory period of isolated cardiac tissue, including atrial, ventricular, Purkinje fiber, and sinoatrial node tissues, without slowing conduction, blocking sodium channels, or decreasing sinus node phase 4 depolarization.[80-82]

In intact subjects, sinus rate slows, presumably due to the increased action potential duration.[82] P-R and Q-T intervals increase without a change in QRS duration. Sotalol prolongs A-H but not H-V intervals[83,84] and increases accessory pathway refractoriness.[85] Sotalol, 4 mg/kg followed by a 1.5 mg/kg/hr infusion, reduces the defibrillation threshold of dogs.[86]

Hemodynamic Effects

In isolated cardiac tissue, sotalol augments contractility by 13 to 15%.[87] Perhaps increased action potential duration, a positive inotropic intervention, reduced the effects of sotalol-induced β-blockade. Intravenous sotalol minimally affects stroke volume but reduces heart rate and, consequently, cardiac output.[88] Yet patients with mildly depressed ejection fractions tolerate oral sotalol well. Although heart rate decreases, a significant increase in filling pressures, secondary to bradycardia, and a decrease in afterload increased stroke volume and maintained cardiac output.[88] Patients with severe ventricular dysfunction may develop refractory heart failure during sotalol therapy.[89]

TABLE 4-4. ELECTROPHYSIOLOGIC EFFECTS OF CLASS III ANTIARRHYTHMIC AGENTS

Parameter	Tissue	Amiodarone	Sotalol	Bretylium
V_{max}	Ventricular muscle	↓	0	0
	Purkinje fibers			
Conduction	Atrioventricular node	↓		
APD	Ventricular muscle	0 (acute) ↑ (chronic)	↑	↑
	Purkinje fibers	↓ (acute) 0 ↑ (chronic)	↑	↑
ERP/APD		0	0	
Normal automaticity	Sinoatrial node	↓	↓	
	Purkinje fibers	↓		
DADs				
EADS			↑	
ERP	AV node	↑	↑	0 ↑
	His-Purkinje	↑	↑	↑
	Atrium	↑	↑	↑
	Ventricle	↑	↑	↑
	Accessory pathway	↑	↑	0
ECG intervals				
RR		↑	↑	0
PR		0 ↑		0 ↑
QRS		↑	0	0
QT		↑	↑	0 ↑
AH		↑	↑	
HV		↑	0	

APD, action potential duration; ERP, effective refractory period; DADs, delayed afterdepolarizations; EADs, early afterdepolarizations, AV, atrioventricular; ECG, electrocardiogram.

↑ variable increased; ↓ variable decreased; 0 no change.

Pharmacokinetics

Table 4-5 presents pharmacokinetic data. The concentration-effect relationship between β-blocking and class III effects of sotalol is interesting, since β-blockade (0.8 µg/ml) is typically seen at lower levels than prolongation of ventricular refractoriness (6.8 µg/ml), a class III effect in anesthetized dogs.[90] Suppression of ventricular arrhythmias in humans occurs at serum levels of 1 to 2 µg/ml, but effective levels vary widely and do not play a role in clinical monitoring.[91]

TABLE 4-5. PHARMACOKINETICS OF SOTALOL

Oral bioavailability	90–100%
Protein binding	Minimal
Volume of distribution	1.5 liters/kg
Elimination half-life	7–18 hr
Metabolism	Minimal, no active metabolites
Excretion	Urine, dose interval varies with creatinine clearance
	normal: twice daily
	30–60 ml/min: once daily
	10–30 ml/min: 36-48 hr
	<10 ml/min: individualized
	Hemodialysis removes sotalol[207]

Clinical Use

Ventricular Arrhythmias

Initial approved indications are limited to the treatment of life-threatening ventricular arrhythmias. In a group of 486 patients with sustained inducible ventricular tachyarrhythmias randomly assigned to receive either sotalol, procainamide, quinidine, mexiletine, propafenone, or pirmenol, sotalol was most effective as assessed by programmed electrical stimulation (36% versus 13% for all other drugs) and by Holter monitoring (41% versus 45% for all other drugs combined).[92] After 2 years of chronic therapy, sotalol had the lowest mortality rate (13% versus 22%), recurrence rate (30% versus 60%), and drug withdrawal rate (38% versus 75%).[92] McGovern and colleagues[93] summarize the results of several clinical trials of sotalol for suppression of ventricular tachyarrhythmias. Doses ranging from 160 to 480 mg/day have efficacy comparable to class Ia agents, with suppression rates between 18 and 67% and recurrence rates of 0 to 38%.[93] Lower doses (250 mg/day), however, may be equally effective.[94] More recently, Kehoe and co-workers[89] found sotalol curative in 106 of 236 patients, while it slowed induced ventricular tachycardia or made induction more difficult in an additional 62 patients. Of 151 patients discharged receiving sotalol (480 mg/day), only 27 (18%) had recurrence of arrhythmia within 1 year.[89]

In 38 patients with inducible ventricular arrhythmias, Schwartz and co-workers[84] infused 2 mg/kg of sotalol intravenously and found that in 18 patients, in whom QT interval increased, ventricular arrhythmias became noninducible. Griffith and co-authors[95] found that 1 mg/kg of sotalol, infused over 5 min, was as ineffective as lidocaine in terminating sustained ventricular tachycardia (36 and 30% termination, respectively). However, 2 of 14 patients receiving sotalol became severely hypotensive.[95]

Supraventricular Arrhythmias

That sotalol, a β-blocker, slows sinus tachycardia and the ventricular response to atrial fibrillation requires no discussion. Although 1.5 mg/kg of sotalol is often recommended for conversion of atrial flutter or fibrillation to sinus rhythm, Teo and co-workers[96] demonstrated conversion rates of 86% after 0.34 mg/kg for atrial fibrillation and 33% after 0.6 mg/kg for atrial flutter. However, sotalol is only moderately effective in maintaining sinus rhythm in patients with paroxysmal atrial fibrillation.[93]

In patients with atrioventricular nodal reentrant tachycardia, 5 of 7 converted to normal sinus rhythm with sotalol, 1.5 to 2.5 mg/kg intravenously.[97] Tachycardia in 4 of 6 was controlled by oral sotalol (427 mg/day) over 14 months.[97] The same study found intravenous sotalol curative in 13 of 18 patients with Wolff-Parkinson-White syndrome and a long-term success of 77%.[97] Compared to placebo, sotalol, 1.5 mg/kg intravenously, converted atrioventricular nodal reentry or atrioventricular reentrant tachycardia via an accessory pathway to sinus rhythm in 83% of 38 patients with compared to 16% conversion after placebo.[98] In patients with accessory pathways who develop atrial fibrillation, sotalol decreased maximal ventricular rate.[99,100]

Sotalol has been used in management of ventricular and supraventricular tachycardias in children.[101]

Adverse Reactions

Both the β-blocking and action potential-prolonging effects of sotalol contribute to its toxicities. Table 4-6 summarizes potential adverse reactions.

TABLE 4-6. SIDE EFFECTS AND TOXICITY OF SOTALOL

Class II (β-blocking) effects
  Bronchospasm
  Congestive heart failure
  Hypotension
  Masks signs of hypoglycemia
  Heart block, symptomatic sinus bradycardia, syncope
  Hyperadrenergic rebound if acutely discontinued
  Fatigue, impotence, weakness, dizziness
  Hyperlipidemia not a problem[208]
Class III effects
  Proarrhythmia, including torsade de pointes (4% of patients)

Torsade de pointes during sotalol therapy occurs in about 4% of patients, especially those with congestive heart failure, hypokalemia, or renal failure or those receiving digitalis therapy.[102] It has been seen in patients without ventricular disease receiving sotalol for supraventricular arrhythmias.[97] Therapy includes withdrawal of sotalol, infusion of isoproterenol, overdrive pacing, and magnesium infusion.[93,103] Hemodialysis can be considered if conventional therapy fails.[104]

McGovern and colleagues[93] outline practical guidelines for the use of sotalol. The drug is indicated in patients with refractory arrhythmias. For patients with reciprocating atrioventricular tachycardia or ventricular tachycardia without severe ventricular dysfunction, sotalol should be tried before amiodarone therapy. The initial dose, given during hospitalization with continuous electrocardiographic monitoring, is 80 mg twice daily in patients with ventricular dysfunction and 160 mg twice daily in those with normal ventricular function. Dosage is increased every 3 days until efficacy is demonstrated, intolerable adverse reactions appear, or QT interval, corrected for heart rate, reaches 550 msec. Adverse effects from excessive β-blockade are common at daily doses exceeding 480 mg.

Intravenous therapy begins with 0.2 mg/kg with doses up to 1.5 mg/kg over 5 min. Obviously, intravenous administration assumes careful monitoring of heart rate and blood pressure.[93]

Table 4-7 summarizes drug interactions. Anesthetic interactions, both caused by β-blocking and QT interval-prolonging effects of sotalol, were discussed in the introduction. Alfentanil has specifically been mentioned as an agent producing frequent bradycardia in patients chronically receiving β-blockers.[105]

Clofilium, similar in structure to sotalol, is a pure class III agent. Anesthesia increases the ability of clofilium to prolong ventricular refractoriness in intact dogs.[106] This is most pronounced during halothane anesthesia and suggests a need for caution when anesthetizing any patient receiving class III drugs.

TABLE 4-7. DRUG INTERACTIONS WITH SOTALOL

Drug	Interaction
Albuterol	Antagonism by β-blocker
Amiodarone	Bradycardia, hypotension
Tricyclic antidepressants	QT prolongation, proarrhythmia
Terfenadine	
Disopyramide	
Anesthetics	
Flecainide	Heart block

Bretylium

Bretylium was originally developed as an antihypertensive drug with selective adrenergic neuronal blocking properties. However, incomplete absorption and a variety of side effects limited its use.[107] Advanced cardiac life support now provides the major indication for bretylium.[108] Bretylium possesses both antiadrenergic and direct electrophysiologic actions. After administration, bretylium concentrates in peripheral adrenergic nerve terminals, initially inducing norepinephrine release. Reserpine depletion of peripheral norepinephrine stores prevents this transient sympathomimetic response. Subsequently, bretylium prevents norepinephrine release and blocks its reuptake by the nerve terminal, an antiadrenergic effect that increases sensitivity to circulating or intravenously administered catecholamines. The magnitude of the antiadrenergic effect depends on the dose and the preexisting adrenergic state of the patient.[107,109]

The contribution of the antiadrenergic effects to the antiarrhythmic action of bretylium is not definitely known. Pretreatment with tricyclic antidepressants, which antagonize the effects of bretylium at the nerve terminal, attenuate the acute antifibrillatory efficacy of bretylium.[110] Atrioventricular node refractory period shortening observed in humans after bretylium infusion represents an additional indirect adrenergic effect.[111]

Direct Electrophysiologic Effects

Bretylium increases action potential duration without altering conduction velocity, upstroke, or membrane responsiveness (Table 4-4). As seen in Table 4-1, these changes reflect block of I_K. Bretylium specifically prolongs shortened action potentials in ischemic regions of the heart, thus reducing the dispersion of refractoriness common in the ischemic heart.[112] Bretylium seldom prolongs the QT interval sufficiently to induce torsade de pointes.[113]

Other Effects

Bretylium binds to cardiac-type high-affinity muscarinic receptors, but has a 10-fold lower affinity for central nervous system type M and L muscarinic receptors.[114] Bretylium also increases prostaglandin levels, important because prostaglandin I_2 prevents ventricular fibrillation in dogs with myocardial infarcts.[115] Administration of cyclooxygenase inhibitors abolishes bretylium-induced increases in ventricular fibrillation threshold.[113]

Hemodynamic Effects

Bretylium produces a biphasic hemodynamic response. An initial transient tachycardic and hypertensive period of 15-min duration reflects catecholamine release. Subsequently, heart rate, blood pressure, and vascular resistance fall, with a maximal effect at 2 hr.[107]

Pharmacokinetics

Table 4-8 describes the pharmacokinetics of bretylium. Plasma concentration does not correlate with the beneficial or toxic effects of bretylium. Unfortunately, data suggest that bretylium may be less effective when administered after another agent than when administered alone.[113] Bernstein and Koch-Weser,[116] for example, found that bretylium, when administered alone initially, prevented ventricular fibrillation more effectively than when given after lidocaine. Similarly, quinidine and bretylium produced opposing effects on action potential duration, while increasing the slowing of atrioventricular conduction.[117] In contrast, Nowak and co-workers[118] found that the combination enhanced survival after cardiac arrest. Lidocaine may synergistically increase fibrillation threshold when given with bretylium.[119]

Clinical Uses

Bretylium is considered primarily an antifibrillatory drug with modest antiectopic activity. Bretylium, in a dose-related manner, markedly raises the fibrillation threshold in canine myocardium[119] and reverses the reduction in fibrillation threshold associated with acute ischemia.[110] Sanna and Arcidiacono[120] demonstrated chemical defibrillation by 5 to 10 mg/kg of bretylium in 5 of 7 patients. Bretylium also decreases the energy required to defibrillate the heart, and the success rate for countershock improves after bretylium.[113]

TABLE 4-8. PHARMACOKINETICS OF BRETYLIUM

Oral bioavailability	12–30%, not used orally
Protein binding	Negligible 1–6%
Volume of distribution	589 liters in normal adults (8 liters/kg)
Elimination half-life	6–10 hr
Metabolism	None
Excretion	Unchanged in urine
	Glomerular filtration rate 10–50 ml/min: reduce dose to 25–50%
	Glomerular filtration rate <10 ml/min: avoid bretylium
	Dialysis removes bretylium[209]

In a canine model of ventricular fibrillation, bretylium was marginally better than lidocaine at preventing recurrence of fibrillation after electrical defibrillation, but no other differences were seen.[121] In humans, however, lidocaine and bretylium were equieffective in resuscitation from ventricular fibrillation.[122] Still, when repeated episodes of fibrillation occur after myocardial infarction despite lidocaine therapy, bretylium may prevent recurrence.[123]

According to the Advanced Adult Cardiac Life Support algorithm, bretylium is used when defibrillation, epinephrine, and lidocaine fail to correct ventricular fibrillation; when ventricular fibrillation has recurred despite lidocaine and epinephrine; when lidocaine and procainamide have failed to control ventricular tachycardia associated with a pulse; or when lidocaine and adenosine have failed to control wide-complex tachycardias.[108] Thus, bretylium is considered a second line drug when others have failed.

Most authorities consider bretylium contraindicated in treatment of ventricular arrhythmias caused by digitalis toxicity where catecholamine release may be arrhythmogenic. Yet, reports exist of successful use of bretylium to treat digitalis toxicity rhythms.[113] Bretylium may successfully treat torsade de pointes,[124] but use of the drug in the treatment of long QT syndrome-induced arrhythmias is controversial.[113] Bretylium, lidocaine, and procainamide are uniformly unsuccessful in correcting polymorphous ventricular tachycardia after myocardial infarction.[125]

Prophylactic administration of bretylium after myocardial infarction prevents arrhythmias. A 10-year study examined 1255 patients after acute myocardial infarction, 843 of whom received bretylium, 10 mg/kg/day, by continuous infusion for 5 to 7 days. The remainder received more conventional antiarrhythmic agents. Primary ventricular fibrillation appeared in 1.3% of bretylium-treated patients and in 5.9% in those not receiving bretylium.[113]

Kirlangitis and colleagues[126] compared bretylium and lidocaine in the prevention of ventricular fibrillation after release of the aortic cross-clamp in coronary revascularization patients. Fibrillation occurred in 91% of saline-treated patients, 64% of lidocaine-treated patients, and only 36% of bretylium-treated patients.[126] However, in dogs, intravenous amiodarone, 5 mg/kg, more effectively prevented reperfusion arrhythmias than bretylium.[127] Bretylium also controls refractory ventricular fibrillation after open heart surgery,[128] but the appearance of hypotension causes concern in these patients. Prophylactically administered bretylium prevents ventricular arrhythmias after open heart surgery.[129]

Bretylium has found a particular role in the therapy of ventricular tachyarrhythmias after accidental intravenous injection of bupivacaine.[130] Prolonged resuscitation attempts should be anticipated, with administration of bretylium for ventricular tachycardia and epinephrine and atropine for mechanical dissociation.[131,132] However, in swine after bupivacaine injection, amiodarone (10 mg/kg) was more effective than bretylium or placebo.[133]

Bretylium is administered as an intravenous bolus of 5 to 10 mg/kg repeated 15 to 30 min later if needed, to a maximum of 30 mg/kg. Therapeutic levels can be maintained with a continuous infusion of 1 to 2 mg/min. Intramuscular injections of 5 to 10 mg/kg every 6 to 8 hr achieve effective drug levels, but produce muscle necrosis. Similar intravenous bolus doses can be given to children.[134] Table 4-9 summarizes the side effects of bretylium and Table 4-10 outlines the potential interactions between bretylium and other drugs.

Amiodarone

Amiodarone, an unusually large iodine-containing benzofuran derivative, was first synthesized as an antianginal coronary vasodilator.[135] Since its antiarrhythmic effects were recognized,[136] amiodarone has become the most effective, most toxic, and most studied antiarrhythmic agent available.

TABLE 4-9. SIDE EFFECTS AND TOXICITY OF BRETYLIUM

Severe nausea and vomiting (intravenous)
Muscle necrosis (intramuscular)
Initial transient hypertension
Inevitable hypotension
Rare bradycardia
Rare proarrhythmia[210]
Parotitis
Hyperthermia
    (may reach 108°F within 30 min of administration[211,212])

TABLE 4-10. DRUG INTERACTIONS WITH BRETYLIUM

Drug	Interaction
Tricyclic antidepressants, guanethidine	Interfere with Bretylium's actions[110]
Quinidine	Antagonism of antiarrhythmic effects[117]
Digitalis	Catecholamine release aggravates arrhythmias

Electrophysiology

The ionic and electrophysiologic effects of amiodarone are depicted in Tables 4-1 and 4-4. Amiodarone has actions that qualify for inclusion in antiarrhythmic classes I, II, III, and IV. Rather than repeat data in the tables, I will focus on the differences between acute and chronic amiodarone administration, the contribution of the monodesethyl metabolite, the role of thyroid suppression in amiodarone's effects, and the significance of additional antiadrenergic actions.

Amiodarone produces homogeneous prolongation of action potential duration and refractoriness in all cardiac tissues. However, these changes begin 1 week after initiation of therapy and increase until a maintenance dose is reached. During chronic therapy, PR, RR, and QT intervals increase.[137] Acutely administered intravenous amiodarone, in contrast, causes little or no change in ventricular refractoriness. Both antegrade and retrograde atrioventricular nodal refractoriness, however, are prolonged, and conduction is slowed.

Acutely administered amiodarone immediately decreased rabbit sinoatrial node rate[136] by decreasing the rate of diastolic depolarization, at least in part due to a calcium channel blocking effect.[137] In Purkinje fibers, acute exposure to amiodarone (5 µg/ml) produced a use-dependent reduction in V_{max}[138] and shortened action potential duration.[137–139] Acute exposure of ventricular muscle to amiodarone similarly reduced V_{max} and shortened action potentials.[137,140] Chronic pretreatment, in contrast, either prolongs[139] or does not change Purkinje fiber action potential duration.[137] Ventricular muscle V_{max} continues to be depressed, but action potential duration increases.[136,140] In atrial tissue, as chronic therapy proceeds, the drug progressively prolongs action potential duration without changing resting potential.[141]

In humans, intravenous amiodarone causes little change in heart rate or mild, transient tachycardia.[142] Sympathetic activation in response to vasodilation offsets antiadrenergic effects and direct suppression of phase 4 depolarization.[143] The AH interval increases 10-20%,[141,142] without significant change in the HV or QRS durations or QTc intervals. The refractory period of the atrium and ventricle do not change, but the atrioventricular nodal and accessory pathway refractory periods increase 15%.[142] Within hours of acute oral administration of 30 mg/kg, amiodarone was effective in suppressing ventricular arrhythmias and atrioventricular conduction of atrial fibrillation, effects which correlated with plasma concentrations.[144]

Chronic oral administration produces distinctly different electrophysiologic effects. A 30% heart rate reduction can yield symptomatic bradycardia or sinus arrest, occasionally requiring permanent pacing to allow the continued use of amiodarone.[145–147] The AH interval increases by 20%, and refractoriness of atrial, atrioventricular nodal, ventricular, and accessory pathway increases by 10 to 45%. Conduction in the His-Purkinje system may decrease, but this is more likely in patients with preexisting bundle branch block.[145,148–150] The electrocardiogram shows a 20% increase in PR interval[145,151] and, invariably, a prolonged QT interval, often accompanied by T-wave abnormalities and prominent U waves.[148] The QRS duration increases by 10 to 20% at resting rates, but demonstrates use-dependent widening up to 40% above pretreatment values at faster paced rates.[142,152] In contrast to the differences between acute and chronic therapy found by others, Nattel[153] observed that both single doses and chronic administration of amiodarone prolonged QT interval and QRS duration of anesthetized rats.

Desethylamiodarone

During chronic therapy, plasma concentrations of desethylamiodarone approach those of the parent drug.[154] Acutely administered amiodarone or desethylamiodarone slow sinus rate similarly in rabbits.[155] Each drug increases action potential duration of atrial muscle without affecting other parameters.[155] Desethylamiodarone is less potent than the parent compound in reducing ventricular upstroke velocity and prolonging action potential duration and refractory period, but causes a similar shortening of action potentials in Purkinje fibers.[155]

Both single doses and chronic administration of either amiodarone or the desethyl metabolite prolonged QT interval and QRS duration of anesthetized rats.[153] Amiodarone prolonged atrioventricular conduction more, but desethylamiodarone had more potent effects on atrial and ventricular conduction.[153] In anesthetized dogs, both drugs caused frequency-dependent slowing of atrioventricular and ventricular conduction. Amiodarone prolonged Wenckebach cycle length more and desethylamiodarone increased QRS duration and refractory periods of the ventricles and atria more.[156] Clearly, the chronic clinical efficacy of amiodarone depends to some degree on accumulation of the desethyl metabolite.

Thyroid Effects

Since the structure of amiodarone is that of a iodinated aromatic ring compound similar to thyroxine (T_4) and triiodothyronine (T_3), it is not surprising that thyroid function in the body is altered.[157] Doses of 300 mg/day represent a marked increase in iodine intake, so that the free iodine produced by drug metabolism frequently causes a transient suppression of T_4 production by the thyroid.[158] Amiodarone also inhibits the activity of 5-deiodinase, the enzyme needed for conversion of T_4 to the metabolically active T_3, resulting in a decrease in serum T_3 and a compensatory increase in T_4 and its conversion to reverse T_3 (rT_3).[157–159] With chronic administration, most patients demonstrate a persistent low normal T_3 level, modestly increased T_4 level, increased rT_3 level, and normal thyroid-stimulating hormone level. Although Singh and Vaughan-Williams[136] prevented an amiodarone-induced increase in action potential duration by giving rabbits 3 weeks of treatment with intraperitoneal T_4, Lambert and co-workers[160] found that amiodarone increased ventricular refractory period to a similar extent whether rats were euthyroid, hypothyroid, or pretreated with T_3 or T_4. Administration of T_3 to patients does not abolish the effects of amiodarone,[161] and the rapid onset of effects in vitro or with intravenous and even oral dosing suggests that thyroid actions are not required for certain actions of amiodarone.

Autonomic Effects

Amiodarone interacts with α- and β-adrenergic adrenoceptors and muscarinic receptors,[55,162,163] contributing to blood pressure and heart rate reductions. However, the drug is only a weak antagonist at α- and β-adrenoceptors[162] and the further reduction in heart rate after administration of propranolol implies additional mechanisms.[6] More importantly, amiodarone reduces β-adrenergic receptor density,[164,165] providing noncompetitive antagonism. Yin et al.[166] found that a euthyroid state was necessary for amiodarone effects on heart rate and that in hypothyroid rats, amiodarone induced no further decrease in the density of β-adrenergic receptors, nor in heart rate. Since hypothyroidism decreases β-adrenergic receptor density[167] and prolongs action potential duration,[168] amiodarone's chronic effects may be in part through altered thyroid regulation of myocardial tissue.

The coronary vasodilator action of amiodarone, by relieving ischemia, could contribute to antiarrhythmic efficacy.

Clinical Use

Ventricular Tachyarrhythmias

Amiodarone is indicated for the treatment of sustained ventricular tachycardia or ventricular fibrillation refractory to other pharmacologic agents. Comparing patients with coronary artery disease without arrhythmias and patients with coronary artery disease and ventricular tachycardia or ventricular fibrillation treated with amiodarone, Kay and co-authors[169] concluded that amiodarone reduced the mortality of the arrhythmia group to that due to the severity of underlying myocardial dysfunction.

Herre and colleagues[170] followed 429 patients discharged from the hospital receiving amiodarone after class I antiarrhythmic drugs failed. The incidence of sudden death was a remarkable 10% at 1 year, 15% at 3 years, and 21% at 5 years. Others have shown a 40 to 100% (mean 70%) incidence of arrhythmia-free survival, with follow-ups ranging from 6 to 36 months.[171–175] The rate of recurrent cardiac arrest while patients received amiodarone therapy ranged from 6 to 24% by 10 to 16 months.[172,173,176]

A report of the effects of amiodarone for prevention of postmyocardial infarction sudden death[177] suggests that amiodarone suppresses ectopy (85% suppression versus 27% for placebo) and improves survival (6% sudden cardiac and 10% overall deaths in the amiodarone-treated group and 14% sudden cardiac and 21% overall deaths in the placebo group). Despite amiodarone doses of 300 to 400 mg/day after a 10 mg/kg/day 3-week load, similar numbers of amiodarone- and placebo-treated patients withdrew, with thyroid stimulating hormone elevation and skin coloration seen in the amiodarone group.[177]

The Role of Electrophysiologic Testing

The efficacy of programmed electrical stimulation to predict the ability of amiodarone to suppress arrhythmias is controversial. Noninducibility in patients receiving amiodarone carries a good prognosis.[178] However, of the 80% of patients with persistently inducible arrhythmias while receiving amiodarone, 85% had good outcomes.[178] Electrophysiologic testing, however, may still be valuable. Both Horowitz and co-workers[178] and Kadish and colleagues[179] agreed that well-tolerated, relatively slow induced arrhythmias presaged nonfatal spontaneous recurrences. Fast or poorly tolerated induced arrhythmias predicted a high incidence of sudden death.

Supraventricular Tachyarrhythmias

Although not yet approved in the United States for the therapy of supraventricular arrhythmias, amiodarone is effective in many situations at doses lower than those administered for refractory ventricular arrhythmias.[180] Amiodarone has successfully corrected atrial fibrillation or flutter in greater than 70% of patients.[172] Rosenbaum and colleagues,[181] for example, reported reversion to sinus rhythm in 29 of 30 patients with atrial fibrillation or flutter. Amiodarone effectively controls atrial tachyarrhythmias in patients with sick sinus syndrome.[180]

Successful therapy of Wolff-Parkinson-White syndrome patients approaches 80% with doses of 200 to 400 mg/day.[181,182] Efficacy is accompanied by a significant increase in the accessory pathway antegrade effective refractory period, with variable effects on the retrograde effective refractory period, and suppression of premature impulses that initiate tachycardia.[149,150] In 76 patients followed for 67 months, 100 to 200 mg/day of amiodarone effectively prevented circus movement tachycardia while higher doses of 200 to 400 mg/day were required in patients with atrial fibrillation.[149]

Effects of Intravenous Amiodarone

Ventricular Tachyarrhythmias

Helmy and colleagues[183] administered 5 mg/kg of intravenous amiodarone followed by 1 g/day to 46 patients with recurrent, drug-refractory, sustained ventricular tachycardia or ventricular fibrillation. Of these patients, 33% responded within 2 hr, an additional 23% by 72 hr, and 1 patient after 72 hr. Another 6 patients (13%) responded after a change to oral amiodarone. Of the 46 patients, 6 had significant adverse effects: 2 patients had hypotension requiring dopamine, 2 had significant bradycardia requiring temporary pacing (1 permanently), and 2 had polymorphous ventricular tachycardia associated with QT prolongation.

Morady et al.[184] administered intravenous amiodarone (5 to 10 mg/kg) to 15 patients with ventricular tachycardia refractory to two or more drugs. Intravenous amiodarone abolished ventricular tachycardia in 12 without proarrhythmic effect.

Supraventricular Tachycardias

Benaim and colleagues[185] treated 100 patients with 5 mg/kg of intravenous amiodarone. Successful conversion occurred in 30% of patients with atrial fibrillation and 65% of those with atrial or junctional tachycardia. In those patients in whom conversion to sinus rhythm was not successful, 47% had slowing of the ventricular response rate. Overall, 80% of patients benefited from intravenous amiodarone.[185] For a more extensive review, see Kadish and Morady.[186]

After coronary revascularization, intravenous amiodarone suppressed both ventricular and supraventricular arrhythmias. Incidence of atrial fibrillation decreased from 21 to 5% in amiodarone-treated patients and nonsustained ventricular tachycardia episodes occurred in 16% of control subjects and 3% of treated patients. Only 2 of 77 patients required drug discontinuation, both for excessively prolonged QT intervals.[187] Compared to oral quinidine, however, intravenous amiodarone was less effective at converting atrial fibrillation or flutter to sinus rhythm in patients after open heart surgery. More side effects occurred in quinidine-treated patients.[188]

Pharmacokinetics

Amiodarone kinetics are characterized by slow onset, huge distribution volume, an incredibly long duration of action, extreme lipophilicity and an active metabolite.[189,190] Table 4-11 summarizes pharmacokinetics of amiodarone.

A single dose produces high serum levels (5 µg/ml or more) that last for 1 to 2 hours. This central compartment fills a peripheral compartment comprising of most body organs over the next 5 days. Continued administration fills an enormous deep compartment, probably fat, over the next 3 to 10 months.[191] At steady state the myocardial concentration is 10 to 50 times the plasma level.[189]

Dose

Typical dosing regimens for the chronic treatment of ventricular tachyarrhythmias begin with a loading phase at 800 to 1600 mg/day in two to three divided doses for 10 to 14 days, followed by 600 to 800 mg/day for 4 to 8 weeks. Maintenance doses of 200 to 600 mg/day produce plasma amiodarone concentrations of 1.45 to 3.8 µg/ml,[154] but the correlation between serum levels and the daily or cumulative amiodarone dose is variable.[172] Serum levels greater than 1.5 µg/ml were noted in those with a therapeutic response on ventricular tachycardia.[172] Ventricular arrhythmias recurred in 9 patients when serum amiodarone concentrations fell below 1 µg/ml.[172] Amiodarone concentrations greater than 2.5 µg/ml predispose toward side effects.[172] However, the correlation between serum levels and either efficacy or toxicity is weak, so measurements of serum amiodarone concentration are of limited value.[154,161,172]

TABLE 4-11. PHARMACOKINETICS OF AMIODARONE

Oral bioavailability	35–65%
Protein binding	>96%
Volume of distribution	60 liters/kg
	Extensive deposition in fat, liver, lung
Elimination half-life	Biphasic; initial 10 days, terminal 107 days for desethylamiodarone 60 days
Metabolism	Extensively metabolized active metabolite desethylamiodarone
Excretion	Biliary, none found in urine
	Neither amiodarone or desethylamiodarone is dialyzable

Drug Toxicity and Side Effects

Toxic side effects are common and are summarized in Table 4-12. Side effects occur in three fourths of all patients and require drug discontinuation in as many as 18%. Toxicity may be dose-related since reactions are seen more frequently in the United States where doses higher than those reported in Europe and South America have been used.[180]

Proarrhythmic Effects

Torsade de pointes occurs in 2 to 5% of patients treated with amiodarone alone or in combination with other antiarrhythmic agents, including digoxin, quinidine, procainamide, disopyramide, propafenone, and mexiletine.[161,180,192] Symptomatic sinus bradycardia, sinus arrest, or, in patients with underlying conduction system disease, advanced atrioventricular block may occur, requiring lowered amiodarone doses or permanent pacemaker implantation.

TABLE 4-12. AMIODARONE TOXICITY

Organ system	Toxicity	Incidence
Lung	Interstitial fibrosis	2–15%, fatal in 10% of these
Thyroid	Hyperthyroidism	2%
	Hypothyroidism	2–10%
		Action of antithyroid drugs may be delayed
		Radioactive iodine therapy contraindicated due to low uptake
Heart	Proarrhythmia	2–5%
	Congestive heart failure	3%
	Bradycardia	2–4%
	Sinoatrial node dysfunction	May require pacer
Nervous system	peripheral neuropathy, ataxia, tremor	20–40%
Skin	Photosensitivity	10%
	Blue gray discoloration	
Eye	Corneal microdeposits	Very common, 10% have blurred vision
Gastrointestinal	Nausea, vomiting elevated liver enzymes	>25%
Various cells	Intracellular multilamellar inclusion bodies[161]	Consist of abnormal phospholipids due to inhibition of phospholipases A1, A2, and C

Pulmonary Toxicity

Pulmonary toxicity is the most serious complication of amiodarone.[193] The syndrome begins with cough and dyspnea, but occasionally as acute respiratory failure.[194] Pulmonary function tests demonstrate a decreased diffusing capacity for carbon monoxide and decreased total lung capacity.[193] Incidence ranges from 2 to 17% and mortality approaches 10% if the drug is not promptly discontinued. Two patterns occur: hypersensitivity pneumonitis and interstitial pneumonitis. The therapeutic role of corticosteroids is uncertain. In hypersensitivity pneumonitis steroid use appears appropriate, but improvement of severe symptoms may occur in interstitial pneumonitis after empirical treatment with 40 to 60 mg of prednisone per day, with tapering of dosage over 2 to 6 months.[161,193]

Thyroid Dysfunction

In addition to the transient alteration in thyroid function caused by amiodarone, 8.6 to 11% of patients demonstrate persistently deranged thyroid function with chronic administration,[158] leading some authorities to suggest that such patients should have thyroid function studies every 6 months.[195] Compilation of various studies suggests that persistent hypothyroidism occurs in 6% of patients (range: 0 to 30%), commonly manifested as atrioventricular block or bradycardia and documented by an elevated thyroid-stimulating hormone level (>20 µUI/ml).[158] The incidence of amiodarone-induced hyperthyroidism or thyrotoxicosis is lower, averaging 2.6% (range: 1 to 15%).[158] Thryotoxicosis is more likely to develop in areas of low iodine intake. In addition to other clinical signs, the hyperthyroid state may become apparent with tachyarrhythmias, including reappearance of that for which the amiodarone was administered and is documented by a decreased serum thyroid-stimulating hormone level. If amiodarone cannot be discontinued, antithyroid therapy may be instituted with propylthiouracil, steroids, and β-adrenergic blockers if needed. If chemotherapy is ineffective or requires too long a period, thyroidectomy has been used as an effective alternative.[196] A desire to avoid general anesthesia has led some to perform thyroidectomy under local anesthesia.[197]

Drug and Anesthetic Interactions

The potential for interaction between amiodarone and other commonly used drugs is phenomenal. Table 4-13 summarizes many important interactions between amiodarone and other drugs. Of particular importance to anesthesiologists is the potential interaction between amiodarone and anesthetic agents.

In isolated guinea pig hearts, Rooney and co-workers[198] found additive cardiac depressant effects of halothane and acutely adminis-

TABLE 4-13. DRUG INTERACTIONS WITH AMIODARONE

Drug	Interaction
Warfarin	Amiodarone inhibits metabolism, half warfarin dose
Digoxin	Serum concentration increased, half dose
Type Ia antiarrhythmics	Serum concentrations of quinidine and procainamide increase, a potential risk for torsade de pointes, reduce dose or discontinue
Aprinidine	Neurologic toxicity
Propafenone	Bradycardia
Type Ic antiarrhythmics	Exacerbation of encainide and flecainide proarrhythmia with QRS widening and atrioventricular block
β-Blockers	Exacerbate negative inotropic and
Calcium antagonists	chronotropic effects of amiodarone
Radiocontrast agents	Acute adult respiratory distress syndrome reported after pulmonary angiography

tered amiodarone. Heart rate and developed left ventricular pressure decreased while atrioventricular conduction slowed. In Purkinje fibers obtained from dogs treated chronically with amiodarone, resting membrane potential was reduced and action potential duration prolonged. Addition of halothane decreased action potential amplitude and upstroke velocity without changing duration. These changes suggested a risk of conduction defects if halothane were administered to patients receiving amiodarone.[2]

Since the 1981 report of perioperative hemodynamic instability in a patient treated with amiodarone,[1] several groups have reported perianesthetic experiences with amiodarone-treated patients. A greater incidence of intraoperative cardiac rhythm disturbances, including atropine-resistant bradycardia, slow nodal rhythm, complete heart block, pacemaker dependency, and increased requirements for perioperative circulatory and respiratory support have been documented in patients receiving amiodarone.[199,200]

Liberman and Teasdale[200] described the anesthetic courses of 16 patients receiving amiodarone during cardiac procedures (12 patients), abdominal aortic surgery (2 patients), pulmonary lobectomy (1 patient) and transurethral prostatectomy (1 patient). Patients had received amiodarone for periods ranging between 3 weeks and 3 years (serum amiodarone concentration averaged 0.9 mg/liter). Compared with 30 coronary revascularization patients with poor left ventricular function not taking amiodarone, three forms of interaction appeared. Nodal rhythm or complete heart block developed in 10 of 15 patients (the other patient had a permanent pacer in place). Inadequate cardiac output required balloon counterpulsation in 6 of 12 cardiopulmonary bypass patients. Finally, apparent α-adrenergic blockade led to low systemic vascular resistance despite α-agonist administration in two patients (Table 4-14).

Kupferschmid and colleagues[201] reviewed factors leading to unanticipated hepatic, cardiac, and pulmonary dysfunction after palliative surgery for obstructive hypertrophic cardiomyopathy in 71 consecutive patients. Fifty-five received β-adrenergic-blockers, calcium entry antagonists, or both, and 16 received amiodarone (Table 4-15). More complications occurred in patients treated with amiodarone, despite discontinuance of amiodarone an average of 91 days before surgery (range: 0 to 457 days). Although the necessary drug-free interval required to reduce perioperative risk could not be determined, discontinuation of amiodarone several months before surgery diminished the probability of major complications.[201] From 1978 to 1983, 129 patients underwent map-

TABLE 4-14. PERIOPERATIVE COMPLICATIONS IN PATIENTS RECEIVING AMIODARONE[200]

Complication	Amiodarone treated (*n* = 16)	Control (*n* = 30)
Bradyarrhythmias	73%*	17%
Atropine-resistant sinus bradycardia, slow nodal rhythm, complete heart block, pacemaker dependency		
Intraaortic balloon pump	50%*	7%
Low vascular resistance	13%*	0%

*$p < 0.05$ versus control patients.

TABLE 4-15. AMIODARONE-INDUCED COMPLICATIONS AFTER OPERATION FOR OBSTRUCTIVE HYPERTROPHIC CARDIOMYOPATHY[201]

Complications	Amiodarone-treated (*n* = 16)	Control (*n* = 55)
Hepatic dysfunction	50%	2%
Pulmonary dysfunction	25%	0%
Low cardiac output	19%	2%
Death	13%	0%

Hepatic dysfunction caused a 10-fold increase in serum glutamic-oxaloacetic and glutamic-pyruvic transaminase concentrations.
Pulmonary dysfunction caused a 4-fold increase in the number of days of ventilatory support.

ping guided endocardial resection at the University of Pennsylvania, of whom 33 received amiodarone to within 25 days of surgery. Despite similar death rates, the conditions of amiodarone-treated patients were less stable, and they required a longer period of mechanical ventilation (Table 4-16).[199]

A series of 67 patients who received an implanted defibrillator or endocardial resection revealed a remarkably high incidence of postoperative adult respiratory distress syndrome in amiodarone-treated patients. Nine of 18 surgical survivors who received amiodarone developed hypoxemia and pulmonary infiltrates, despite normal cardiac output and pulmonary capillary wedge pressure, while none of 44 survivors not receiving amiodarone developed adult respiratory distress syndrome.[194] Noncardiogenic pulmonary edema during or after surgery has also been reported.[202,203]

To avoid potential general anesthetic complications, Kopacz[204] described continuous spinal anesthesia for colon resection. Avoidance of postoperative respiratory complications, slow elevation of anesthetic level with ability to compensate for hemodynamic changes, and avoidance of high serum local anesthetic concentrations were suggested as benefits by the advocate.[204]

Not all investigators have found such reactions in amiodarone-treated patients. Chassard and others[205] compared 10 patients who received 10 g of amiodarone electively before valvular surgery to 9 control patients. No increased complications were observed in amiodarone-treated patients, and amiodarone proved an effective perioperative antiarrhythmic agent. However, the elective, short-term use of the drug immediately prior to surgery differentiates this study from those previously mentioned. Elliott and her colleagues[206] similarly found no differences in perioperative course between 21 patients receiving amiodarone and matched control subjects.

TABLE 4-16. AMIODARONE AND ENDOCARDIAL RESECTION[199]

Complications	Amiodarone-treated (n = 33)	Control (n = 96)
Epinephrine infusion after bypass	87%*	33%
Balloon counterpulsation	39%	25%
Respiratory complications	21%*	4%
Death	9%	12%

*$p < 0.05$ versus control patients.

SUMMARY

Amiodarone, bretylium and sotalol are potent antiarrhythmic agents that act by prolonging action potential duration, cardiac refractory period, and QT interval. Each drug has additional effects. Most notably, each is an antiadrenergic agent, although by very different pathways: sotalol by β-adrenergic blockade; bretylium by norepinephrine depletion; and amiodarone by decreasing β-adrenergic receptor synthesis (apparently via an antithyroid action). In addition to their direct actions on ionic fluxes, these actions may contribute to their efficacy and potential for drug interactions. While the indications for each drug limit use to serious or desperate situations, these agents may find increased application in patients with various arrhythmias. The anesthetist must understand the uses and indications for each agent and be ready to diagnose and treat possible interactions, recognizing that amiodarone may be present months after its discontinuation.

References

1. Gallagher JD, Lieberman RW, Meranze J, Spielman SR, Ellison N. Amiodarone-induced complications during coronary artery surgery. Anesthesiology 1981;55:186.
2. Gallagher JD. The electrophysiologic effects of amiodarone and halothane on canine Purkinje fibers. Anesthesiology 1991;75:106.
3. Frumin H, Kerin NZ, Rubenfire M. Classification of antiarrhythmic drugs. J Clin Pharmacol 1989;29:387.
4. Schmeling WT, Warltier DC, McDonald DJ, Madsen KE, Atlee JL, Kampine JP. Prolongation of the QT interval by enflurane, isoflurane, and halothane in humans. Anesth Analg 1991;72:137.
5. Levine JH, Morganroth J, Kadish AH. Mechanisms and risk factors for proarrhythmia with type 1a compared with 1c antiarrhythmic drug therapy. Circulation 1989;80:1063.
6. Charlier R. Cardiac actions in the dog of a new antagonist of adrenergic excitation which does not produce competitive blockade of adrenoceptors. Br J Pharmacol 1970;39:674.
7. Antonoccio MJ, Gomoll A. Pharmacology, pharmacodynamics and pharmacokinetics of sotalol. Am J Cardiol 1990;65:12A.
8. Carmeliet E. Slow inactivation of the sodium current in rabbit cardiac Purkinje fibers. Pflugers Arch 1987;408:18.
9. Patlack JB, Ortiz M. Slow currents through single sodium channels of the adult rat heart. J Gen Physiol 1985;86:89.
10. Nilius B. Modal gating behavior of cardiac Na channels in cell-free membrane patches. Biophys J 1988;53:857.
11. Follmer CH, Aomine M, Yeh JZ, Singer DH. Amiodarone-induced block of sodium current in isolated cardiac cells. J Pharmacol Exp Ther 1987; 243:187.

12. Kohlhardt M, Fichtner H. Block of single cardiac Na⁺ channels by antiarrhythmic drugs: the effects of amiodarone, propafenone and diprafenone. J Membr Biol 1988;102:105.
13. Aomine M. Inhibition of tetrodotoxin-sensitive plateau sodium current by amiodarone in guinea pig cardiac muscles. Gen Pharmacol 1989;20:653.
14. Sheldon RS, Hill RJ, Cannon NJ, Duff HJ. Amiodarone: biochemical evidence for binding to a receptor for class I drugs associated with the rat cardiac sodium channel. Circ Res 1989;65:477.
15. Varro A, Nanasi PP, Lathrop DA. Effect of sotalol on transmembrane ionic currents responsible for repolarization in cardiac ventricular myocytes from rabbit and guinea pig. Life Sci 1991;49:PL7.
16. Nilius B, Hess P, Lansman JB, Tsien RW. A novel type of cardiac calcium channel in ventricular cells. Nature (Lond) 1985;316:443.
17. Lee KS, Tsien RW. Mechanism of calcium channel blockade by verapamil, D600, diltiazem and nitrendipine in single dialyzed heart cells. Nature (Lond) 1983;302:790.
18. Kameyama M, Hoffman F, Trautwein W. On the mechanism of betaadrenergic regulation of the Ca channel in the guinea-pig heart. Pflugers Arch 1985;405:285.
19. Hagiwara N, Irisawa H, Kameyama M. Contribution of two types of calcium currents to the pacemaker potentials of rabbit sino-atrial node cells. J Physiol (Lond) 1988;395:233.
20. Mitchell MR, Powell T, Terrar DA, Twist VW. Characteristics of the second inward current in cells isolated from rat ventricular muscle. Proc R Soc Lond [Biol] 1983;219:447.
21. Isenberg G, Klockner U. Calcium currents of isolated bovine ventricular myocytes are fast and of large amplitude. Pflugers Arch 1982;395:30.
22. Nishimura M, Follmer CH, Singer DH. Amiodarone blocks calcium current in single guinea pig ventricular myocytes. J Pharmacol Exp Ther 1989;251:650.
23. Cohen CJ, Spires S, Van Skiver D. Block of T-type Ca channels in guinea pig atrial cells by antiarrhythmic agents and Ca channel antagonists. J Gen Physiol 1992;100:703.
24. Varro A, Surawicz B. Effect of antiarrhythmic drugs on membrane channels in cardiac muscle. In: Fisch C, Surawicz B, eds. Cardiac electrophysiology and arrhythmias. New York: Elsevier, 1991:277.
25. Matsuda H, Saigusa A, Irisawa H. Ohmic conductance through the inwardly rectifying K channel and blocking by internal Mg²⁺. Nature (Lond) 1987;325:156.
26. Luo C, Rudy Y. A model of the ventricular cardiac action potential: depolarization, repolarization, and their interaction. Circ Res 1991;68:1501.
27. Wasserstrom JA, Ten Eick RE. Electrophysiology of mammalian ventricular muscle. In: Dangman KH, Miura DS, ed. Electrophysiology and pharmacology of the heart. New York: Marcel Dekker, 1991:199.
28. Pancrazio JJ, Frazer MJ, Lynch C III. Barbiturate anesthetics depress the resting K⁺ conductance of myocardium. J Pharmacol Exp Ther 1993;265:358.
29. Sanguinetti MC, Jurkiewicz NK. Two components of cardiac delayed rectifier K⁺ current: differential sensitivity to block by class III antiarrhythmic agents. J Gen Physiol 1990;96:195.
30. Sato R, Hisatome I, Singer DH. Amiodarone blocks the inward rectifier K⁺ channel in guinea pig ventricular myocytes. Circulation 1987;76(suppl IV):151. Abstract.

31. Carmeleit E. Electrophysiologic and voltage clamp analysis of the effects of sotalol on isolated cardiac muscle and Purkinje fibers. J Pharmacol Exp Ther 1985;232:817.
32. Berger F, Borchard U, Hafner D. Effects of (+)- and (±)-sotalol on repolarizing outward currents and pacemaker current in sheep cardiac Purkinje fibres. Naunyn Schmiedebergs Arch Pharmacol 1989;340:696.
33. Komeichi K, Tohse N, Nagaya H, Shimizu M, Zhu MY, Kanno M. Effects of N-acetylprocainamide and sotalol on ion currents in isolated guinea-pig ventricular myocytes. Eur J Pharmacol 1990;187:313.
34. Satoh H. Class III antiarrhythmic drugs (amiodarone, bretylium and sotalol) on action potentials and membrane currents in rabbit sinoatrial node preparations. Naunyn Schmiedebergs Arch Pharmacol 1991; 344:674.
35. Argentieri TM, Sullivan ME, Wiggins JR. Class III antiarrhythmic agents. In: Dangman KH, Miura DS, eds. Electrophysiology and pharmacology of the heart. New York: Marcel Dekker, 1991:677.
36. Noble D, Tsien RW. Outward membrane currents activated in the plateau range of potentials in cardiac Purkinje fibres. J Physiol (Lond) 1969;200:205.
37. Balser JR, Bennett PB, Hondeghem LM, Roden DM. Suppression of time-dependent outward current in guinea pig ventricular myocytes. Actions of quinidine and amiodarone. Circ Res 1991;69:519.
38. Bkaily G, Payet MD, Benabderrazik M, et al. Intracellular bretylium blocks Na^+ and K^+ currents in heart cells. Eur J Pharmacol 1988;151:389.
39. Snyders DJ, Tamkun MM, Bennett PB. A rapidly activating and slowly inactivating potassium channel cloned from human heart: functional analysis after stable mammalian cell culture expression. J Gen Physiol 1993;101:513.
40. Coraboeuf E, Carmeliet E. Existence of two transient outward currents in sheep cardiac Purkinje fibers. Pflugers Arch 1982;392:352.
41. Siegelbaum SA, Tsien RW. Calcium-activated transient outward current in calf cardiac Purkinje fibers. J Physiol (Lond) 1980;299:485.
42. Callewaert G, Vereecke J, Carmeliet E. Existence of a calcium-dependent potassium channel in the membrane of cow cardiac Purkinje cells. Pflugers Arch 1986;406:424.
43. Tohse N. Calcium-sensitive delayed rectifier potassium current in guinea pig ventricular cells. Am J Physiol 1990;27:H1200.
44. Noma A. ATP-regulated K^+ channels in cardiac muscle. Nature (Lond) 1983;305:147.
45. Kakei M, Noma A. Adenosine-5'-triphosphate-sensitive single potassium channel in the atrioventricular node cell of the rabbit heart. J Physiol (Lond) 1984;352:265.
46. Gettes LS. The electrophysiology of acute ischemia. In: Fisch C, Surawicz B, eds. Cardiac electrophysiology and arrhythmias. New York: Elsevier, 1991:13.
47. Brooks CM, Gilbert JL, Greenspan ME, Lange G, Mazella HM. Excitability and electrical response of ischemic heart muscle. Am J Physiol 1960; 198:1143.
48. Elharrar V, Zipes DP. Cardiac electrophysiologic alterations during myocardial ischemia: afterdepolarizations and ventricular arrhythmias. Am J Physiol 1977;233:H329.
49. Yao Z, Cavero I, Gross GJ. Activation of cardiac KATP channels: an endogenous protective mechanism during repetitive ischemia. Am J Physiol 1993;264:H495.

50. Haworth RA, Goknur AB, Berkoff HA. Inhibition of ATP-sensitive potassium channels of adult rat heart cells by antiarrhythmic drugs. Circ Res 1989;65:1157.

51. Egan TM, Noble D, Noble SJ, Powell T, Spindler AJ, Twist VN. Sodium-calcium exchange during the action potential in guinea-pig ventricular cells. J Physiol (Lond) 1989;411:639.

52. Powell T, Noble D. Calcium movements during each heart beat. Mol Cell Biochem 1989;89:103.

53. Gadsby DC. The Na/K pump of cardiac myocytes. In: Zipes DP, Jalife J, eds. Cardiac electrophysiology: from cell to bedside. Philadelphia: WB Saunders, 1990:35.

54. Noble D. Ionic mechanisms in normal cardiac activity. In: Zipes DP, Jalife J, eds. Cardiac electrophysiology: from cell to bedside. Philadelphia: WB Saunders, 1990:163.

55. Aomine M. Suggestive evidence for inhibitory action of amiodarone on Na^+, K^+-pump activity in guinea pig heart. Gen Pharmacol 1989;20:491.

56. Dzimiri N, Almotrefi AA. Inhibition of myocardial Na^+-K^+-ATPase activity by bretylium: role of potassium. Arch Int Pharmacodyn Ther 1992;318:76.

57. Dzimiri N, Almotrefi AA. Interaction of bretylium tosylate with guinea-pig myocardial Na^+-K^+-ATPase. Gen Pharmacol 1991;22:935.

58. The Cardiac Arrhythmia Suppression Trial (CAST) Investigators. Preliminary report: effect of encainide and flecainide on mortality in a randomized trial of arrhythmia suppression after myocardial infarction. N Engl J Med 1989;321:406.

59. Task Force of the Working Group on Arrhythmias of the European Society of Cardiology. CAST and beyond: implications of the Cardiac Arrhythmia Suppression Trial. Circulation 1990;81:1123.

60. Hondeghem LM, Snyders DJ. Class III antiarrhythmic agents have a lot of potential but a long way to go: reduced effectiveness and dangers of reverse use dependence. Circulation 1990;81:686.

61. Colatsky TJ, Follmer CH, Starmer CF. Channel specificity in antiarrhythmic drug action: mechanism of potassium channel block and its role in suppressing and aggravating cardiac arrhythmias. Circulation 1990;82:2235.

62. Funck-Brentano C, Kibleur Y, Le Coz F, Poitier J-M, Mallet A, Jaillon P. Rate dependence of sotalol-induced prolongation of ventricular repolarization during exercise in humans. Circulation 1991;83:536

63. Lathrop DA. Electromechanical characterization of the effects of racemic sotalol and its optical isomers in isolated canine ventricular trabecular muscles and Purkinje strands. Can J Physiol Pharmacol 1985;63:1506.

64. Anderson KP, Walker R, Dustman T, et al. Rate-related electrophysiologic effects of long-term administration of amiodarone on canine ventricular myocardium in vivo. Circulation 1989;79:948.

65. Spach MS, Miller WT III, Geselowitz DB, Barr RC, Kootsey JM, Johnson EA. The discontinuous nature of propagation in normal canine cardiac muscle: evidence for recurrent discontinuities of intracellular resistance that affect the membrane currents. Circ Res 1981;48:39.

66. Balke CW, Lesh MD, Spear JF, Kadish A, Levine JH, Moore EN. Effects of cellular uncoupling on conduction in anisotropic canine ventricular myocardium. Circ Res 1988;63:879.

67. Spach MS, Dolber PC, Heidlage JF. Influence of the passive anisotropic properties on directional differences in propagation following modification of the sodium conductance in human atrial muscle. A model of reentry based on anisotropic discontinuous propagation. Circ Res 1988;62:811.
68. Osaka T, Kodama I, Tsuboi N, Toyama J, Yamada K. Effects of activation sequence and anisotropic cellular geometry on the repolarization phase of action potential of dog ventricular muscles. Circulation 1987;76:226.
69. Biagetti MO, de Forteza E, Quinteiro RA. Differential effects of amiodarone on V_{max} and conduction velocity in anisotropic myocardium. J Cardiovasc Pharmacol 1990;15:918.
70. Wysocka E, Wysocki H, Siminiak T, Szczepanik A. Effect of selected antiarrhythmic drugs on the superoxide anion production by polymorphonuclear neutrophils in vitro. Cardiology 1989;76:264.
71. Siminiak T, Wysocki H, Veit A, Maurer HR. The effect of selected antiarrhythmic drugs on neutrophil free oxygen radicals production measured by chemiluminescence. Basic Res Cardiol 1991;86:355.
72. Silver PJ, Connell MJ, Dillon KM, Cumiskey WR, Volberg WA, Ezrin AM. Inhibition of calmodulin and protein kinase C by amiodarone and other class III antiarrhythmic agents. Cardiovasc Drugs Ther 1989;3:675.
73. Tohse N, Kameyama M, Sekiguchi K, Shearman MS, Kanno M. Protein kinase C activation enhances the delayed rectifier potassium current in guinea-pig heart cells. J Mol Cell Cardiol 1990;22:725.
74. The Multicenter Postinfarction Research Group. Risk stratification and survival after myocardial infarction. N Engl J Med 1983;309:331.
75. May GS, Furberg CD, Eberlein KA, Geraci BJ. Secondary prevention after myocardial infarction: a review of short-term acute phase trials. Prog Cardiovasc Dis 1983;25:335.
76. Beta-Blocker Heart Attack Trial Research Group. A randomized trial of propranolol in patients with acute myocardial infarction. JAMA 1982; 247:1707.
77. Malliani A, Schwartz PJ, Zanchetti A. Neural mechanisms in life-threatening arrhythmias. Am Heart J 1980;100:705.
78. Patterson E, Holland K, Eller BT, Lucchesi BR. Ventricular fibrillation resulting from ischemia at a site remote from previous myocardial infarction. A conscious canine model of sudden coronary death. Am J Cardiol 1982;50:1412.
79. Uprichard AGC, Lucchesi BR. Antifibrillatory drugs. In: Dangman KH, Miura DS, eds. Electrophysiology and pharmacology of the heart: a clinical guide. New York: Marcel Dekker, 1991:723.
80. McComb JM, McGowan JB, McGovern BA, Ruskin GH. Comparison of the electrophysiologic properties of d- and dl-sotalol. J Am Coll Cardiol 1985;5:438. Abstract.
81. Kato R, Ikeda N, Yabek SM, Kannan R, Singh BN. Electrophysiologic effects of the levo- and dextrorotatory isomers of sotalol in isolated cardiac muscle and their in vivo pharmacokinetics. J Am Coll Cardiol 1986;7:116.
82. Beyer T, Brachmann J, Aidonidis I, Rizos I, KÅbler W. Effects of d-sotalol on electrophysiological parameters in rabbit AV-nodal preparations: demonstration of acute class II actions. J Mol Cell Cardiol 1986; 18(suppl):279. Abstract.
83. McComb JM, McGovern B, McGowan JB, Ruskin JN, Garan H. Electrophysiologic effects of d-sotalol in humans. J Am Coll Cardiol 1987;10:211.

84. Schwartz J, Crocker K, Wynn J, Somberg JC. The antiarrhythmic effects of d-sotalol. Am Heart J 1987;114:539.

85. Huikuri HV, Koistinen MJ, Takkunen JT. Efficacy of intravenous sotalol for suppressing inducibility of supraventricular tachyarrhythmias at rest and during isometric exercise. Am J Cardiol 1992;69:498.

86. Wang M, Dorian P. DL and D sotalol decrease defibrillation energy requirements. PACE Pacing Clin Electrophysiol 1989;12:1522.

87. Parmley WW, Rabinowitz B, Chuck L, Bonorris G, Katz JP. Comparative effects of sotalol and propranolol on contractility of papillary muscles and adenyl cyclase activity of myocardial extracts of cat. J Clin Pharmacol 1972;12:127.

88. Mahmarian JJ, Verani MS, Pratt CM. Hemodynamic effects of intravenous and oral sotalol. Am J Cardiol 1990;65:28A.

89. Kehoe RF, Zheutlin TA, Dunnington CS, Mattioni TA, Yu G, Spangenberg RB. Safety and efficacy of sotalol in patients with drug-refractory sustained ventricular tachyarrhythmias. Am J Cardiol 1990;65:58A.

90. Nattel S, Feder-Elituv R, Matthews C, Nayebpour M, Talajic M. Concentration dependence of class III and beta-adrenergic blocking effects of sotalol in anesthetized dogs. J Am Coll Cardiol 1989;13:1190.

91. Follath F. The utility of serum drug level monitoring during therapy with class III antiarrhythmic agents. J Cardiovasc Pharmacol 1992;20(suppl 2):S41.

92. The ESVEM Investigators. The ESVEM trial. Electrophysiologic study versus electrocardiographic monitoring for selection of antiarrhythmic therapy of ventricular tachyarrhythmias. Circulation 1989;79:1354.

93. McGovern BA, Ruskin JN, Garan H. Sotalol. In: Horowitz LN, ed. Current management of arrhythmias. Philadelphia: BC Decker, 1991:368-371.

94. Jordaens LJ, Palmer A, Clement DL. Low-dose oral sotalol for monomorphic ventricular tachycardia: effects during programmed electrical stimulation and follow-up. Eur Heart J 1989;10:218.

95. Griffith MJ, Linker NJ, Garratt CJ, Ward DE, Camm AJ. Relative efficacy and safety of intravenous drugs for termination of sustained ventricular tachycardia. Lancet 1990;336:670.

96. Teo KK, Harte M, Horgan JH. Sotalol infusion in the treatment of supraventricular tachyarrhythmias. Chest 1985;87:113.

97. Millar RN. Efficacy of sotalol in controlling reentrant supraventricular tachycardias. Cardiovasc Drugs Ther 1990;4(suppl 3):625.

98. Jordaens L, Gorgels A, Stroobandt R, Temmerman J. Efficacy and safety of intravenous sotalol for termination of paroxysmal supraventricular tachycardia. The Sotalol Versus Placebo Multicenter Study Group. Am J Cardiol 1991;68:35.

99. Brugada P, Smeets JL, Brugada J, Farre J. Mechanism of action of sotalol in supraventricular arrhythmias. Cardiovasc Drugs Ther 1990;4(suppl 3):619.

100. Madrid AH, Moro C, Marin Huerta EM, Novo L, Mestre JL. Atrial fibrillation in Wolff-Parkinson-White syndrome: reversal of isoproterenol effects by sotalol. PACE Pacing Clin Electrophysiol 1992;15:2111.

101. Maragnes P, Tipple M, Fournier A. Effectiveness of oral sotalol for treatment of pediatric arrhythmias. Am J Cardiol 1992;69:751.

102. Soyka LF, Wirtz C, Spangenberg RB. Clinical safety profile of sotalol in patients with arrhythmias. Am J Cardiol 1990;65:74A.

103. Arstall MA, Hii JT, Lehman RG, Horowitz JD. Sotalol-induced torsade de pointes: management with magnesium infusion. Postgrad Med J 1992; 68:289.
104. Singh SN, Lazin A, Cohen A, Johnson M, Fletcher RD. Sotalol-induced torsades de pointes successfully treated with hemodialysis after failure of conventional therapy. Am Heart J 1991;121:601.
105. Reitz JA. Alfentanil in anesthesia and analgesia. Drug Intell Clin Pharm 1986;20:335.
106. Sullivan ME, Steinberg MI. Cardiovascular and electrocardiographic effects of clofilium in conscious or anesthetized dogs. Pharmacologist 1981;23:209. Abstract.
107. Heissenbuttel RH, Bigger JTJ. Bretylium tosylate: a newly available antiarrhythmic drug for ventricular arrhythmias. Ann Intern Med 1979; 91:229.
108. Adult advanced cardiac life support. JAMA 1992;268:2199.
109. Koch-Weser J. Bretylium. N Engl J Med 1978;300:473.
110. Bacaner MB. Quantitative comparison of bretylium with other antifibrillatory drugs. Am J Cardiol 1968;21:504.
111. Anderson JL, Brondine WN, Patterson E, Marshall HW, Allison SD, Lucchesi BR. Serial electrophysiologic effects of bretylium in man and their correlation with plasma concentration. J Cardiovasc Pharmacol 1982;4:871.
112. Cardinal R, Sasyniuk BI. Electrophysiological effects of bretylium tosylate on subendocardial Purkinje fibers from infarcted canine hearts. J Pharmacol Exp Ther 1978;204:159.
113. Reyes W, Milstein S, Goldstein MA, Benditt DG. Bretylium. In: Horowitz LN, ed. Current management of arrhythmias. Philadelphia: BC Decker, 1991:308.
114. Gillard M, Brunner F, Waelbroeck M, Svoboda M, Christophe J. Bretylium tosylate binds preferentially to muscarinic receptors labelled with [₃H]oxotremorine M (SH or 'high affinity' receptors) in rat heart and brain cortex. Eur J Pharmacol 1989;160:117.
115. Parratt JR, Coker SJ, Wainwright CL. Eicosanoids and susceptibility to ventricular arrhythmias during myocardial ischemia and reperfusion. J Mol Cell Cardiol 1977;19(suppl 5):55.
116. Bernstein JG, Koch-Weser J. Effectiveness of bretylium tosylate against refractory ventricular arrhythmias. Circulation 1972;45:1024.
117. De Azedo IM, Watanabe Y, Dreifus LS. Electrophysiologic antagonism on quinidine and bretylium tosylate. Am J Cardiol 1974;33:633.
118. Nowak RM, Bodnar TJ, Dronen S, Gentzkow G, Tomlanovich MC. Bretylium tosylate as the initial treatment for cardiopulmonary arrest presenting to the emergency department: a randomized comparison with placebo. Ann Emerg Med 1981;10:404.
119. Hanyok JJ, Chow MS, Kluger J, Fieldman A. Antifibrillatory effects of high dose bretylium and a lidocaine-bretylium combination during cardiopulmonary resuscitation. Crit Care Med 1988;16:691.
120. Sanna G, Arcidiacono R. Chemical defibrillation of the human heart with bretylium tosylate. Am J Cardiol 1973;32:982.
121. Vachiery JL, Reuse C, Blecic S, Contempre B, Vincent JL. Bretylium tosylate versus lidocaine in experimental cardiac arrest. Am J Emerg Med 1990;8:492.

122. Haynes RE, Chinn TL, Copass MK, Cobb LA. Comparison of bretylium tosylate and lidocaine in management of out of hospital ventricular fibrillation: a randomized clinical trial. Am J Cardiol 1981;48:353.
123. Dhurandhar RW, Pickron J, Goldman AM. Bretylium tosylate in the management of recurrent ventricular fibrillation complicating acute myocardial infarction. Heart Lung 1980;9:265.
124. Smith WM, Gallagher JJ. "Les torsades de pointes": an unusual ventricular arrhythmia. Ann Intern Med 1980;93:578.
125. Wolfe CL, Nibley C, Bhandari A, Chatterjee K, Scheinman M. Polymorphous ventricular tachycardia associated with acute myocardial infarction. Circulation 1991;84:1543.
126. Kirlangitis J, Middaugh R, Knight R, et al. Comparison of bretylium and lidocaine in the prevention of ventricular fibrillation after aortic cross-clamp release in coronary artery bypass surgery. J Cardiothorac Anesth 1990;4:582.
127. Rosalion A, Snow NJ, Horrigan TP, Noon DL, Mostow ND. Amiodarone versus bretylium for suppression of reperfusion arrhythmias in dogs. Ann Thorac Surg 1991;51:81.
128. Santora AH, Finucane BT. Bretylium tosylate therapy for ventricular fibrillation after cardiopulmonary bypass. South Med J 1983;76:1197.
129. Casteneda AR, Bacaner MB. Effect of bretylium tosylate on the prevention and treatment of postoperative arrhythmias. Am J Cardiol 1970;25:461.
130. Feldman HS, Arthur GR, Pitkanen M, Hurley R, Doucette AM, Covino BG. Treatment of acute systemic toxicity after the rapid intravenous injection of ropivacaine and bupivacaine in the conscious dog. Anesth Analg 1991;73:373.
131. Kasten GW, Martin ST. Bupivacaine cardiovascular toxicity, comparison of treatment with bretylium and lidocaine. Anesth Analg 1985;64:911.
132. Kasten GW, Martin ST. Successful cardiovascular resuscitation after massive intravenous bupivacaine overdosage in anesthetized dogs. Anesth Analg 1985;64:491.
133. Haasio J, Pitkanen MT, Kytta J, Rosenberg PH. Treatment of bupivacaine-induced cardiac arrhythmias in hypoxic and hypercarbic pigs with amiodarone or bretylium. Reg Anesth 1990;15:174.
134. Zaritsky A. Pediatric resuscitation pharmacology. Members of the Medications in Pediatric Resuscitation Panel. Ann Emerg Med 1993;22:445.
135. Vastesaeger M, Gillot P, Rason G. Etude clinique d'une nouvelle médication anti-angoreuse. Acta Cardiol 1967;22:483.
136. Singh BN, Vaughan-Williams EM. The effect of amiodarone, a new antianginal drug, on cardiac muscle. Br J Pharmacol 1970;39:657.
137. Yabek SM, Kato R, Singh BN. Acute effects of amiodarone on the electrophysiologic properties of isolated neonatal and adult cardiac fibers. J Am Coll Cardiol 1985;5:1109.
138. Varro A, Nakaya Y, Elharrar V, Surawicz B. Use-dependent effects of amiodarone on V_{max} in cardiac Purkinje and ventricular muscle fibers. Eur J Pharmacol 1985;112:419.
139. Gallagher JD, Bianchi J, Gessman LJ. A comparison of the electrophysiologic effects of acute and chronic amiodarone administration on canine Purkinje fibers. J Cardiovasc Pharmacol 1989;13:723.
140. Aomine M. Multiple electrophysiological actions of amiodarone on guinea pig heart. Naunyn Schmiedebergs Arch Pharmacol 1988;338:589.

141. Ikeda N, Nademanee K, Kannaw R, Singh BN. Electrophysiologic effects of amiodarone: experimental and clinical observations relative to serum and tissue drug concentrations. Am Heart J 1984;108:890.
142. Wellens HJJ, Brugada P, Abdollah H. A comparison of the electrophysiologic effects of intravenous and oral amiodarone in the same patient. Circulation 1984;69:120.
143. Goupil N, Lenfant J. The effects of amiodarone on the sinus node activity of the rabbit heart. Eur J Pharmacol 1976;39:23.
144. Escoubet B, Coumel P, Poirier J-M, et al. Suppression of arrhythmias within hours after a single oral dose of amiodarone and relation to plasma and myocardial concentrations. Am J Cardiol 1985;55:696.
145. Waxman HL, Groh WC, Marchlinski FE, et al. Amiodarone for control of sustained ventricular tachycardia: clinical and electrophysiologic effects in 51 patients. Am J Cardiol 1982;50:1066.
146. McGovern B, Garan H, Ruskin JN. Sinus arrest during treatment with amiodarone. Br Med J 1982;284:160.
147. Lee TH, Friedman PL, Goldman L, Stone PH, Antman EM. Sinus arrest and hypotension with combined amiodarone-diltiazem therapy. Am Heart J 1985;109:163.
148. Rosenbaum MB, Chiale PA, Ryba D, Elizari M. Control of tachyarrhythmias associated with Wolff-Parkinson-White syndrome by amiodarone hydrochloride. Am Heart J 1974;34:215.
149. Wellens HJJ, Lie KI, Bär FW, et al. Effect of amiodarone in the Wolff-Parkinson-White syndrome. Am J Cardiol 1976;38:189.
150. Rowland E, Krikler DM. Electrophysiological assessment of amiodarone in the treatment of resistant supraventricular arrhythmias. Br Heart J 1980;44:82.
151. Heger JJ, Prystowski EN, Jackman WM, et al. Amiodarone: clinical efficacy and electrophysiology during long-term therapy for recurrent ventricular tachycardia or ventricular fibrillation. N Engl J Med 1981;305:539.
152. Morady F, DiCarlo LA, Krol RB, Baerman JM, de Buitleir M. Acute and chronic effects of amiodarone on ventricular refractoriness, intraventricular conduction and ventricular tachycardia induction. J Am Coll Cardiol 1986;7:148.
153. Nattel S. Pharmacodynamic studies of amiodarone and its active N-desethyl metabolite. J Cardiovasc Pharmacol 1986;8:771.
154. Heger JJ, Prystowsky EN, Zipes DP. Relationships between amiodarone dosage, drug concentrations, and adverse side effects. Am Heart J 1983;106:931.
155. Yabek SM, Kato R, Singh BN. Effects of amiodarone and its metabolite, desethylamiodarone, on the electrophysiologic properties of isolated cardiac muscle. J Cardiovasc Pharmacol 1986;8:197.
156. Talajic M, DeRoode MR, Nattel S. Comparative electrophysiologic effects of intravenous amiodarone and desethylamiodarone in dogs: evidence for clinically relevant activity of the metabolitee. Circulation 1987;75:265.
157. Figge HL, Figge J. The effects of amiodarone on thyroid hormone function: a review of the physiology and clinical manifestations. J Clin Pharmacol 1991;30:588.
158. Nademanee K, Piwonka RW, Singh BN, Hershman JH. Amiodarone and thyroid function. Prog Cardiovasc Dis 1989;31:427.

159. Kennedy RL, Griffiths H, Gray TA. Amiodarone and the thyroid. Clin Chem 1989;35:1882.
160. Lambert C, Cardinal R, Vermeulen M. Lack of relation between the ventricular refractory period prolongation by amiodarone and the thyroid state in rats. J Pharmacol Exp Ther 1987;242:320.
161. Katzung BG, Lee MA, Langberg JJ. Amiodarone. In: Dangman KH, Miura DS, eds. Electrophysiology and pharmacology of the heart: a clinical guide. New York: Marcel Dekker, 1991:637.
162. Polster P, Broekhuysen J. The adrenergic antagonism of amiodarone. Biochem Pharmacol 1976;25:131.
163. Cohen-Armon M, Schreiber G, Sokolovsky M. Interaction of the antiarrhythmic drug amiodarone with the muscarinic receptor in the rat heart and brain. J Cardiovasc Pharmacol 1984;6:1148.
164. Nokin P, Clinet M, Schoenfeld P. Cardiac beta-adrenoceptor modulation by amiodarone. Biochem Pharmacol 1983;32:2473.
165. Venkatesh N, Padbury JF, Singh BN. Effects of amiodarone and desethylamiodarone on rabbit myocardial beta-adrenoceptors and serum thyroid hormones—absence of relationship to serum and myocardial drug concentrations. J Cardiovasc Pharmacol 1986;8:989.
166. Yin Y-L, Perret GY, Nicolas P, Vassy R, Uzzan B, Tod M. In vivo effects of amiodarone on cardiac β-adrenoceptor density and heart rate require thyroid hormones. J Cardiovasc Pharmacol 1992;19:541.
167. Williams TL, Lefkowitz RJ. Thyroid hormone regulation of β-adrenergic receptors number. J Biol Chem 1977;88:2787.
168. Freedberg AS, Papp JG, Vaughan-Williams EM. The effect of altered thyroid state on atrial intracellular potentials. J Physiol (Lond) 1970;207:357.
169. Kay GN, Pryor DB, Lee KL, et al. Comparison of survival of amiodarone treated patients with coronary artery disease and malignant ventricular arrhythmias with that of a control group with coronary artery disease. J Am Coll Cardiol 1987;9:877.
170. Herre JM, Sauve MJ, Malone P, et al. Long-term results of amiodarone therapy in patients with recurrent sustained ventricular tachycardia or ventricular fibrillation. J Am Coll Cardiol 1989;13:442.
171. Rosenbaum MB, Chiale PA, Haedo A, Lazzari JO, Elizari MV. Ten years experience with amiodarone. Am Heart J 1983;106:957.
172. Haffajee CI, Love JC, Alpert JS, Asdourian GK, Sloan KC. Efficacy and safety of long-term amiodarone in treatment of cardiac arrhythmias: dosage experience. Am Heart J 1983;106:935.
173. Morady F, Sauve MJ, Malone P, et al. Long-term efficacy and toxicity of high-dose amiodarone therapy of ventricular tachycardia or ventricular fibrillation. Am J Cardiol 1983;52:975.
174. Kaski JC, Girotti LA, Messuti H, Rutitzky B, Rosenbaum MB. Long-term management of sustained, recurrent, symptomatic ventricular tachycardia with amiodarone. Circulation 1981;64:273.
175. Mason JW, the Amiodarone Toxicity Study Group. Toxicity of amiodarone. Circulation 1985;72:III-272. Abstract.
176. DiCarlo LA, Morady F, Sauve MJ, et al. Cardiac arrest and sudden death in patients treated with amiodarone for sustained ventricular tachycardia or ventricular fibrillation: risk stratification based on clinical variables. Am J Cardiol 1985;55:372.

177. Cairns JA, Connolly SJ, Gent M, Roberts R. Post-myocardial infarction mortality in patients with ventricular premature depolarizations. Canadian Amiodarone Myocardial Infarction Arrhythmia Trial Pilot Study. Circulation 1991;84:550.

178. Horowitz LN, Greenspan AM, Spielman SR, et al. Usefulness of electrophysiologic testing in evaluation of amiodarone therapy for sustained ventricular tachyarrhythmias associated with coronary heart disease. Am J Cardiol 1985;55:367.

179. Kadish AH, Buxton AE, Waxman HL, Flores B, Josephson ME, Marchlinski FE. Usefulness of electrophysiologic study of determine the clinical tolerance of arrhythmia recurrences during amiodarone therapy. J Am Coll Cardiol 1987;10:90.

180. Waters JB, Haffajee CI. Amiodarone. In: Horowitz JN, ed. Current management of arrhythmias. Philadelphia: BC Decker, 1991:294-301.

181. Rosenbaum MB, Chiale PA, Halpern MS, et al. Clinical efficacy of amiodarone as an antiarrhythmic agent. Am J Cardiol 1976;38:934.

182. Kopelman HA, Horowitz LN. Efficacy and toxicity of amiodarone for the treatment of supraventricular tachyarrhythmias. Prog Cardiovasc Dis 1989;31:355.

183. Helmy I, Herre JM, Gee G, et al. Use of intravenous amiodarone for emergency treatment of life-threatening ventricular arrhythmias. J Am Coll Cardiol 1988;12:1015.

184. Morady F, Scheinman MM, Shen E, Shapiro W, Sung RJ, DiCarlo L. Intravenous amiodarone in the acute treatment of recurrent symptomatic ventricular tachycardia. Am J Cardiol 1983;51:156.

185. Benaim R, Denizeau JP, Melon J. Les effets antiarythmiques de l'amiodarone injectable: a propos de 100 cas. Arch Mal Coeur 1976;69:513.

186. Kadish A, Morady F. The use of intravenous amiodarone in the acute therapy of life-threatening tachyarrhythmias. Prog Cardiovasc Dis 1989;31:281.

187. Hohnloser SH, Meinertz T, Dammbacher T, et al. Electrocardiographic and antiarrhythmic effects of intravenous amiodarone: results of a prospective, placebo-controlled study. Am Heart J 1991;121:89.

188. McAlister HF, Luke RA, Whitlock RM, Smith WM. Intravenous amiodarone bolus versus oral quinidine for atrial flutter and fibrillation after cardiac operations. J Thorac Cardiovasc Surg 1990;99:911.

189. Holt DW, Tucker GT, Jackson PR, Storey GCA. Amiodarone pharmacokinetics. Am Heart J 1983;106:840.

190. Chatelain P, Laruel R. Amiodarone partitioning with phospholipid bilayers and erythrocyte membranes. J Pharm Sci 1985;74:783.

191. Zipes DP, Prystowsky EN, Heger JJ. Amiodarone: electrophysiologic actions, pharmacokinetics and clinical effects. J Am Coll Cardiol 1984;4:1059.

192. Bajaj BP, Baig MW, Perrins EJ. Amiodarone-induced torsades de pointes: the possible facilitatory role of digoxin. Int J Cardiol 1991;33:335.

193. Rakita L, Sobol SM, Mostow N, Vrobel T. Amiodarone pulmonary toxicity. Am Heart J 1983;106:906.

194. Greenspon AJ, Kidwell GA, Hurley W, Mannion J. Amiodarone-related postoperative adult respiratory distress syndrome. Circulation 1991;84:III407.

195. Wilson JS, Podnil PJ. Side effects of amiodarone. Am Heart J 1991;121:158.

196. Farwell AP, Abend SL, Huang S, Patwardhan NA, Braverman WE. Thyroidectomy for amiodarone-induced thyrotoxicosis. JAMA 1990;263:1526.
197. Mehra A, Widerhorn J, Lopresti J, Rahimtoola SH. Amiodarone-induced hyperthyroidism: thyroidectomy under local anesthesia. Am Heart J 1991;122:1160.
198. Rooney RT, Stowe DF, Marijic J, Kampine JP, Bosnjak ZJ. Additive depressant effects of amiodarone given with halothane in isolated hearts. Anesth Analg 1991;72:474.
199. Feinberg BI, LaMantia KR, Levy WJ. Amiodarone and general anesthesia—a retrospective analysis. Eighth Annual Meeting of the Society of Cardiovascular Anesthesiologists. Phoenix, AZ, 1986:137. Abstract.
200. Liberman BA, Teasdale SJ. Anaesthesia and amiodarone. Can Anaesth Soc J 1985;32:629.
201. Kupferschmid JP, Rosengart TK, McIntosh CL, Leon MB, Clark RE. Amiodarone-induced complications after cardiac operation for obstructive hypertrophic cardiomyopathy. Ann Thorac Surg 1989;48:359.
202. Herndon JC, Cook AO, Ramsay MAE, Swygert TH, Capehart J. Postoperative unilateral pulmonary edema: possible amiodarone pulmonary toxicity. Anesthesiology 1992;76:308.
203. Satz AK, Stoelting RK, Gibbs PS. Intraoperative pulmonary edema in a patient being treated with amiodarone. Anesth Analg 1991;73:821.
204. Kopacz DJ. Continuous spinal anaesthesia for abdominal surgery in a patient receiving amiodarone. Can J Anaesth 1991;38:341.
205. Chassard D, George M, Guiraud M, et al. Relationship between preoperative amiodarone treatment and complications observed during anaesthesia for valvular cardiac surgery. Can J Anaesth 1990;37:251.
206. Elliott PL, Schauble JF, Rogers MC, Reid PR. Risk of decompensation during anesthesia in presence of amiodarone. Circulation 1983;68:III-280. Abstract.
207. Blair AD, Burgess ED, Maxwell BM. Sotalol kinetics in renal insufficiency. Clin Pharmacol Ther 1981;29:457.
208. Frisk Holmberg M, Juhlin Dannfelt A, Kaiser P, Rîssner S, Elliasson K, Hylander B. Therapeutic and metabolic effects of sotalol. Clin Pharmacol Ther 1984;36:174.
209. Adir J, Narang PK, Josselson J. Nomogram for bretylium dosing in renal failure. Ther Drug Monit 1985;7:265.
210. Anderson JL, Popat KD, Pitt B. Paradoxical ventricular tachycardia and fibrillation after intravenous bretylium therapy. report of two cases. Arch Intern Med 1981;141:801.
211. Thibault J. Hyperthermia associated with bretylium tosylate injection. Clin Pharmacol 1989;8:145.
212. Perlman PE, Adams WGJ, Ridgeway NA. Extreme pyrexia during bretylium administration. Postgrad Med 1989;85:111.

David E. Haines

Diagnosis and Treatment of Supraventricular Tachycardia

5

Supraventricular tachycardia (SVT) is a broad term that includes a wide variety of atrial, atrioventricular (AV) nodal, and AV reciprocating tachycardias. Our knowledge and understanding of the arrhythmia mechanisms and their associated anatomic substrates has increased dramatically in the past 10 years, as have the number of treatment options that are available. New antiarrhythmic compounds and new catheter ablation techniques have allowed us to suppress or entirely eliminate many arrhythmias in symptomatic patients, which were refractory to therapy in the past. Unfortunately, therapeutic challenges still remain in the treatment of some SVTs, particularly atrial fibrillation. The goal of this chapter is to introduce a framework for SVT diagnosis by arrhythmia mechanism, then present various acute and chronic treatment strategies based upon these diagnoses.

SUPRAVENTRICULAR TACHYCARDIA MECHANISMS

In the acute and chronic management of patients with SVT, it is imperative to consider the anatomic basis and electrophysiologic mechanism of each rhythm in order to optimize management. In most cases, careful analysis of a 12-lead electrocardiogram when combined with knowl-

Clinical Cardiac Electrophysiology: Perioperative Considerations
Edited by Carl Lynch III. J.B. Lippincott Company, Philadelphia, PA ©1994

edge of the patient's anatomic substrate can lead to the proper diagnosis of the arrhythmia mechanism. However, in some cases, invasive electrophysiologic testing may be required. In general, arrhythmias may originate from any structure (e.g., sinoatrial node, AV node, and atrium) and may be reentrant or automatic in mechanism. Reentrant rhythms are most common and are characterized by a sudden (paroxysmal) onset and offset.[1] In the setting of the electrophysiology laboratory, these rhythms may typically be induced and terminated with appropriately timed paced premature beats. Automatic rhythm mechanisms include normal automaticity (e.g., sinus tachycardia in response to physiologic stress) and abnormal automaticity (e.g., atrial tachycardia caused by digitalis intoxication). These rhythms are characterized by a "warm-up" and "cool-down" phenomenon and will momentarily suppress, but usually not terminate, with overdrive pacing[2] (Fig. 5-1). A final category of arrhythmogenesis, triggered automaticity, has clinical characteristics of both reentrant and abnormal automatic rhythms.[3] Although this mechanism is well established in the in vitro setting, its contribution as a mechanism of SVT in the clinical setting is less certain.

Reentrant Atrial Tachycardias

The reentrant atrial arrhythmias include atrial fibrillation, atrial flutter, and intraatrial reentrant tachycardia.[4] Although the typical form of each of these rhythms is unique in the characteristics of the patterns of

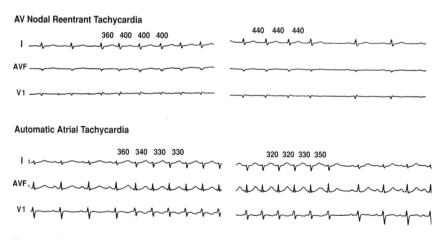

FIGURE 5-1. The onset and offset of two mechanisms of supraventricular tachycardia are illustrated. Atrioventricular nodal reentrant tachycardia (*top panel*) is initiated paroxysmally with a premature atrial beat and also terminates suddenly. In contrast, automatic atrial tachycardia (*bottom panel*) shows evidence of a "warm-up" and a "cool-down" phenomenon.

reentry, many intermediate forms occur, resulting in a continuum of arrhythmias ranging from very rapid and disorganized fibrillation to relatively slow and very organized intraatrial reentry. A common feature among patients with reentrant atrial arrhythmias is the high prevalence of structural abnormalities of the atria including enlargement and hypertrophy. Atrial fibrillation is a disorganized rhythm in which wavelets of reentry occur throughout both atria.[5] Electrical recordings from discreet atrial sites cycle between organized "flutter-like" activity and totally disorganized fibrillation. The critical determinants of atrial fibrillation propagation are regional conduction velocity and refractoriness (or "wavelength").[5] If atrial dimensions exceed the wavelength, then the arrhythmia may propagate. Since the AV node is bombarded with variably organized electrical impulses at high rates during atrial fibrillation, the resultant ventricular response follows an irregularly irregular pattern. Because the AV node has decremental conduction properties, only a portion of the atrial beats will penetrate the node and be propagated as a ventricular beat. Although more organized forms of atrial fibrillation (i.e., atrial fibrillation/flutter) are frequently observed, the ventricular response is typically irregularly irregular, and their response to therapy is similar to typical atrial fibrillation.

Atrial flutter and intraatrial reentrant tachycardia probably represent a similar arrhythmia mechanism with differing rates.[6] The pathway of intraatrial electrical conduction has most commonly been mapped as a macroreentrant circuit, which uses the caval orifices and the tricuspid valve as anatomic barriers. In addition, a critical zone of slowed conduction most likely exists in the region of the low interatrial septum. Typical atrial flutter has atrial rates of 250 to 350 beats/min whereas intraatrial reentrant tachycardia has rates below 250 beats/min. The surface electrocardiogram demonstrates organized atrial activity manifest by discreet flutter waves or P-waves. The AV node is stimulated at a regular rate and usually conducts with a regular ratio ranging from 4:1 to 1:1 (Fig. 5-2).

ATRIOVENTRICULAR NODAL REENTRY

AV nodal reentrant tachycardia is an arrhythmia that originates entirely from the AV node and perinodal atrial tissue. Classically, the reentrant circuit is described as consisting of a slow AV nodal pathway as the anterograde limb of the circuit and a fast AV nodal pathway as the retrograde limb[7] (Fig. 5-2). The uncommon form of AV nodal reentry utilizes the fast pathway as the anterograde limb and the slow pathway as the retrograde limb of the reentry circuit. Recently, AV nodal reentrant tachycardias have been described where

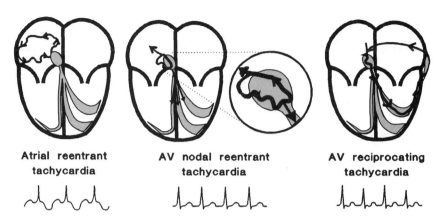

Atrial reentrant
tachycardia

AV nodal reentrant
tachycardia

AV reciprocating
tachycardia

FIGURE 5-2. Schematic diagrams of the pathway of reentry and the resultant surface electrocardiographic patterns in three common forms of reentrant supraventricular tachycardia. (Reproduced with permission from Haines DE, DiMarco JP. Curr Prob Cardiol 1992;17:411.)

both limbs of the reentrant circuit manifest prolonged conduction times in comparison to the fast pathway conduction. This indicates the presence of multiple slow pathways in selected patients. These arrhythmias have been termed "slow-slow" AV nodal reentry.

The era of highly focal ablative therapy with radiofrequency catheter ablation has yielded considerable insight into the anatomic correlates to the variants of AV nodal reentrant tachycardia. Selective ablation in the region of an unusually late and discrete low right posteroseptal atrial electrogram, designated by some as the "slow pathway potential," has frequently led to successful elimination of all slow pathway conduction (essentially returning the AV node to entirely normal function).[8] However, its presence does not appear to be a prerequisite for successful slow pathway ablation, and it frequently persists despite successful ablation, suggesting that it is a marker for the site of the slow pathway, but not the slow pathway potential per se. Conversely, selective ablation in the anterior-cranial aspect of the interatrial septum at the apex of the triangle of Koch (the presumed site of the compact zone of the AV node) yields fast pathway modification or conduction block. The "anterior approach" carries a 5 to 10% risk of complete heart block.[9] Thus, the "fast pathway" seems to represent the normal AV nodal conduction pathway, whereas the "slow pathways" may be branching points of atrial input into the node. Although this anatomic relationship is typical, several variations in the point of atrial insertion of slow and fast AV nodal pathways have been reported.

Because antegrade conduction time over the His-Purkinje network is similar to the retrograde conduction time over the fast pathway, the atrial and ventricular activations are usually simultaneous. This typically leads to superimposition of the P-waves on the QRS complexes. In the atypical form of AV nodal reentrant tachycardia, the PR interval is normal, hence the term "long R-P' tachycardia."

AV Reciprocating Tachycardia

With exception of the normal AV conduction system, the atria are electrically isolated from the ventricles by the annulus fibrosis of the AV valves in normal patients. Occasionally, an accessory AV conducting fiber is congenitally present. The common type of these fibers is variably referred to as accessory pathways, bypass tracts, or bundles of Kent.[10] If they are capable of propagating a depolarizing electrical wavefront in an anterograde fashion, the ventricle will be depolarized prematurely, resulting in the abnormal conduction pattern called the Wolff-Parkinson-White syndrome. If the pathways can propagate impulses retrogradely, then patients may experience orthodromic AV reciprocating tachycardia. In this common form of paroxysmal SVT, an impulse conducts retrograde up the accessory pathway through the atrium, travels anterograde through the AV conduction system, traverses the ventricle, then reenters in the accessory pathway (Fig. 5-2). This arrhythmia may occur whether the pathway is overt and conducts anterogradely (Wolff-Parkinson-White) or is concealed and conducts in a retrograde fashion only (concealed bypass tract).

The appearance of AV reciprocating tachycardia is similar to that of AV nodal reentrant tachycardia except that it tends to be somewhat faster and is frequently associated with visible retrograde P-waves immediately following the QRS in the ST segment. If the atypical form of this tachycardia occurs where the anterograde limb of the reentrant circuit is the accessory pathway and the retrograde limb the AV conduction system, then the tachycardia is called an antidromic AV reciprocating tachycardia and has a totally "preexcited" wide QRS morphology. Finally, unusual forms of accessory pathways exist, such as the decrementally conducting posteroseptal pathways of the permanent form of junctional reciprocating tachycardias[11] and the decremental right anteroseptal or free wall pathways that are termed Mahaim fibers.[12]

Automatic Supraventricular Tachycardias

Automatic rhythms may originate from any supraventricular structure, leading to SVT. SVT which has a P-wave identical to that found in sinus rhythm is either physiologic or "inappropriate" sinus tachycardia.[13] The latter diagnosis is one of exclusion after other etiologies such as anemia or hyperthyroidism have been eliminated. The source of inappropriate sinus tachycardia appears to be a diffuse region of enhanced automaticity rather than a highly focused origin. In contrast, automatic (ectopic) atrial tachycardia is presumed to be caused by abnormal automaticity originating from a very localized source in the right or left atrium, yielding a P-wave morphology which differs from that of sinus rhythm. The tachycardia may be intermittent or incessant and occasionally has been implicated as the cause of a cardiomyopathy.[14] Automatic tachycardia arising from the AV node or His bundle is termed junctional ectopic tachycardia. It is found most frequently in the pediatric population or in the setting of digitalis intoxication in adults. On an electrocardiogram it appears similar to AV nodal reentrant tachycardia, although it is nonparoxysmal. Multifocal tachycardia is an atrial rhythm which, by definition, has three or more P-wave morphologies.[15] The mechanism of this rhythm is unknown, but there is indirect evidence to suggest that it is triggered automaticity, including enhancement of the tachycardia with catecholamines and slight slowing with calcium channel blockers.

ACUTE EVALUATION OF SUPRAVENTRICULAR TACHYCARDIA

Differential Diagnosis

When the clinician is faced with a patient who has a narrow-complex tachycardia, diagnosis of the proper arrhythmia mechanism and understanding the patient's predisposing anatomic and physiologic conditions will allow institution of optimal acute and chronic therapy. In the diagnosis of SVT mechanisms, the most valuable tools are an excellent medical history and a high-quality 12-lead electrocardiogram acquired during tachycardia. The medical history should address questions such as the characteristics of SVT onset and offset, the duration of arrhythmia symptoms, and the presence of past cardiac illness. This, supplemented by a careful cardiac examination and, in some cases echocardiography, will lead the clinician toward a more limited range of possible diagnoses. For example, an older patient with a history of significant valvular disease is likely to have a reentrant atrial arrhythmia, whereas a patient with no cardiac difficulties except for a long history of paroxysmal palpitations that terminate with vagal maneuvers most certainly has AV node reentrant tachycardia or AV reciprocating tachycardia.

When the clinician examines the electrocardiograms of patients with tachycardia, the pattern of arrhythmia onset and offset is useful information. Note that wide complex tachycardia in patients with structural heart disease is much more likely to be ventricular tachycardia than SVT. Also, rhythms that appear to be narrow-complex on a single-channel monitor recording may clearly be wide-complex when evaluated with multiple electrocardiogram leads. Thus, obtaining a 12-lead electrocardiogram during tachycardia is strongly advised whenever possible. Rhythms that abruptly start and stop with premature beats are most likely reentrant in mechanism, whereas a slowly increasing and decreasing rate at onset and offset suggests an automatic mechanism (Fig. 5-1).

The pattern of ventricular response to the SVT is important. A regular rhythm with a 1:1 A-V relationship could be observed with many SVT mechanisms, but any evidence of AV nodal block during SVT, such as Wenckebach or 2:1 block, implies that the rhythm does not involve the AV node as a requisite component of the tachycardia circuit (Fig. 5-3). The P-wave morphology during tachycardia will identify the arrhythmia origin as high atrium (upright P-wave in leads II, III, and aVF) or low atrium/AV junction (inverted P-wave in inferior leads). Simultaneous P-waves and QRS complexes are most common in AV nodal arrhythmias and inverted P-waves following the QRS onset by more than 80 msec suggest an AV reciprocating or slow-slow AV nodal reentrant tachycardia. Since the AV relationship during atrial tachycardia is dependent upon tachycardia rate and the dromotropic properties of the AV node, the PR interval may be normal or markedly prolonged in this setting.

Diagnostic Maneuvers

The differentiation among SVT mechanisms is greatly facilitated by the use of provocative maneuvers. Since determining the contribution of the AV node to arrhythmia propagation is a useful branch point in the SVT diagnosis, the response of SVT to a temporary slowing or block of AV nodal conduction should be observed. This may be evoked with a variety of vagal maneuvers such as the Valsalva maneuver or carotid sinus massage.[16] Unfortunately, patients in acute care settings often are in a parasympatholytic state and are unresponsive to vagal maneuvers. In this setting, intravenous adenosine may be used successfully to accomplish the same goal.[17] Tachycardias that persist despite temporary induced high-grade AV block are invariably atrial in origin (Fig. 5-4). During the period of AV block, the P-wave morphology and rate may be examined more closely to determine its axis and determine if there is a 1:1 or 2:1 AV conduction ratio at baseline. Abrupt termination of SVT with adenosine indicates that the sinoatrial or AV node is a critical limb of a reentrant circuit. Finally, gradual slowing and then acceleration of the arrhythmia suggests that it originates from a site of automaticity in the sinoatrial or AV node. More sophisticated rhythm analysis may be achieved by direct measurement of atrial and ventricular electrograms using either transesophageal or transvenous electrodes. Ultimately, complete delineation of the arrhythmia mechanism may be obtained by performing a multicatheter invasive electrophysiologic study.

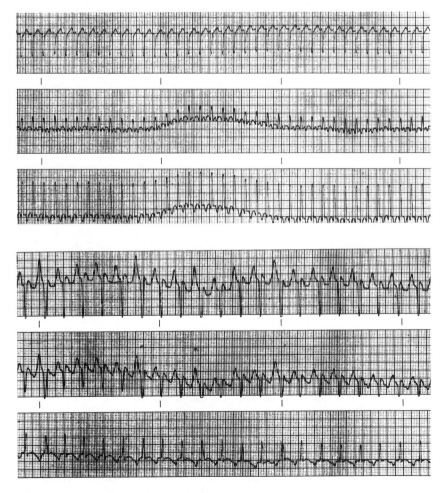

FIGURE 5-3. Example of a surface three-lead electrocardiogram from a patient during sustained intraatrial reentrant tachycardia. The patient had 1:1 atrioventricular conduction (*top panel*). After administration of 10 mg of intravenous verapamil, the patient had pseudotermination of the tachycardia, but in fact merely had slowing of atrioventricular conduction to 2:1 (*bottom panel*). The underlying P-waves at twice the ventricular rate may be discerned on closer examination.

A

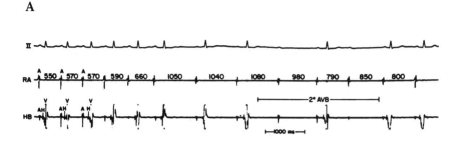

B

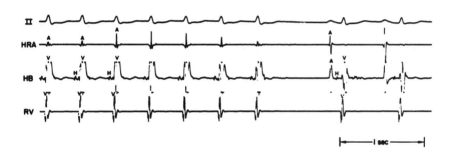

C

FIGURE 5-4. Response of supraventricular tachycardias to bolus administration of intravenous adenosine. A, surface electrocardiogram lead II and intracardiac tracings from the high right atrium (RA or HRA) and the vicinity of the bundle of His (HB) during sinus tachycardia. There is gradual slowing of sinus node automaticity and a brief episode of atrioventricular node block during peak adenosine effect. B, similar surface and intracardiac electrograms including a recording from the right ventricular apex (RV) in a patient with atrioventricular nodal tachycardia. Adenosine causes transient block in the antegrade limb of the reentrant circuit (the "slow" atrioventricular nodal pathway) resulting in abrupt tachycardia termination. C, typical episode of atrial flutter. Despite induction of increased atrioventricular nodal block by adenosine, the atrial tachycardia rate is unchanged. (Reproduced with permission from DiMarco JP, et al. Circulation 1983;68:1254, and DiMarco JP, et al. J Am Coll Cardiol 1985;6:417.)

DRUG THERAPY

A wide variety of antiarrhythmic drugs are now available for use in the treatment of SVT. In order to better understand their action, many of these agents may be grouped into classes as proposed by Vaughn-Williams.[18] Although this proposed schema has some limitations, generally it is may be used to predict how an individual arrhythmia will respond to a specific drug. It must be noted that, although the drugs listed below are commonly prescribed for the treatment of SVT, many of them have not received Food and Drug Administration approval for this indication. Thus, it is incumbent upon the prescribing physician to have a clear understanding of the risks and benefits of therapy with these agents before their use.

Class I Drugs

The class I antiarrhythmic agents are designated as such because they all demonstrate local anesthetic properties by competitive inhibition of the fast sodium channel. These drugs may be subclassified into Ia, Ib, and Ic categories.[18] The Ia agents such as quinidine, procainamide, and disopyramide have moderate use-dependent sodium channel-blocking capabilities and uniformly delay repolarization.[19] They effectively slow conduction in atria, ventricles, and accessory pathways. The Ib agents shorten action potential durations and have a net effect that results in little alteration of conduction. As a rule, they are not effective in the treatment of SVT. Class Ic drugs have powerful sodium channel-blocking effects and slow use dependence kinetics. They impart little change in refractoriness.[18] The currently available drugs in this category are flecainide and propafenone. All class I antiarrhythmic compounds have a variety of toxicities and side effects, those of most concern being a significant potential for arrhythmia aggravation.

Class II Drugs

The class II antiarrhythmic agents include those with a primary action of β-adrenergic antagonism.[18] The antiarrhythmic effects of these drugs are mediated by their effects of slowing conduction and decreasing automaticity in adrenergically sensitive tissue—most notably the sinoatrial node and atrioventricular node. Little change in atrial conduction or refractoriness is observed. Accessory pathway conduction is either unchanged or slightly slowed (unlike digoxin and verapamil, which may accelerate accessory pathway conduction), which allows for empiric therapy of patients with Wolff-Parkinson-White syndrome with β-blockers without being concerned about accelerating ventricular response to atrial fibrillation. Many drugs that are not primarily categorized as class II have important β-adrenergic blocking activity, including propafenone,[20] sotalol,[21] and amiodarone.[22] Drugs that have additional actions such as intrinsic sympathomimetic activity or α-adrenergic blockade offer little incremental gain over the pure β-blockers. The β-blockers are very safe and have moderate efficacy in the treatment of a wide range of SVTs. They are particularly useful for the treatment of arrhythmias that occur in settings of high endogenous catecholamine levels such as exercise-induced SVT or for rate control in patients with postoperative atrial fibrillation.

Class III Drugs

The electrophysiologic effect that characterizes class III antiarrhythmic drugs is action potential prolongation with resultant prolongation of refractoriness. This is probably mediated by potassium channel blockade.[23] The drugs in this class that are prescribed for the treatment of SVTs are amiodarone and sotalol. Amiodarone has many actions including sodium channel blockade and noncompetitive β-adrenergic blockade.[22] It has powerful effects of conduction slowing, lengthening refractoriness, and decreasing automaticity.[24] The only factors that prevent amiodarone from being an "ideal" antiarrhythmic drug are its unusual pharmacokinetic properties with an extremely long half-life and large volume of distribution and its significant toxicities including pulmonary, hepatic, and skin toxicity.[24]

Sotalol is a potent class III agent that has recently received Food and Drug Administration approval for the treatment of ventricular arrhythmias. Whereas the *d*-sotalol enantiomer has relatively pure class III effects, the *l*-sotalol enantiomer contributes the property of significant β-adrenergic blockade to the racemic compound.[21] This combination of effects makes it well suited for the treatment of a variety of SVTs. The class III effects of sotalol demonstrate a reverse use dependence,[25] which, in conjunction with the sinus bradycardia often observed during initiation of therapy, can result in the proarrhythmic complication of torsade de points.

Class IV Drugs

These drugs that are calcium channel antagonists have a role in the treatment of SVT which is similar to that of the β-blockers. Their predominant sites of action are SA and AV nodes. Although the dihydropyridine agents have powerful smooth muscle effects, their electrophysiologic effects are minimal.[26] Hence, verapamil and diltiazem are the agents available for the treatment of SVT.[27] These drugs are relatively well tolerated and have moderate efficacy in the control of selected arrhythmias. Tachycardias that are dependent upon high endogenous catecholamine tone are treated more effectively with β-blockers than calcium channel blockers, although the latter are useful when β-blockers are relatively contraindicated.

Cardiac Glycosides

Digitalis and other glycosides have historically been first-line therapy for many SVTs. Their action is effected by a centrally mediated vagotonia,[28] with little direct electrophysiologic effect on the heart in therapeutic doses. Because of this mechanism of action, digitalis loses its efficacy in the setting of spontaneous vagolysis, such as during exercise or other physiologic stress. The onset of the digitalis effect is slow (up to 5 hr),[29] so its efficacy in the acute management of symptomatic arrhythmias is limited. Digitalis intoxication is an important cause of automatic atrial and junctional tachycardias.

Purinergic Agonists

Adenosine is an endogenous nucleoside that can profoundly slow SA and AV nodal conduction. This effect is mediated by the activation of the A_1 adenosine receptor that increases potassium conductance and hyperpolarizes the nodal cells, thus slowing or completely blocking conduction.[30] Adenosine's plasma half-life is less than 6 sec, so rapid bolus administration is required to observe an effect.[30] Continuous infusion of adenosine does not achieve therapeutic concentrations for arrhythmia therapy and, in fact, causes a peripheral vasodilation that can result in a reflex increase in sympathetic tone and tachycardia exacerbation. Thus, adenosine is a useful agent for termination of reentrant rhythms utilizing the SA or AV node, but is not a good choice for the continued slowing of ventricular response to an atrial tachycardia. Dipyridamole inhibits adenosine deaminase and will enhance adenosine effects. Methylxanthines are competitive inhibitors of the adenosine receptors and will block adenosine's effects.[17]

ABLATION THERAPY

Catheter ablation is one of the most significant advances in the therapy of arrhythmias to occur in the past decade. The rationale for ablation therapy is that, for every arrhythmia, a critical anatomic substrate exists that allows for impulse formation or propagation of that rhythm. If that anatomic region is irreversibly damaged or destroyed, then the arrhythmia should no longer occur spontaneously or with provocation. Historically, this has been accomplished with an open surgical approach. However, since 1981 investigators have been refining techniques for creating focal injury by way of a transvenous or transarterial catheter. Initial attempts at catheter ablation used high-energy direct current shocks through conventional electrode catheters to impart injury.[31] Unfortunately, this approach was associated with significant barotrauma and a high prevalence of reversible injury. When radiofrequency (RF) ablation was first introduced, it was apparent that small discrete endocardial lesions could be reliably produced with minimal risk to the patient. However, the small size of the lesions initially led to unacceptably low success rates and prolonged procedure durations. With improvement in our understanding of the pathoanatomic relationships of arrhythmias, coupled with improvement in catheter design, RF catheter ablation has become the mainstay of therapy for a variety of arrhythmias.[32,33]

The dominant mechanism of tissue destruction during RF catheter ablation appears to be thermal.[34] The tools used in cardiac RF catheter ablation today are merely refinements of the standard electrocautery instruments. RF current is passed through the conducting wire to the tip electrode of an electrode catheter. The current then passes into the tissue that is directly contiguous to the electrode, which is grounded to a large surface area dispersive electrode applied to the skin surface (Fig. 5-5). In comparison to the metal electrode, the impedance of the tissue is high, and the electrical energy is therefore rapidly dissipated as a resistive loss into the tissue in the form of heat.[34] The resistive heating is proportional to the RF power density in the tissue. Therefore, since the surface area of the ablation electrode is small compared to that of the dispersive electrode, heating occurs in

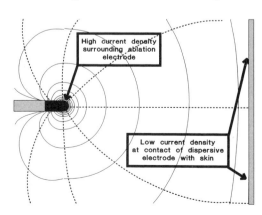

High current density surrounding ablation electrode

Low current density at contact of dispersive electrode with skin

FIGURE 5-5. Schematic representation of the current density and field lines during radiofrequency catheter ablation.

the former but not the latter location. The frequency of RF current used in ablation is typically 300 to 1000 kHz, which is high enough so that no muscle stimulation occurs, but low enough so that the mechanism of power transmission through the catheter is predominantly electrical. Although a number of electrical factors will affect the efficiency of energy transmission from the generator to the catheter tip, the major factors in successful lesion formation are the successful electrical coupling of the catheter to the tissue and heat transfer from the tissue surface into deeper levels.

The mode of tissue heating from RF current is conductive heating from the narrow rim of resistively heated tissue at the catheter contact point. The lesion size is proportional to the RF power as well as to the electrode-tissue interface temperature. Lesion size may be maximized by increasing the power until the temperature approaches 100°C. However, once this temperature reaches or exceeds 100°C, boiling at the electrode tip may occur.[35] This is usually associated with a sudden rise in electrical impedance and formation of "coagulum" and thrombus at the catheter tip. Thus, for any given electrode geometry, a theoretical maximum lesion size exists, which is bounded by the 100°C limit. Lesion size may be maximized by using large tipped electrodes, since lesion size is proportional to the electrode-tissue contact area.[36] Finally, the importance of close electrode-tissue contact must be underscored. Since most RF catheter ablative lesions are created on the endocardial surface within the cardiac chambers, the contact point is subject to significant convective heat loss into the circulating blood pool. Any degree of catheter movement during energy delivery will markedly decrease the efficiency of tissue heating and may result in inadequate lesion formation.

The success rate of catheter ablation has exceeded the expectations of most clinicians, and it has completely supplanted open surgical techniques in the treatment of SVT. The safety profile of this technique is excellent, but like any invasive procedure, complications may occur. The risk of major complications is approximately 1%, including stroke, significant valve trauma, and pericardial tamponade.[32,33] Most of these complications are attributable to the catheterization rather than to the ablations, per se. Although myocardial infarction has been reported as a result of inadvertent delivery of RF current within the coronary artery, no other coronary arterial complications have been seen. This may be due to the significant convective cooling around high-flow vessels, which serves to protect the vascular wall. The magnitude of myocardial injury resulting from the RF ablations appears to be small and is usually associated with a low increase or no increase in serum creatine kinase level. However, this measure may not be a reliable indicator of volume of myocardial injury in this setting. The risk of procedure-related death appears to be approximately 0.1%. Finally, some of these procedures may be prolonged with fluoroscopy times exceeding 1 hr, resulting in a small but definable increased risk of malignancy and birth defects.[37]

APPROACH TO THE PATIENT WITH ATRIAL FIBRILLATION AND FLUTTER

The reentrant atrial arrhythmias are the most prevalent among the various SVTs. There is a wide spectrum of clinical presentations and associated pathologic conditions that predispose patients to this arrhythmia. Diseases that result in atrial hypertension or hypertrophy are most commonly implicated, including mitral and tricuspid valve disease, hypertension, dilated, ischemic, and hypertrophic cardiomyopathies, inflammatory diseases, and congenital heart disease. Noncardiac etiologies for atrial arrhythmias include hyperthyroidism, alcohol consumption, pneumonia, and pulmonary embolism. Approximately 3% of atrial fibrillation patients have a single episode and no obvious cardiac pathologic condition.[38] A high prevalence of atrial fibrillation is found in patients with Wolff-Parkinson-White syndrome. This rhythm is of concern since the accessory pathway can conduct the tachycardia to the ventricles at high rates and precipitate ventricular fibrillation.

Acute Management

The goal of treatment of patients with atrial fibrillation or flutter is to first stabilize the patient's hemodynamics, then follow with a complete evaluation and optimization of chronic therapy. If a patient shows evidence of hemodynamic embarrassment with a rapid ventricular response to the atrial tachycardia, the clinician should have a low threshold for proceeding with electrical cardioversion under heavy sedation or brief general anesthesia. Transthoracic shocks should be administered in a synchronized fashion with an initial energy selected for atrial fibrillation of ≥200 J or 50 to 100 J for atrial flutter.

Medical rate control is best achieved with a β-blocker such as propranolol administered at a rate of 1 mg/min up to a loading dose of 0.1 mg/kg. If β-blockers are contraindicated, intravenous calcium channel blockers (e.g., verapamil or diltiazem) may be used.[39] Adenosine is frequently useful as a diagnostic tool to temporarily achieve high-grade AV block and expose the underlying atrial rhythm, but its very short half-life precludes its use for rate control. Although digoxin is frequently the first drug selected in this setting, it is usually ineffective since most patients will have very low vagal tone, and the major effect of digoxin is vagotonic in this setting. In addition, the peak effect of digoxin is delayed up to 5 hr after intravenous administration. Digoxin is no more effective in the termination of atrial fibrillation than placebo[40] (Fig. 5-6). After the ventricular response to atrial fibrillation and flutter is controlled, tachycardia in many patients will spontaneously convert to sinus rhythm. Intravenous or oral administration of a class I drug such as quinidine or procainamide will be associated with successful conversion to sinus rhythm in approximately 50 to 60% of patients.[41–43] Care must be taken when these drugs are used in the treatment of atrial flutter because the slowing of the flutter rate and the vagolytic effect of some of these drugs can lead to a shift from 2:1 to 1:1 AV conduction and acceleration of the ventricular response rate.

Chronic Medical Management

The chronic treatment of atrial fibrillation is currently one of the most challenging areas in arrhythmia management. The decision whether or not to initiate chronic treatment with antiarrhythmic drugs is complex and must include the factors of the symptomatic state of the patient, the risk of thromboembolic events during chronic atrial fibrillation,[44,45] and the risk of antiarrhythmic drug therapy. No prospective trials have evaluated the relative merits of different strategies; thus, the clinician must extrapolate the available data and individualize the therapy for each patient.

The class I and class III drugs have all shown efficacy in the chronic suppression of atrial fibrillation and flutter. Class Ia drugs have traditionally been used as first-line therapy and have a 50 to 60% long-term suppression rate. The class Ic agents may have a slightly higher

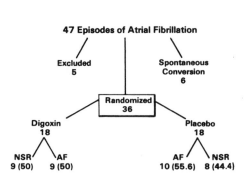

FIGURE 5-6. Course of 47 episodes of new onset atrial fibrillation in patients considered for random selection of intravenous digoxin or placebo therapy. Numbers in parentheses represent percentages of patients receiving either digoxin or placebo. NSR, normal sinus rhythm; AF, atrial fibrillation. (Reproduced with permission from Falk RH, Knowlton AA, Bernard SA, et al. Digoxin for converting recent-onset atrial fibrillation to sinus rhythm. Ann Intern Med 1987;106:503.)

efficacy than the class Ia drugs or may be complementary in their action and therefore appropriate for use in patients in whom class Ia drugs fail.[46] Because disopyramide, flecainide, and propafenone significantly depress myocardial contractility, they probably should not be used in patients with impaired left ventricular function. Sotalol has an efficacy rate in the long-term suppression of refractory atrial arrhythmias that is similar to that of quinidine, but it appears to be better tolerated.[47] Amiodarone has a similar reported efficacy of 50 to 80%,[48,49] but these statistics apply to patients whose conditions have been refractory to multiple antiarrhythmic regimens, which suggests that this therapy delivered de novo might have a higher total efficacy rate.

All antiarrhythmic drugs have a potential for arrhythmia aggravation (or "proarrhythmia"). The occurrence of proarrhythmia may range from 5 to 15%, and may be manifest by an occurrence of or exacerbation in the increase in density of premature ventricular beats, nonsustained ventricular tachycardia, sustained ventricular tachycardia, or torsade de points. The Cardiac Arrhythmia Suppression Trial demonstrated a 3-fold higher mortality in patients treated with encainide and flecainide who had a history of prior myocardial infarction and frequent premature ventricular beats.[50] While these data cannot be routinely extrapolated to the atrial tachycardia population, the majority of patients with atrial fibrillation or flutter have significant structural heart disease and hence may be at increased risk for proarrhythmia. A meta-analysis of six published trials comparing quinidine to placebo in the treatment of atrial fibrillation showed successful suppression of the arrhythmia but also identified a 3-fold excess mortality in the group treated with active drug.[51] Thus, a decision to prescribe chronic suppressive antiarrhythmic drugs must be made with an understanding of the risk inherent in this approach.

Although atrial flutter is an unstable rhythm that should be terminated, it is an acceptable alternative to withhold suppressive drug therapy in patients with chronic atrial fibrillation and control the ventricular response with AV nodal blocking agents or an AV junction ablation and permanent pacemaker. The major risk of this approach is thromboembolic phenomena. This risk may be reduced to an acceptable level if low-dose therapy with coumadin for anticoagulation is used.[44,45]

Ablative Therapy

The use of AV junctional catheter ablation in patients with refractory atrial arrhythmias is an effective modality when drugs are either ineffective or have limiting side effects. Because the junctional escape rhythm that is present after induction of AV block is usually slow (<50 beats/min), this procedure is invariably accompanied by placement of a permanent ventricular pacemaker. In patients in whom the arrhythmia is intermittent, dual chamber pacing may be used in the DDIR mode (i.e., rate-responsive dual chamber pacing, but no ventricular tracking of spontaneous atrial rhythms) so that AV synchrony is maintained when the patient is in sinus rhythm. The success rate of this procedure is now 95%, and the complication rate is below 1%.[52]

The use of ablative therapy as a primary "curative" therapy for atrial arrhythmias has been limited. A surgical procedure called the "maze" procedure has been developed in which a series of atrial incisions render the atria incapable of sustaining the complex pattern of wavelets of reentry that sustain atrial fibrillation.[53] Initial publications report a >90% success rate in highly selected patients with maintenance of atrial transport function. The long-term outcome of these patients, particularly with regard to the risk of thromboembolic events, remains to be determined. Since classical atrial flutter follows a more discrete path of reentry within the right atrium, it appears to be amenable to catheter ablative techniques[54] (Fig. 5-7). Preliminary studies report an acute success rate of approximately 75%, and a chronic success rate of 75% in initial responders.

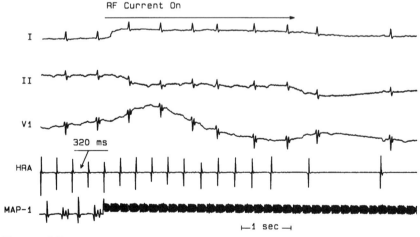

FIGURE 5-7. Example of radiofrequency catheter ablation of sustained intraatrial reentrant tachycardia. Surface electrocardiographic leads I, II, and V₁ and intracardiac electrograms from the high right atrium (HRA) and the mapping/ablation catheter (MAP-1) are shown. The electrogram from the ablation catheter shows a 2:1 pattern of fractionation, suggesting that it is positioned in a region of intraatrial conduction slowing and block. Three seconds after onset of radiofrequency current delivery, the tachycardia is terminated.

APPROACH TO THE PATIENT WITH PAROXYSMAL SUPRAVENTRICULAR TACHYCARDIA

Paroxysmal supraventricular tachycardia is a category of arrhythmias characterized by a regular narrow-QRS complex tachycardia at rates that are typically 150 to 250 beats/min. By definition, these rhythms have a sudden onset and termination. The most common mechanisms for these arrhythmias are AV nodal reentrant tachycardia and AV reciprocating tachycardia. Occasionally, atrial tachycardias have a similar electrocardiographic appearance, and temporary induction of AV block with vagal maneuvers or adenosine is necessary to expose the underlying atrial rhythm. Retrograde atrial activity evidenced by inverted P-waves in surface electrocardiographic leads II, III, and aVF is usually observed within the ST segment in AV reciprocating tachycardia, but is concealed by the simultaneous QRS complex in typical AV nodal reentry. Patients with paroxysmal supraventricular tachycardia are frequently young and usually have no structural heart disease, although a high prevalence of accessory pathways are observed in patients with Epstein's anomaly. The group with Wolff-Parkinson-White syndrome is an important subset of the paroxysmal SVT population. The most prevalent arrhythmia in this group is AV reciprocating tachycardia, but these patients also have a high prevalence of atrial fibrillation, which can lead to a rapid ventricular response rate in this setting and possible cardiac arrest.[55]

Acute Management

Since most mechanisms of paroxysmal supraventricular tachycardia use the AV node as a requisite limb of the reentry circuit, maneuvers and drugs that temporarily slow AV nodal conduction will frequently terminate these tachycardias. Popular vagal maneuvers include the Valsalva manuever, carotid sinus massage, and cold facial immersion (the diving reflex).[16] Although these methods are often successful if self-administered at the onset of the tachycardia, the success rate falls once the tachycardia persists, and the patient has a spontaneous vagolytic and sympathetic response.

Acute pharmacologic termination of paroxysmal supraventricular tachycardia is best achieved with a rapid intravenous bolus of adenosine.[56] This drug is an endogenous nucleoside that has powerful slowing effects on AV nodal conduction. Because it has an ultra-short half-life, the effect of adenosine will be enhanced by central venous administration and diminished by a slow rate of administration or conditions that prolong the central circulation time such as central shunts and low cardiac output states. In unusual cases, very frequent spontaneous reinitiation of the tachycardia occurs. In this setting, a drug with a longer half-life such as verapamil is a better choice for arrhythmia termination. Recent studies suggest that intravenous diltiazem may have an efficacy rate that is comparable to verapamil, but at this time the data are limited.[57]

Chronic Medical Management

Since AV nodal reentrant tachycardia and AV reciprocating tachycardia both utilize the AV node as a critical limb of the reentrant circuit, AV nodal blocking agents are generally used as first tier medical therapy. The β-blockers and calcium channel blockers have advantages of once or twice daily dosing and excellent safety profiles.[58,59] Although well tolerated by most patients, the use of β-blockers is sometimes limited by central nervous system side effects. Historically, digitalis has been used as a first-line therapy in paroxysmal supraventricular tachycardia, despite the absence of objective data to support this strategy. Class Ic drugs such as flecainide and propafenone are highly effective because of their actions of slowing conduction in the AV node as well as extranodal accessory pathways. Although the risk of proarrhythmia must always be considered when using these drugs, their safety profile has been good when administered to a population of patients with structurally normal hearts.[60] The class III drugs, sotalol and amiodarone, have efficacy rates similar to the class Ic drugs. Because of the significant organ toxicity associated with its use, amiodarone should be reserved only for patients with severe refractory tachycardia who are not candidates for nonpharmacologic therapeutic approaches.

ABLATIVE THERAPY

Surgical therapy for patients with the Wolff-Parkinson-White syndrome, tachycardias mediated by accessory pathways, and AV nodal reentrant tachycardia has been a valuable therapeutic modality, but is of historical interest for the most part at this time. The invaluable lessons learned from surgical ablation were that an anatomic approach to arrhythmias was necessary to optimize their management and that arrhythmia cure was a realizable goal. Initial efforts at RF catheter ablations of these arrhythmias were frustrated by poor catheter performance and inadequate pathoanatomic understanding by many operators. As knowledge and tools have advanced, experienced ablation centers routinely report >95% success rates in these patients.[52]

The basic technique for accessory pathway ablation usually involves placement of multiple intracardiac catheters in order to anatomically localize the pathway and document its properties. The ablation catheter is generally a 6 or 7 Fr catheter with a deflectable distal segment, 1 to 3 ring electrodes, and a tip electrode that is 4 mm in length. The catheter is positioned under fluoroscopic guidance such that the tip is very close to the site of atrial or ventricular insertion of the accessory pathway. RF power is delivered while antegrade or retrograde conduction is continuously monitored. If conduction block is achieved within 10 sec, the RF pulse is then continued for a total of 30 to 60 sec (Fig. 5-8). Accessory pathways are approached with transvenous catheter placement for right-sided locations and transseptal or retrograde transaortic placement for left-sided locations.

AV nodal reentrant tachycardia may be treated with an RF ablation technique called AV nodal modification. In this procedure, the His bundle is identified with an electrode catheter in the high anterior right interatrial septum. The mapping/ablation catheter is placed inferior to this catheter along the tricuspid anulus in the midseptal or right posteroseptal region. A large amplitude high-frequency potential that is timed before the His electrogram but after the local atrial electrogram is often identified and has been called by some the "slow pathway potential." Delivery of RF energy at this location most frequently results in selective elimination of slow AV nodal pathway conduction.[8] Ablation in a more anterior/superior septal position can eliminate AV nodal

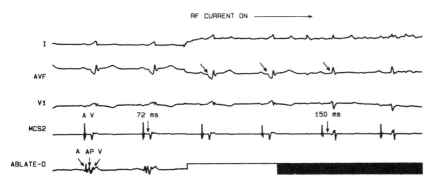

FIGURE 5-8. Example of radiofrequency catheter ablation in a patient with the Wolff-Parkinson-White syndrome. Surface electrocardiographic leads I, II, and V_1 and intracardiac electrograms from the second bipolar pair of an octapolar coronary sinus catheter (MCS2) and the distal bipole of the mapping/ablation catheter (ABLATE-D) are shown. A closely timed local atrial (A) and ventricular (V) electrogram with an intervening accessory pathway electrogram (AP) is recorded from the ablation catheter. The first two heartbeats after onset of ablation show the characteristic delta-wave pattern (arrows) that is permanently ablated on the third beat.

reentrant tachycardia, but will result in damage to the "fast" pathway, which probably anatomically represents the normal AV node. This approach also carries a 5 to 10% risk of complete heart block.

Currently, catheter ablation of paroxysmal supraventricular tachycardia has a reported success rate of >90% when performed by experienced operators. The procedure may be extremely technically challenging in a minority of patients, which accounts for the variability of success rates. Operators who are willing to persist in procedures for 12 or more hours may approach a 100% success rate, but patients' tolerance of prolonged procedure times may limit this approach.

APPROACH TO PATIENTS WITH AUTOMATIC TACHYCARDIAS

Automatic atrial tachycardia (also called ectopic atrial tachycardia) is an arrhythmia that is usually caused by a specific focus of abnormal automaticity in the left or right atrium. It shows typical features of gradual acceleration and termination, although it is sometimes extinguished in a few beats, thus mimicking a reentrant atrial arrhythmia. This uncommon rhythm is most commonly observed in patients with dilated cardiomyopathy.[61] A variety of drugs including β-blockers, calcium channel blockers, class I,[62] and class III drugs have some activity against this resistant arrhythmia, but efficacy data are lacking. Recently, anecdotal reports have indicated a high arrhythmia suppression rate (>85%) with RF catheter ablation. Although these results are very promising, long-term follow-up on these highly selected patients is not available.

Junctional ectopic tachycardia is an arrhythmia most commonly seen in neonates with congestive heart failure[63] and in adults with digitalis toxicity. The mainstay of therapy is to treat the underlying disease. Amiodarone therapy may be moderate successful in suppression of this rhythm. In patients in whom the arrhythmia is incessant and drug therapy ineffective, catheter ablation of the AV junction may be the only viable therapeutic alternative.

Multifocal atrial tachycardia is an arrhythmia most commonly seen in patients with atrial disease and hypoxemia (e.g., cor pulmonale).[15] Although the mechanism of this arrhythmia is unknown, it is thought to be triggered automaticity. The mainstay of treatment of multifocal atrial tachycardia is to improve oxygenation and avoid excessive doses of methylxanthine agents. Although β-blockers are effective therapeutic agents, the usual presence of bronchospasm in these patients limits their use. Verapamil decreases atrial rate marginally and may be useful in selected patients.

SUMMARY

Evaluation and treatment of patients with supraventricular tachycardia require a careful evaluation of the patient's history, physical examination, and appropriate supplemental diagnostic studies to identify the arrhythmia substrate. The 12-lead electrocardiogram must be analyzed with particular emphasis on the P-wave location, morphology, and association to the ventricular activity. Analysis of arrhythmia onset and offset and response to vagal maneuvers or adenosine administration should yield a diagnosis of the arrhythmia mechanism in almost all cases. Once the mechanism has been elucidated, a cogent approach for acute and chronic management may be pursued. Specific drugs should be selected that approach the discrete sources of the arrhythmias.

References

1. Wit AL, Cranefield PF, Hoffman BF. Slow conduction and reentry in the ventricular conducting system. II. Single and sustained circus movement in networks of canine and bovine Purkinje fibers. Circ Res 1972;30:11.
2. Wit AL. Cellular electrophysiologic mechanisms of cardiac arrhythmias. Cardiol Clin 1990;8:393.
3. Wu D, Kou HC, Hung JS. Exercise-triggered paroxysmal ventricular tachycardia. A repetitive rhythmic activity possibly related to afterdepolarization. Ann Intern Med 1981;95:410.
4. Waldo AL. Mechanism of atrial fibrillation, atrial flutter, and ectopic atrial tachycardia—a brief review. Circulation 1987;75:III-37.
5. Allessie MA, Rensma PL, Brugada J, et al. Pathophysiology of atrial fibrillation. In Zipes DP, Jalife J, eds. Cardiac electrophysiology. Philadelphia, WB Saunders, 1990:548.
6. Haines DE, DiMarco JP. Sustained intraatrial reentrant tachycardia. Clinical, electrocardiographic and electrophysiologic characteristics and long-term follow-up. J Am Coll Cardiol 1990;15:1345.
7. Akhtar M, Damato AN, Ruskin JN, et al. Antegrade and retrograde conduction characteristics in three patterns of paroxysmal atrioventricular junctional reentrant tachycardia. Am Heart J 1978;95:22.
8. Jackman WM, Beckman KJ, McClelland JH, et al. Treatment of supraventricular tachycardia due to atrioventricular nodal reentry by radiofrequency catheter ablation of slow pathway for AV nodal reentrant tachycardia. N Engl J Med 1992;327:313.
9. Lee MA, Morady F, Kadish A, et al. Catheter modification of the atrioventricular junction with radiofrequency energy for control of atrioventricular nodal reentry tachycardia. Circulation 1991;83:827.
10. Gallagher JJ, Pritchett ELC, Sealy WC, et al. The preexcitation syndromes. Prog Cardiovasc Dis 1978;20:285.
11. Critelli G, Gallagher JJ, Monda V, et al. Anatomic and electrophysiologic substrate of the permanent form of junctional reciprocating tachycardia. J Am Coll Cardiol 1984;4:601.
12. Gallagher JJ, Smith WM, Kassell JH, et al. Role of Mahaim fibers in cardiac arrhythmias in man. Circulation 1981;64:176.

13. Bauernfeind RA, Amat-Y-Leon F, Dhingra RC, et al. Chronic nonparoxysmal sinus tachycardia in otherwise healthy persons. Ann Intern Med 1979; 91:702.
14. Coleman HN, Taylor RR, Pool PE, et al. Congestive heart failure following chronic tachycardia. Am Heart J 1971;81:790.
15. Kastor JA. Multifocal atrial tachycardia. N Engl J Med 1990;322:1713.
16. Mehta D, Wafa S, Ward DE, et al. Relative efficacy of various physical manoeuvres in the termination of junctional tachycardia. Lancet 1988;1:1181.
17. DiMarco JP, Miles W, Akhtar M, et al. Adenosine for paroxysmal supraventricular tachycardia: dose ranging and comparison with verapamil in placebo-controlled, multicenter trials. Ann Intern Med 1990;113:104.
18. Vaughan-Williams EM. A classification of antiarrhythmic actions reassessed after a decade of new drugs. J Clin Pharmacol 1984;24:129.
19. Campbell TJ. Kinetics of onset of rate-dependent effects of class 1 antiarrhythmic drugs are important in determining their effects on refractoriness in guinea-pig ventricle, and provide a theoretical basis for their subclassification. Cardiovasc Res 1983;17:344.
20. Duke IS, Vaughn-Williams EM. The multiple modes of action of propafenone. Eur Heart J 1984;5:115.
21. Nattel S, Feder-Elituv R, Matthews C, et al. Concentration dependence of class III and beta-adrenergic blocking effects of sotalol in anesthetized dogs. J Am Coll Cardiol 1989;13:1190.
22. Polster P, Broeckhuysen J. The adrenergic antagonism of amiodarone. Biochem Pharmacol 1976;25:133.
23. Singh BN, Vaughan-Williams EM. Effect of altering potassium concentration on the action of lidocaine and diphenylhydantoin on rabbit atrial and ventricular muscle. Circ Res 1971;29:286.
24. Mason JW. Drug therapy: amiodarone. N Engl J Med 1987;316:455.
25. Funck-Brentano C, Kibleur Y, Le Coz F, et al. Rate dependence of sotalol-induced prolongation of ventricular repolarization during exercise in humans. Circulation 1991;83:536.
26. Kawai C, Konishi T, Matsuyama E, et al. Comparative effects of three calcium antagonists, diltiazem, verapamil and nifedipine, on the sinoatrial and atrioventricular nodes. Experimental and clinical studies. Circulation 1981;63:1035.
27. Lathrop DA, Valle-Aguilera JR, Millard RW, et al. Comparative electrophysiologic and coronary hemodynamic effects of diltiazem, nisoldipine and verapamil on myocardial tissue. Am J Cardiol 1982;49:613.
28. Chai CY, Wang HH, Hoffman BF, et al. Mechanisms of bradycardia induced by digitalis substances. Am J Physiol 1967;212:26.
29. Antman EM, Smith TW. Pharmacokinetics of digitalis glycosides. In Smith TW, ed. Digitalis glycosides. Orlando, FL: Grune and Stratton, 1985:45.
30. Belardinelli L, Linden J, Berne RM. The cardiac effects of adenosine. Prog Cardiovasc Dis 1989;32:73.
31. Scheinman MM, Morady F, Hess DS, et al. Catheter-induced ablation of the atrioventricular junction to control refractory supraventricular arrhythmias. JAMA 1982;248:851.
32. Jackman WM, Wang XZ, Friday KJ, et al. Catheter ablation of accessory atrioventricular pathways (Wolff-Parkinson-White syndrome) by radiofrequency current. N Engl J Med 1991;324:1605.

33. Calkins H, Sousa J, el-Atassi R, et al. Diagnosis and cure of the Wolff-Parkinson-White syndrome or paroxysmal supraventricular tachycardias during a single electrophysiologic test. N Engl J Med 1991;324:1612.

34. Haines DE, Watson DD. Tissue heating during radiofrequency catheter ablation: a thermodynamic model and observations in isolated perfused and superfused canine right ventricular free wall. PACE Pacing Clin Electrophysiol 1989;12:962.

35. Haines DE, Verow AF. Observations on electrode-tissue interface temperature and effect on electrical impedance during radiofrequency ablation of ventricular myocardium. Circulation 1990;82:1034.

36. Haines DE, Watson DD, Verow AF. Electrode radius predicts lesion radius during radiofrequency energy heating. Validation of a proposed thermodynamic model. Circ Res 1990;67:124.

37. Calkins H, Niklason L, Sousa J, et al. Radiation exposure during radiofrequency catheter ablation of accessory atrioventricular connections. Circulation 1991;84:2376.

38. Kopecky SL, Gersh BJ, McGoon MD, et al. The natural history of lone atrial fibrillation. A population-based study over three decades. N Engl J Med 1987;317:669.

39. Platia EV, Michelson EL, Porterfield JK, et al. Esmolol versus verapamil in the acute treatment of atrial fibrillation or atrial flutter. Am J Cardiol 1989;63:925.

40. Falk RH, Knowlton AA, Bernard SA, et al. Digoxin for converting recent-onset atrial fibrillation to sinus rhythm. Ann Intern Med 1987;106:503.

41. Suttorp MJ, Kingma JH, Lie-A-Huen L, et al. Intravenous flecainide versus verapamil for acute conversion of paroxysmal atrial fibrillation or flutter to sinus rhythm. Am J Cardiol 1989;63:693.

42. Suttorp MJ, Kingma JH, Jessurun ER, et al. The value of class 1c antiarrhythmic drugs for acute conversion of paroxysmal atrial fibrillation or flutter to sinus rhythm. J Am Coll Cardiol 1990;16:1722.

43. Borgeat A, Goy J, Maendly R, Kaufmann U, Grbic M, Sigwart U. Flecainide versus quinidine for conversion of atrial fibrillation to sinus rhythm. Am J Cardiol 1986;58:496.

44. Stroke Prevention in Atrial Fibrillation Study Group Investigators. Stroke Prevention in Atrial Fibrillation Study. Final results. Circulation 1991;84:527.

45. Boston Area Anticoagulation Trial for Atrial Fibrillation Investigators. The effect of low-dose warfarin on the risk of stroke in patients with non-rheumatic atrial fibrillation. N Engl J Med 1990;323:1505.

46. Anderson JL, Jolivette DM, Fredell PA. Summary of efficacy and safety of flecainide for supraventricular arrhythmias. Am J Cardiol 1988; 62:62D.

47. Juul-Moller S, Edvardsson N, Rehnqvist-Ahlberg N. Sotalol versus quinidine for the maintenance of sinus rhythm after direct current conversion of atrial fibrillation. Circulation 1990;82:1932.

48. Graboys TB, Podrid PJ, Lown B. Efficacy of amiodarone for refractory supraventricular tachyarrhythmias. Am Heart J 1983;106:870.

49. Horowitz LN, Spielman SR, Greenspan AM, et al. Use of amiodarone in the treatment of persistent and paroxysmal atrial fibrillation resistant to quinidine therapy. J Am Coll Cardiol 1985;6:1402.

50. Echt DS, Liebson PR, Mitchell LB, et al. Mortality and morbidity in patients receiving encainide, flecainide, or placebo. The Cardiac Arrhythmia Suppression Trial. N Engl J Med 1991;324:781.
51. Coplen SE, Antman EM, Berlin JA, et al. Efficacy and safety of quinidine therapy for maintenance of sinus rhythm after cardioversion. A meta-analysis of randomized control trials. Circulation 1990;82:1106.
52. Scheinman MM, Laks MM, DiMarco J, et al. Current role of catheter ablative procedures in patients with cardiac arrhythmias. A report for health professionals from the Subcommittee on Electrocardiography and Electrophysiology, American Heart Association. Circulation 1991;83:2146.
53. Cox JL, Boineau JP, Schuessler RB, et al. Successful surgical treatment of atrial fibrillation. Review and clinical update. JAMA 1991;266:1976.
54. Cosio FG, Lopez-Gil M, Goicolea, A, Arribas F, Barroso JL. Radiofrequency ablation of the inferior vena cava-tricuspid valve isthmus in common atrial flutter. Am J Cardiol 1993;71:705.
55. Sharma AD, Yee R, Guiraudon G, et al. Sensitivity and specificity of invasive and non-invasive testing for risk of sudden death in Wolff-Parkinson-White syndrome. J Am Coll Cardiol 1987;10:373.
56. DiMarco JP, Miles W, Akhtar M, et al. Adenosine for paroxysmal supraventricular tachycardia: dose ranging and comparison with verapamil in placebo-controlled, multicenter trials. Ann Intern Med 1990;113:104.
57. Huycke EC, Sung RJ, Dias VC, et al. Intravenous diltiazem for termination of reentrant supraventricular tachycardia: a placebo-controlled, randomized, double-blind, multicenter study. J Am Coll Cardiol 1989;13:538.
58. Chang MS, Sung RJ, Tai TY, et al. Nadolol and supraventricular tachycardia: an electrophysiologic study. J Am Coll Cardiol 1983;2:894.
59. Mauritson DR, Winniford MD, Walker WS, et al. Oral verapamil for paroxysmal supraventricular tachycardia. Ann Intern Med 1982;96:409.
60. Henthorn RW, Waldo AL, Anderson JL, et al. Flecainide acetate prevents recurrence of symptomatic paroxysmal supraventricular tachycardia. The Flecainide Supraventricular Tachycardia Study Group. Circulation 1991; 83:119.
61. Gelb BD, Garson A Jr. Noninvasive discrimination of right atrial ectopic tachycardia from sinus tachycardia in "dilated cardiomyopathy." Am Heart J 1990;120:886.
62. Kuck KH, Kunze KP, Schltuter M, et al. Encainide versus flecainide for chronic atrial and junctional ectopic tachycardia. Am J Cardiol 1988; 62:37L.
63. Bharati S, Moskowitz WB, Scheinman M, et al. Junctional tachycardias: anatomic substrate and its significance in ablative procedures. J Am Coll Cardiol 1991;18:179.

William G. Stevenson

6 | Diagnosis and Treatment of Ventricular Arrhythmias

Over the past two decades there has been substantial progress in elucidating mechanisms and refining therapies for ventricular arrhythmias. This chapter will focus on mechanisms and management of sustained ventricular tachycardias.

MECHANISMS OF ARRHYTHMIAS

Reentry

Reentrant arrhythmias are caused by continuous propagation of an excitation wave through myocardial tissue. The excitation wave may travel repeatedly over the same path producing ordered reentry or the reentry path may change continually producing "random reentry." The reentry paths may be anatomically defined (Fig. 6-1A), for example, reentry around an area of inexcitable tissue, or may be determined by functional properties of the tissue, such as refractoriness[1,2] (Fig. 6-1B). Repeated collision of opposing excitation wavefronts may also produce an area of conduction block defining a reentry circuit (Fig. 6-1C). In some situations a rotating wavefront may form spiral-shaped patterns or rotors around a small central refractory region (Fig. 6-1D).[3] The tachycardia rate is determined by the revolution time through the circuit. For

Clinical Cardiac Electrophysiology: Perioperative Considerations
Edited by Carl Lynch III. J.B. Lippincott Company, Philadelphia, PA ©1994

Types of Reentry Circuits

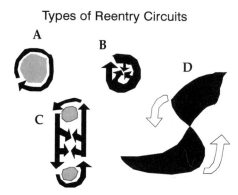

FIGURE 6-1. Four types of reentry circuits are shown. In all panels excitation wavefronts are shown in *black* and the direction of propagation is shown by the *arrows*. In A, reentry occurs around a fixed anatomic obstacle (*gray area*). In B, reentry occurs around a central area which is maintained refractory by continual bombardment from the circulating wavefront. In C, reentry occurs around two anatomic obstacles (*gray areas*) and a central area of conduction block is maintained by continued collision of the reentry wavefronts. In D, spiral wave reentry occurs around a central point (see text). (Reproduced with permission from Stevenson WG, Middlekauff HR, Saxon LA. Ventricular arrhythmias and sudden death in coronary artery disease. J Arrhyth Management Summer 1992:11.)

reentry to continue, each point in the circuit must have sufficient time after depolarization to recover excitability before arrival of the next excitation wavefront. Thus, slow conduction, which increases the revolution time through the circuit, and short recovery times (refractory periods) promote reentry. To initiate reentry, a propagating excitation wavefront encounters refractory tissue in the reentry circuit, propagates around this refractory area of conduction block, and then reenters the initial area of conduction block from another direction. Disparities in local recovery times (heterogeneous refractoriness) facilitate reentry by promoting conduction block. In some situations, the excitation waves produced by an ordered reentry circuit may fractionate into multiple random reentry wavelets causing ventricular fibrillation.

Reflection is a related arrhythmia mechanism in which a propagated impulse encounters a region of depressed excitability, the tissue beyond the depressed region is depolarized by electrotonic current spread, and the depolarization of the "distant" tissue generates an electrotonic current reflecting the excitation back into the tissue from which it arose.[4] Reflection can be demonstrated in some in vitro preparations, but its clinical relevance is unknown.

Automaticity

Automaticity is the generation of action potentials by spontaneous membrane depolarizations. Automaticity is normally present in the sinus node, atrioventricular (AV) node, and His-Purkinje system. Abnormal automaticity occurs in partially depolarized tissue, such as ischemic myocardium. Abnormal automaticity in the infarct border zone may be the mechanism of premature ventricular contractions and accelerated idioventricular rhythms commonly observed during the first few days after acute myocardial infarction.[5,6]

Triggered Automaticity

A second type of automaticity is caused by afterdepolarizations.[7–9] These are membrane depolarizations that occur during or following an action potential (Fig. 6-2). If an afterdepolarization gives rise to a second action potential, this is referred to as triggered automaticity. There are two types of afterdepolarizations. Early afterdepolarizations occur before the completion of repolarization, interrupting the downslope of the action potential. Early afterdepolarizations are produced by many agents that prolong action potential duration, such as the potassium channel blocker cesium, quinidine, sotalol, and N-acetyl-procainamide (NAPA). Slow heart rates and in some situations α-adrenergic stimulation facilitate early afterdepolarizations. Triggered automaticity from early afterdepolarizations is a likely cause of the rapid polymorphic ventricular tachycardia, torsade de pointes, although this has not been definitely established for humans.[8]

Delayed afterdepolarizations occur after completion of repolarization.[7] Delayed afterdepolarizations are often produced by interventions that increase intracellular calcium, such as digitalis, sympathetic stimulation, and rapid pacing. Delayed afterdepolarizations may also

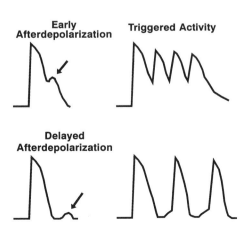

FIGURE 6-2. Schematic diagrams of action potentials illustrating two types of afterdepolarizations and triggered activity are shown. The *top left tracing* illustrates an early afterdepolarization (*arrow*) appearing as a transient membrane depolarization prior to repolarization. The *top right panel* shows a schematic diagram of triggered activity resulting from early afterdepolarizations. The *bottom left panel* shows a delayed afterdepolarization (*arrow*) appearing as a depolarization after return of the membrane potential to its resting potential. The *bottom right panel* shows a schematic diagram of triggered activity from delayed afterdepolarizations.

contribute to arrhythmias observed in the first few days after myocardial infarction. Delayed and early afterdepolarizations are more easily elicited in hypertrophied ventricular myocardium.[10]

A Clinical Classification of Sustained Ventricular Arrhythmias

A practical classification of sustained ventricular arrhythmias based on the electrocardiographic appearance is suggested in Table 6-1. For the purposes of this discussion, sustained ventricular tachycardia is defined as one that has adverse hemodynamic effects or requires an intervention for termination. Ventricular tachycardia has discernible QRS complexes in the surface electrocardiogram. In contrast, ventricular fibrillation displays only chaotic electrical activity on the electrocardiogram. Ventricular tachycardia may be monomorphic with each QRS complex resembling the previous QRS complex or polymorphic with the QRS complexes varying from beat to beat or over several beats.

SUSTAINED MONOMORPHIC VENTRICULAR TACHYCARDIA

In ventricular tachycardias having a consistent beat-to-beat morphology, the ventricles are activated in a repetitive fashion. This suggests the presence of a stable arrhythmia focus such as an area of prior myocardial infarction.

TABLE 6-1. CLASSIFICATIONS OF VENTRICULAR TACHYCARDIA

Sustained Ventricular Tachycardia. Ventricular tachycardia that causes adverse hemodynamic effects or requires an intervention for termination.

Monomorphic Sustained Ventricular Tachycardia. Ventricular tachycardia that has consistently similar QRS complexes from beat to beat. The ventricles are activated repetitively in the same sequence.

Sinusoidal Ventricular Tachycardia. Ventricular tachycardia that has a sinusoidal configuration due to rapid rate (also called ventricular flutter) or prolonged QRS duration.

Polymorphic Ventricular Tachycardia. Ventricular tachycardia in which QRS complexes change from beat to beat, suggesting chaotic ventricular activation.

Ventricular Fibrillation. Chaotic ventricular activation that does not have discernible QRS complexes.

Monomorphic Ventricular Tachycardia Late After Myocardial Infarction

The most common cause of sustained monomorphic ventricular tachycardia is reentry within an area of infarct scar. After surviving the acute phase of myocardial infarction and being discharged home, 4 to 10% of patients suffer an episode of sustained ventricular tachycardia.[11–15] Ventricular tachycardia may degenerate to ventricular fibrillation, causing sudden death. It may produce hypotension and syncope following which it may terminate spontaneously. Alternatively, ventricular tachycardia may present with palpitations, angina, dyspnea, or other cardiac symptoms without syncope. The clinical presentation depends on the rate of the tachycardia, the severity of left ventricular dysfunction, tachycardia-induced ischemia, and the response of the autonomic nervous system to the tachycardia.[16–19]

Pathophysiology

During the healing phase after acute myocardial infarction necrotic myocytes are replaced by fibrous scar tissue. Some surviving ventricular myocyte bundles persist within the scar and become encased in fibrous tissue.[15,20–23] The distortion in architecture of the myocyte bundles is associated with increased resistance between the myocytes, which slows conduction velocity through the bundles. In some instances depressed membrane responses in surviving myocytes slow conduction. This slow conduction is the substrate for reentrant ventricular tachycardia. The resulting reentry circuits can have complex geometric configurations. For example, the circulating reentry wavefronts may propagate through the scar, exit out into the surrounding myocardium, propagate along the border of the scar, and then turn back into the scar. Often the complex nature of the scar allows an individual patient to have several different QRS morphologies of ventricular tachycardia.[22]

Risk Factors for Ventricular Tachycardia (Table 6-2)

The larger the infarct, the greater the likelihood of ventricular tachycardia.[12,20] As left ventricular ejection fraction falls below 0.4, the risk of sudden death or ventricular tachycardia late after myocardial infarction increases. Failure to achieve early reperfusion during the acute phase of myocardial infarction also appears to increase the risk of ventricular tachycardia. Because these tachycardias are the result of a relatively fixed anatomic substrate (slow conduction through an infarct scar), a patient who suffers a spontaneous episode of ventricular tachycardia has a >30% likelihood of suffering recurrent episodes dur-

TABLE 6-2. RISK FACTORS FOR SUSTAINED MONOMORPHIC VENTRICULAR TACHYCARDIA

1. Prior episode of sustained monomosphic ventricular tachycardia
2. Prior myocardial infarction with:
   A. Left ventricular ejection fraction <0.4
   B. History of syncope
   C. Bundle branch block
   D. Abnormal signal-averaged ECG
   E. Inducible ventricular tachycardia on electrophysiology study
   F. Failure to reperfuse the infarct artery during the acute infarction

ing the following 2 to 3 years despite antiarrhythmic drug therapy.[24,25] The electrophysiologic effects of many antiarrhythmic drugs can be largely reversed by β-adrenergic stimulation, possibly increasing the risk of recurrences during periods of stress.[26] Self-terminating runs of ventricular tachycardia often present as syncope. Patients with prior infarction and a history of unexplained syncope should be evaluated for the possibility of ventricular tachycardia as the cause.

Several methods can be used to assess the risk of ventricular tachycardia in postmyocardial infarction patients. Programmed electrical stimulation can be used to detect the presence of a potential reentry circuit by providing the trigger for tachycardia initiation and forms the basis for electrophysiologic testing.[12-14] When performed within 1 month of myocardial infarction, programmed electrical stimulation initiates sustained monomorphic ventricular tachycardia in 6 to 40% of patients. Patients with inducible ventricular tachycardia that is slower than 260 beats/min have approximately a 20% risk of spontaneous sustained ventricular tachycardia or sudden death within 2 years. Interestingly, inducible ventricular fibrillation and monomorphic ventricular tachycardia faster than 260 beats/min are nonspecific responses and do not predict an increased risk of arrhythmic events during follow-up. In patients without inducible ventricular tachycardia, the risk is <5%.

Slow conduction in the infarct scar, the substrate for postmyocardial infarction ventricular tachycardia, can also be detected noninvasively.[11,13-15] Slow conduction results in portions of the scar being activated late relative to the QRS complex. The slow conduction produces low-amplitude electrical signals, which are not detectable in the standard electrocardiogram. However, signal averaging the electrocardiogram to reduce noise, followed by amplification and filtering (time domain analysis) allows detection of the slow conduction as "late potentials." Late potentials are detectable in 20 to 40% of myocardial infarct survivors. Approximately 20% of patients with late potentials suffer spontaneous sustained ventricular tachycardia or sudden death within 2 years of their acute infarction. Signal-averaged electrocardiograms can also be subjected to spectral analysis to assess high-frequency signal content. Time domain analysis is most widely used and has been most extensively validated, but is unreliable in patients with bundle branch block. Frequency domain analysis should theoretically be more accurate in patients with bundle branch block but has technical limitations and is still evolving.

Ambient ventricular ectopy has not been a clinically useful predictor of risk for sustained ventricular tachycardia. The presence of >6 ventricular ectopic beats/hr on an ambulatory electrocardiogram or complex ventricular ectopy is associated with an increased risk of sudden death in myocardial infarction survivors, but the predictive accuracy is low.[13,15] Suppression of ventricular ectopic beats with class I antiarrhythmic agents does not reduce sudden death and in fact may increase the risk of death.[27]

Although programmed electrical stimulation and signal-averaged electrocardiography reveal the substrate for reentry in up to 40% of infarct survivors, only a minority of these patients will actually suffer a spontaneous sustained arrhythmia. The reasons for this are not entirely clear. Spontaneous initiation of ventricular tachycardia probably involves a complex interplay of autonomic tone, heart rate, and ectopic activity. Combined analysis of heart rate variability, ambient ventricular ectopy, and signal averaged electrocardiography may improve prediction of spontaneous occurrences of ventricular tachycardia in patients with clinically latent reentry circuits. Farrell and coworkers[13] found that patients who had both late potentials and diminished heart rate variability had a 33% risk of ventricular tachycardia or sudden death. Risk factors for ventricular tachycardia in patients who have survived a myocardial infarction are summarized in Table 6-2.

Diagnostic Features

Reentrant tachycardias typically have a sudden onset (Fig. 6-3, *A* and *B*). The first few beats may display a variable QRS morphology before stabilizing to a repetitive monomorphic QRS. When ventricular tachycardia is relatively slow (e.g., <180 beats/min), it may be hemodynamically well tolerated. This often leads to misdiagnosis as supraventricular tachycardia with aberrancy. Of patients who have a regular wide QRS complex tachycardia the most useful diagnostic feature in distinguishing paroxysmal supraventricular tachycardia with aberrancy from ventricular tachycardia is the past medical history. If there is a history of prior myocardial infarction, the likelihood that the rhythm is ventricular tachycardia is >90% even if it is hemodynamically well tolerated.[28]

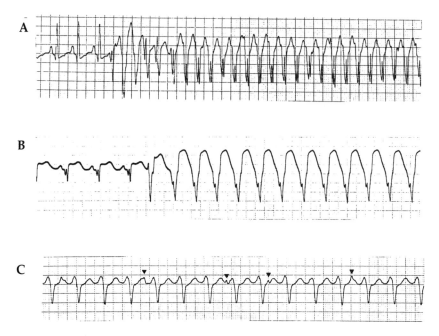

FIGURE 6-3. Three examples of sustained monomorphic ventricular tachycardia are shown. A shows initiation of rapid ventricular tachycardia. The first 5 beats of tachycardia are polymorphic before it stabilizes to a monomorphic tachycardia having a rate of 220 beats/min. B shows initiation of sustained monomorphic ventricular tachycardia. The fourth beat from the left is a premature ventricular contraction followed by monomorphic ventricular tachycardia at a rate of 120 beats/min. C shows a sustained monomorphic ventricular tachycardia which became incessant but slow (100 beats/min) after treatment with amiodarone. Dissociated P-waves are indicated by *arrows*. (Reproduced with permission from Stevenson WG, Stevenson LW. Evaluation and management of arrhythmias in heart failure. In: Parmley WW, Chatterjee K, eds. Cardiology. Philadelphia: JB Lippincott, 1993.)

When a slower wide QRS complex tachycardia produces hemodynamic collapse, the rhythm is also usually ventricular tachycardia. Electrocardiographic features useful for distinguishing ventricular tachycardia from supraventricular tachycardia with aberrancy have been recently reviewed.[28,29] The QRS duration ranges from as little as 0.10 sec in rare cases to >0.2 sec and is of limited value in distinguishing ventricular tachycardia from supraventricular tachycardia. During ventricular tachycardia impulses may conduct from the ventricle back up the His-Purkinje system, through the AV node to the atrium. Hence the presence of a 1:1 relation between the ventricles and atria does not exclude ventricular tachycardia. AV dissociation strongly favors a diagnosis of ventricular tachycardia and is detectable in approximately 30% of cases as P-waves deforming the ST segments or T-waves at a rate slower than that of QRS complexes (Fig. 6-3C) or as fusion beats. When tachycardia is well tolerated and supraventricular tachycardia is suspected, administration of intravenous adenosine many either terminate supraventricular tachycardia or produce transient AV block and allow P-waves of atrial tachycardia or atrial flutter to be observed although often only for a few seconds. Adenosine rarely terminates ventricular tachycardia, although occasionally AV dissociation becomes more obvious. If tachycardia is well tolerated such that a 12-lead electrocardiogram can be obtained, several QRS morphology criteria can be applied to attempt to establish the origin of the tachycardia.[28,29] Morphologies that favor ventricular tachycardia include absence of an RS complex in all precordial leads (from V_1 to V_6) or presence of an RS complex with a duration from the onset of the R-wave to the nadir of the S-wave of >100 msec.

Management

Sustained ventricular tachycardia that causes loss of consciousness should be treated with immediate cardioversion. Although ventricular tachycardia can often be terminated with relatively low energies, such as 10 W-sec, cerebral hypoperfusion mandates initial attempts with higher energy to increase the likelihood of immediate success. In adults, external cardioversion with 200 W-sec is recommended and if unsuccessful is followed immediately with 360 J. Ideally the shock should be synchronized with the QRS complex. A shock that occurs during the T-wave, when the ventricles are in various states of repolarization, is more likely to precipitate ventricular fibrillation than a QRS synchronous shock. Immediate delivery of an asynchronous shock is preferable, however, to a delay in administering a synchronous shock if there is difficulty achieving adequate QRS detection by the defibrillator.

If ventricular tachycardia is hemodynamically well tolerated, a bolus of lidocaine (generally 1.5 mg/kg) should be considered. If this fails to terminate tachycardia, a bolus of procainamide (10 to 15 mg/kg intravenously at a rate of 25 to 50 mg/min) with careful monitoring for hypotension due to the vasodilating effects of procainamide may be tried. Bretylium can also be considered, but the electrophysiologic effects may not be apparent for over 20 min following intravenous administration. Sedation followed by QRS-synchronous cardioversion should be immediately performed if there is evidence of hemodynamic compromise, impaired cerebral perfusion, or angina pectoris. One approach to the patient with hemodynamically well tolerated wide QRS tachycardia is shown in Figure 6-4.

Potential precipitating factors should be sought and removed. These include hypokalemia, hypoxia, sympathetic stimulation, ectopy induced by intracardiac catheters or pacing wires, and myocardial ischemia.

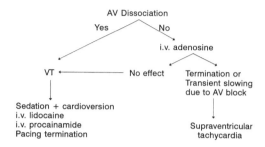

- Hemodynamically tolerated
- Regular

AV Dissociation

Yes / No

i.v. adenosine

VT ← No effect | Termination or Transient slowing due to AV block

Sedation + cardioversion
i.v. lidocaine
i.v. procainamide
Pacing termination

Supraventricular tachycardia

FIGURE 6-4. A schematic diagram for one approach to the management of wide QRS tachycardia in a hemodynamically stable patient is shown. See text for discussion.

It is important to recognize that antiarrhythmic drugs may occasionally precipitate ventricular tachycardia or render tachycardia more difficult to manage.[30] For example, a patient may have suffered one episode of ventricular tachycardia, resulting in hospitalization. Antiarrhythmic drug therapy is initiated, and after several doses ventricular tachycardia recurs, but is slower and more difficult to terminate. Sometimes tachycardia becomes incessant, and cardioversion is followed after a few sinus rhythm beats by resumption of tachycardia. This is a form of drug-induced proarrhythmia, which can occur with many antiarrhythmic agents, but is most frequent with the class Ic antiarrhythmic drugs (e.g., flecainide, encainide, and propafenone), which markedly slow conduction velocity through the myocardium. These tachycardias may have a sinusoidal configuration (see below). It is important in this situation to avoid further administration of class I antiarrhythmic drugs, which slow conduction and may make the situation worse rather than better. Hemodynamic support and multiple electrical cardioversions may be required until the drug is excreted. In some cases β-adrenergic blocking agents or lidocaine has successfully terminated incessant ventricular tachycardias caused by class Ic antiarrhythmic drugs.[31-33] The action of lidocaine may be related to its ability to bind rapidly to the sodium channel, which then protects the site from the more tightly bound class Ic agents. Administration of sodium bicarbonate may reverse some of the electrophysiologic effects.[33]

As noted above, an episode of sustained monomorphic ventricular tachycardia late after myocardial infarction indicates the presence of a chronic arrhythmia focus and a high risk for arrhythmia recurrences. Chronic prophylactic therapy with antiarrhythmic drugs is generally considered. For patients who have hemodynamically poorly tolerated ventricular tachycardia, an implantable cardioverter defibrillator is often warranted. Uncommonly, ventricular tachycardia becomes incessant in the early postoperative period after coronary artery bypass surgery or implantation of a automatic cardioverter-defibrillator.[34] In our experience patients who have had an episode of ventricular tachycardia but who have had no recurrences while receiving medications have had a low risk of arrhythmia recurrence during noncardiac surgery and in the postoperative period. We advise, however, careful monitoring and attention to maintaining antiarrhythmic therapy throughout the hospital course.

Other Causes of Sustained Monomorphic Ventricular Tachycardia

Sustained monomorphic ventricular tachycardia may also occur, although less commonly in a variety other cardiac disorders. In general, the prognostic significance and management are similar to for those tachycardias caused by chronic myocardial infarct scars.

Chagas disease caused by the myocardial effects of *Trypanosoma cruzii* infection is a common cause of myocardial scarring and ventricular tachycardia in South and Central America.[35] This diagnosis should be suspected, particularly in patients from endemic areas who have unexplained bundle branch block (often right bundle branch block with left or right axis deviation) or cardiomegaly.

Right ventricular dysplasia is a rare disorder characterized by replacement of portions of the right ventricle with adipose and fibrous tissue.[36] Development of ventricular tachycardia is common. This diagnosis should be suspected in patients with ventricular arrhythmias and right ventricular wall motion abnormalities, usually detected on echocardiography.

Progressive systemic sclerosis (scleroderma) can involve the myocardium, producing regions of scar.[37] Sustained ventricular tachycardia is uncommon but does occur.

Sarcoidosis produces areas of myocardial infiltration with noncaseating granuloma. Ventricular tachycardia occasionally occurs.[38] Many, but not all, patients with cardiac involvement also have evidence of pulmonary sarcoidosis.

Occasionally ventricular tachycardia will develop in a patient with valvular heart disease or nonischemic, idiopathic cardiomyopathy. These tachycardias may be related to areas of ventricular scar in some cases. In approximately one third of patients with ventricular tachycardia caused by nonischemic cardiomyopathy the etiology is macroreentry within the bundle branches. The circulating reentry wavefronts propagate up one bundle branch, down the contralateral bundle branch, and back through the septum to complete the circuit.[39] This is important to recognize because the tachycardia is easily curable by catheter ablation targeting the right bundle branch, which is a component of the reentry circuit. Bundle branch macroreentry should be suspected in any patient with sustained, monomorphic ventricular tachycardia who has not had a prior myocardial infarction. Most of these patients have some evidence of conduction system disease, typically an electrocardiographic pattern of left ventricular conduction delay.

Rarely, sustained monomorphic ventricular tachycardia occurs in structurally normal hearts.[40] These tachycardias often present as palpitations in young healthy individuals and tend to fall into one of two distinct groups. The most common idiopathic ventricular tachycardia originates in the right ventricular outflow tract below the pulmonary valve. Tachycardias have a left bundle branch block configuration and an inferior and right-ward frontal plane QRS axis. Often they are provoked by exercise or administration of β-adrenergic agonists. Some terminate with administration of adenosine, wrongly suggesting that they are supraventricular in origin.[41] The second group originate in the left ventricle. The tachycardia has a right bundle branch block configuration, often with a superior or left-ward frontal plane QRS axis. Some of these tachycardias terminate with administration of intravenous verapamil, again wrongly suggesting a supraventricular origin. Usually these tachycardias are hemodynamically well tolerated. Although arrhythmia recurrences are common, sudden death is rare. β-Adrenergic blocking agents are usually the first line of therapy. If these are not successful, calcium channel blocking agents can be tried. Many idiopathic ventricular tachycardias respond to class I antiarrhythmic agents as well. Endocardial catheter ablation is often curative.[42]

POLYMORPHIC VENTRICULAR TACHYCARDIAS

Occasionally, sustained polymorphic ventricular tachycardia is caused by reentry in a chronic myocardial scar. In our experience polymorphic ventricular tachycardia is due much more commonly to torsade de pointes or acute ischemia.

Torsade de Pointes

It is extremely important to consider the possibility of torsade de pointes (translated from French as "twisting about the points") for any sustained polymorphic ventricular tachycardia (Fig. 6-5). This arrhythmia is preceded by QT interval prolongation in the sinus rhythm QRS complexes, reflecting a delayed repolarization of the cardiac action potential, has a characteristic presentation, can be aggravated or caused by antiarrhythmic drugs, and responds dramatically to some therapies

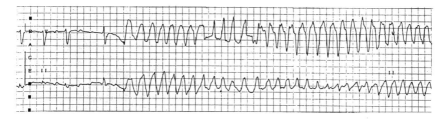

FIGURE 6-5. An example of torsade de pointes caused by procainamide and N-acetyl-procainamide is shown. The underlying rhythm is atrial flutter with variable AV block. Following the third supraventricular QRS there is a pause created by increased AV block. The fourth supraventricular QRS has a markedly prolonged QT interval evident in the *top* electrocardiogram lead. The T-wave is interrupted by the first beat of the polymorphic ventricular tachycardia. See text for discussion. (Reproduced with permission from Stevenson WG, Stevenson LW. Evaluation and management of arrhythmias in heart failure. In: Parmley WW, Chatterjee K, eds. Cardiology. Philadelphia: JB Lippincott, 1993.)

that are likely to aggravate other types of ventricular tachycardia.[9,43,44] If not recognized and treated, it is likely to be fatal.

The mechanism of torsade de pointes is not definitively known, but studies in animal models link it to early afterdepolarizations in the setting of QT prolongation.[8] There are two clinical forms, recently reviewed by Jackman and co-workers.[9] The most common is the pause-dependent, acquired form. The congenital long QT syndrome is rare. Both present with polymorphic ventricular tachycardia. Following termination or immediately prior to initiation, the sinus rhythm QT interval is prolonged, usually >0.46 sec and often >0.56 sec.

Acquired Torsade de Pointes—Long QT Syndrome

Torsade de pointes can be caused by anything that delays ventricular repolarization, thereby prolonging the QT interval (Table 6-3). Antiarrhythmic drugs are the most frequent cause. It is estimated that the risk with quinidine or sotalol is 0.8 to 2%. It occurs, but with a less defined incidence, with disopyramide and procainamide. The procainamide metabolite N-Acetylprocainamide (NAPA) is often implicated as a cause in patients with renal insufficiency. Although procainamide is hepatically metabolized, NAPA is entirely excreted via the urine. When procainamide is administered to patients with renal failure, NAPA progressively accumulates, having a half-life of 48 hr. Interestingly, amiodarone, which can markedly prolong the QT interval, appears to rarely cause torsade de pointes.[9,44] Phenothiazines, erythromycin, pentamidine, and terfenadine have also been implicated in anecdotal cases. Hypokalemia prolongs the QT interval and may precipitate torsade de pointes by itself or when it occurs in com-

TABLE 6-3. CAUSES OF ACQUIRED TORSADE DE POINTES

Antiarrhythmic drugs	Other drugs
Quinidine	Erythromycin
Sotalol	Terfenadine
Procainamide/N-acetylprocainamide	Probucol
Disopyramide	Tricyclic and tetracyclic
Amiodarone (rare)	antidepressants
Electrolyte abnormalities	Phenothiazines
Hypokalemia	Organophosphate insecticides
Hypocalcemia/hypomagnesiemia	Severe bradycardia
	Subarachnoid hemorrhage

bination with a previously tolerated drug that prolongs the QT interval. Hypocalcemia and hypomagnesemia may also theoretically contribute to QT prolongation and torsade de pointes. Initiation is favored by slow heart rates or pauses that further prolong the QT interval.

Diagnostic Features

The sinus rhythm QT interval is usually >0.46 sec and often markedly prolonged. Importantly, QT prolongation may be observed only in the beats immediately preceding the episode. Initiation of tachycardia often occurs with a characteristic sequence of events (Fig. 6-5). Premature ventricular contractions or variations in ventricular response to atrial fibrillation produce a sudden lengthening of the R-R interval as compared to preceding beats. The QT interval following this long R-R interval may be dramatically prolonged and the T-wave is interrupted by a premature ventricular beat, which is the first beat of the ventricular tachycardia. This characteristic initiation following a "pause" has led some to designate this tachycardia "pause-dependent."[9] The tachycardia itself is polymorphic, often having a characteristic waxing and waning QRS amplitude in one or more ECG leads. Episodes may spontaneously terminate or degenerate to ventricular fibrillation. Occasionally it may appear monomorphic for several beats in a single electrocardiogram lead.

Management

Sustained episodes that degenerate to ventricular fibrillation should be treated with immediate defibrillation. All precipitating factors should be corrected or removed. Magnesium sulfate often prevents recurrences and should be administered as a bolus of 1 g intravenously over 1 to 2 min.[44] Ventricular ectopy and runs of nonsustained ventricular tachycardia are good guides to the effects of therapy, and the magnesium bolus can be immediately repeated if ventricular ectopy persists. A continuous infusion of magnesium sulfate of 3 to 20 mg/min may be required with monitoring for neuromuscular depression especially in patients with renal impairment. Magnesium administration is often effective even if serum magnesium levels are normal. If magnesium alone is ineffective, isoproterenol infusion can be used to shorten the QT interval through direct β-adrenergic effects and by increasing the heart rate. Isoproterenol should be titrated to a tolerated dose, which suppresses ventricular ectopy. Overdrive ventricular or atrial pacing should be added next and may allow isoproterenol therapy to be discontinued. Pacing should be at a rate that suppresses ventricular ectopy, often between 100 and 120 beats/min.

If the diagnosis is uncertain and the patient has not responded to magnesium therapy, a major concern is ventricular tachycardia caused by acute ischemia (see below), or rarely, reentry in an old myocardial infarct scar. Drugs that may be useful in this circumstance without aggravating torsade de pointes include lidocaine, nitroglycerin, and possibly bretylium (Table 6-4).

Congenital Long QT Syndrome.

The congenital long QT syndromes are rare disorders. The Jervell and Lange-Nielson syndrome is associated with congenital deafness and has an autosomal recessive pattern of inheritance. The Romano Ward

TABLE 6-4. ACQUIRED TORSADE DE POINTES VERSUS ISCHEMIC POLYMORPHIC VENTRICULAR TACHYCARDIA

	Torsade de Pointes	Ischemic Polymorphic Ventricular Tachycardia
$MgSO_4$	+	+
Lidocaine	+/−	+ or +/−
Isoproterenol	+	−
Overdrive pacing	+	+/− or −
β-Adrenergic blockers	−	+
Bretylium	+/−	+
Amiodarone	+/− or −	+
Sotalol	−	+
Class IA*	−	+ or −
Nitroglycerin/intraaortic balloon counterpulsation	?	+

+ = beneficial; − = detrimental, +/− = no effect.
* Quinidine, procainamide, disopyramide.

syndrome is inherited as autosomal dominant.[9,45] Patients typically have an episode of VT and have syncope or cardiac arrest before the age of 30 years. When the patient is in sinus rhythm, the electrocardiogram demonstrates QT prolongation: the corrected QT interval (QT/square root of the R-R) is typically >0.46. However, it can be normal and may become prolonged only immediately before initiation of the arrhythmia. Drugs and hypokalemia, which further prolong the QT interval, may precipitate attacks and are contraindicated. The congenital long QT syndromes differ from the acquired form in that sympathetic stimulation may aggravate the arrhythmia. In some patients episodes tend to occur during excitement. In contrast to the acquired form, therefore, sympathomimetic agents should be avoided. Intravenous magnesium has been useful in some cases. β-Adrenergic blockade and removal of precipitating factors are the first line of therapy.[9] Sustained ventricular tachycardia/ventricular fibrillation responds to electrical defibrillation. Relatively little information is available regarding surgical risk. In our experience, patients who are receiving β-adrenergic blockers tolerate general anesthesia well. Attention to avoiding electrolyte abnormalities and careful perioperative monitoring are prudent.

Polymorphic Ventricular Tachycardia Caused by Acute Ischemia

The second major cause of polymorphic ventricular tachycardia is acute myocardial ischemia. During severe ischemia cellular potassium loss increases extracellular potassium and causes partial depolarization of myocytes.[15,20] This occurs, for example, in the infarct border zone during acute myocardial infarction. Conduction velocity through these regions is consequently slowed, and refractoriness is altered in a heterogeneous manner. This sets the stage for reentry. The reentry path is not anatomically fixed, and reentrant impulses rapidly fractionate into multiple wavelets, producing ventricular fibrillation. The likelihood of ventricular fibrillation during acute ischemia is increased by high sympathetic tone, hypokalemia, and preexisting ventricular hypertrophy.[10,15,20] It is reduced by high parasympathetic tone, β-adrenergic blocking agents, and amiodarone.[46-49]

Diagnostic Features

This polymorphic ventricular tachycardia often degenerates rapidly to ventricular fibrillation (Fig. 6-6).[50] In contrast to torsade de pointes, it is not always preceded by QT prolongation. The QT interval may be mildly prolonged by ischemia, however. In contrast to acquired, "pause-

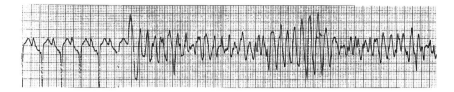

FIGURE 6-6. Initiation of rapid polymorphic ventricular tachycardia caused by acute myocardial ischemia. The initial rhythm is sinus tachycardia. A premature ventricular contraction initiates rapid polymorphic ventricular tachycardia. See text for discussion. (Reproduced with permission from Stevenson WG, Stevenson LW. Evaluation and management of arrhythmias in heart failure. In: Parmley WW, Chatterjee K, eds. Cardiology. Philadelphia: JB Lippincott, 1993.)

dependent" torsade de pointes, ischemic ventricular fibrillation is more likely to occur when sympathetic tone is high and the heart rate is rapid. It may also be preceded by long short R-R interval sequences but less commonly then torsade de pointes. Finally, it is unlikely to respond to measures that increase heart rate (e.g., isoproterenol and overdrive pacing), as these would be expected to increase ischemia.

Management

Sustained episodes respond to defibrillation. During acute myocardial infarction a single promptly defibrillated episode is not necessarily followed by repeated episodes as cells in the infarct border zone become inexcitable as the infarct progresses. Recurrent episodes suggest ongoing ischemia. Administration of magnesium, β-adrenergic blocking drugs, and lidocaine may be beneficial (Table 6-4). Aggressive therapy for ischemia with intravenous nitroglycerin and intraaortic balloon counterpulsation may be useful for stabilization to allow coronary angiography followed by therapeutic interventions to improve coronary blood flow and alleviate ischemia.

SINUSOIDAL VENTRICULAR TACHYCARDIA AND VENTRICULAR FLUTTER

When the QRS complex has a duration similar to the ST segment and T-wave, the electrocardiogram has a sinusoidal appearance, without clearly distinguishable QRS complexes. This occurs when either the tachycardia rate is very rapid (Fig. 6-7) or the QRS duration is markedly prolonged due to slow conduction through the ventricle (Fig. 6-8).[30-33]

Causes for rapid sinusoidal ventricular tachycardias (ventricular flutter) are the same as for sustained monomorphic ventricular tachy-

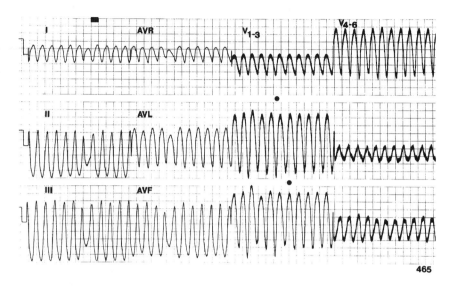

FIGURE 6-7. An example of sinusoidal ventricular tachycardia (ventricular flutter) caused by reentry in an area of right ventricular scar is shown. See text for discussion.

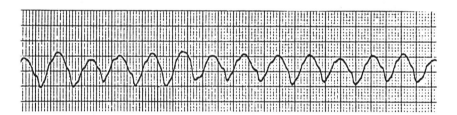

FIGURE 6-8. Sinusoidal ventricular tachycardia caused by hyperkalemia is shown. Tachycardia rate is 140 beats/min and the QRS duration is markedly prolonged. See text for discussion.

cardia. Tachycardia is usually poorly tolerated hemodynamically and requires immediate cardioversion. QRS synchronization of the direct current shock may be problematic as the cardioverter-defibrillator may select the peak of a T-wave rather than the peak of the QRS complex.

Ventricular tachycardia with a markedly prolonged QRS duration (e.g., >0.3 sec) and slower rate (usually <150 beats/min) suggests that ventricular conduction has been slowed by hyperkalemia or the toxic effects of class I antiarrhythmic drugs, tricyclic antidepressants, or phenothiazines (Fig. 6-8). In the setting of hyperkalemia hemodynamic collapse is imminent. Administration of intravenous calcium gluconate or calcium chloride improves conduction, shortens the QRS duration and improves hemodynamics within minutes, as does sodium bicarbonate.[50] The two should not be administered concomitantly through the same intravenous line because they will precipitates as calcium carbonate. Immediate arrangements to correct acidosis and possibly to remove potassium with dialysis or potassium-binding agents administered via the gastrointestinal tract should be considered.

Slow ventricular tachycardia caused by sodium channel-blocking agents such as class I antiarrhythmic drugs, phenothiazines, and tricyclic antidepressants also responds to administration of hypertonic sodium bicarbonate or sodium lactate.[33,34] Hemodynamic support with inotropic agents and repeated cardioversions may be required until the drug is excreted.

References

1. Hoffman BF, Rosen MR. Cellular mechanisms for cardiac arrhythmias. Circ Res 1981;49:1.
2. Allessie MA, Bonke FIM, Schopman FJC. Circus movement in rabbit atrial muscle as a mechanism of tachycardia. II. The "leading circle" concept: a new model of circus movement in cardiac tissue without the involvement of an anatomic obstacle. Circ Res 1977;41:9.
3. Jalife J, Davidenko JM, Michaels DC. A new perspective on the mechanisms of arrhythmias and sudden cardiac death: spiral waves of excitation in heart muscle. J Cardiovasc Electrophysiol 1991;2(suppl)S133.
4. Antzelevitch C, Jalife J, Moe GK. Characteristics of reflection as a mechanism of reentrant arrhythmias and its relationship to parasystole. Circulation 1980;61:182.
5. Vos MA, Gorgels A, Leunissen J, et al. Programmed electrical stimulation and drugs identify two subgroups of ventricular tachycardias occurring 16–24 hours after occlusion of the left anterior descending artery. Circ Res 1992;85:747.
6. El-Sherif N, Mehra R, Gough WB, Zeiler RH. Ventricular activation patterns of spontaneous and induced ventricular rhythms in canine one-day-old myocardial infarction: evidence for focal and reentrant mechanisms. Circ Res 1982;51:152.
7. Rosen MR, Reder RF. Does triggered activity have a role in the genesis of cardiac arrhythmias? Ann Intern Med 1981;94:794.
8. Levine JH, Spear JF, Guarnieri T, Weisfledt ML, DeLAngen CDJ, Becker LC, et al. Circulation 1985;72:1092.
9. Jackman WM, Friday KJ, Anderson JL, et al. The long QT syndromes: a critical review, new clinical observations and a unifying hypothesis. Prog Cardiovasc Dis 1988;31:115.
10. Aronson RS. Mechanisms of arrhythmias in ventricular hypertrophy. J Cardiovasc Electrophysiol 1991;2:249.
11. Kuchar DL, Thoburn CW, Sammel NL. Prediction of serious arrhythmic events after myocardial infarction: signal-averaged electrocardiogram, Holter monitoring, and radionuclide ventriculography. J Am Coll Cardiol 1987;9:531.

12. Bourke JP, Richards DAB, Ross DL, Wallace EM, McGuire MA, Uther JB. Routine programmed electrical stimulation in survivors of acute myocardial infarction for prediction of spontaneous ventricular tachyarrhythmias during follow-up results, optimal stimulation protocol and cost-effective screening. J Am Coll Cardiol 1991;18:780.

13. Farrell TG, Bashir Y, Cripps T, Malik M, Poloniecki J, Bennett ED. Risk stratification for arrhythmic events in postinfarction patients based on heart rate variability, ambulatory electrocardiographic variables and the signal-averaged electrocardiogram. J Am Coll Cardiol 1991;18:687.

14. Denniss AR, Richards DA, Cody DV, et al. Prognostic significance of ventricular tachycardia and fibrillation induced at programmed stimulation and delayed potentials detected on the signal averaged electrocardiogram of survivors of acute myocardial infarction. Circulation 1986;74:731.

15. Weiss JN, Nademanee K, Stevenson WG, Singh B. Ventricular arrhythmias in ischemic heart disease. Ann Intern Med 1991;114:784.

16. Stevenson WG, Brugada P, Waldecker B, Zehender M, Wellens HJJ. Clinical, angiographic and electrophysiologic findings in patients with aborted sudden death as compared with patients with sustained ventricular tachycardia after myocardial infarction. Circulation 1985;71:1146.

17. Adhar GC, Larson LW, Bardy GH, Greene HL. Sustained ventricular arrhythmias: differences between survivors of cardiac arrest and patients with recurrent sustained ventricular tachycardia. J Am Coll Cardiol 1988; 12:159.

18. Feldman T, Carroll JD, Munkenbeck F, et al. Hemodynamic recovery during simulated ventricular tachycardia: role of adrenergic receptor activation. Am Heart J 1988;115:576.

19. Smith ML, Ellenbogen KA, Beightol LA, Eckberg DL. Sympathetic neural responses to induced ventricular tachycardia. J Am Coll Cardiol 1991; 18:1015.

20. Janse MJ, Wit AL: Electrophysiological mechanisms of ventricular arrhythmias resulting from myocardial ischemia and infarction. Physiol Rev 1989;69:1049.

21. Bolick DR, Hackel DB, Reimer KA, Ideker RE. Quantitative analysis of myocardial infarct structure in patients with ventricular tachycardia. Circulation 1986;74:1266.

22. Stevenson WG, Weiss JN, Wiener I, Nademanee K. Slow conduction in the infarct scar: relevance to the occurrence, detection, and ablation of ventricular reentry circuits resulting from myocardial infarction. Am Heart J 1989;117:452.

23. de Bakker JM, van Capelle FJL, Janse MJ, et al. Reentry as a cause of ventricular tachycardia in patients with chronic ischemic heart disease: electrophysiologic and anatomic correlation. Circulation 1988;77:589.

24. Poole JE, Mathiesen TL, Kudenchuk PJ, et al. Long-term outcome in patients who survive out of hospital ventricular fibrillation and undergo electrophysiologic studies: evaluation by electrophysiologic subgroups. J Am Coll Cardiol 1990;16:657.

25. Willems AR, Tijussen JG, Van Cappelle FJL, et al. Determinants of prognosis in symptomatic ventricular tachycardia or ventricular tachycardia or ventricular fibrillation late after myocardial infarction. J Am Coll Cardiol 1990;16:521.

26. Markel ML, Miles WM, Luck JC, Klien LS, Prystowsky EN. Differential effects of isoproterenol on sustained ventricular tachycardia before and during porcainamide and quinidine antiarrhythmic drug therapy. Circulation 1993;87:783.

27. The cardiac arrhythmia suppression trial (CAST) investigators. CAST mortality and morbidity. Treatment versus placebo. N Engl J Med 1991;324:781.

28. Akhtar M, Shenasa M, Jazzayeri, Caceres J, Tchou PJ. Wide QRS complex tachycardia: reappraisal of a common clinical problem. Ann Intern Med 1988;109:905.

29. Brugada P, Brugada J, Mont L, Smeets J, Andries EW, et al. A new approach to the differential diagnosis of a regular tachycardia with a wide QRS complex. Circulation 1991;83:1649.

30. Herre JM, Titus C, Oeff M, et al. Inefficacy and proarrhythmic effects of flecainide and encainide for sustained ventricular tachycardia and ventricular fibrillation. Ann Intern Med 1990;113:671.

31. Myerburg RJ, Kessler KM, Cox MM, et al. Reversal of proarrhythmic effects of flecainide acetate and encainide hydrochloride by propranolol. Circulation 1989;80:1571.

32. Winkelmann BR, Leinberger H. Life-threatening flecainide toxicity. Ann Intern Med 1987;106:807.

33. Bajaj AK, Woosley RL, Roden DM. Acute electrophysiologic effects of sodium administration in dogs treated with O-desmethyl encainide. Circulation 1989;80:994.

34. Kim SG, Sisher JD, Choue CW, et al. Influence of left ventricular function on outcome of patients treated with implantable defibrillators. Circulation 1992;85:1304.

35. Mendoza I, Camardo J, Moleiro F. Sustained ventricular tachycardia in chronic Chagasic myocarditis: electrophysiologic and pharmacologic characteristics. Am J Cardiol 1986;57:423.

36. Marcus FI, Fontaine GH, Guiraudon G, et al. Right ventricular dysplasia: a report of 24 adult cases. Circulation 1982;65:384.

37. Moser DK, Stevenson WG, Woo MA, Weiner SR, Clements PJ, Suzuki SM, et al. Frequency of late potentials in systemic sclerosis. Am J Cardiol 1991;67:541.

38. Winters SL, Cohen M, Greenburg S, et al. Sustained ventricular tachycardia associated with sarcoidosis: assessment of the underlying cardiac anatomy and the prospective utility of programmed ventricular stimulation, drug therapy, and an implantable antitachycardia device. J Am Coll Cardiol 1991;18:937.

39. Caceres J, Jazzayeri M, McKinnie J, et al. Sustained bundle branch reentry as a mechanism of clinical tachycardia. Circulation 1989;79:256.

40. Akhtar M. Clinical spectrum of ventricular tachycardia. Circulation 1990; 82:1561.

41. Lerman BB. Response of nonreentrant catecholamine-mediated ventricular tachycardia to endogenous adenosine and acetylcholine: evidence for myocardial receptor-mediated effects. Circulation 1993;87:382.

42. Klein LS, Shih HT, Hackett K, Zipes DP, Miles WM. Radiofrequency catheter ablation of ventricular tachycardia in patients without structural heart disease. Circulation 1992;85:1666.

43. Tzivoni D, Banai S, Schuger C, Benhorn J, Kereu A, Gottlieb S, Stern S, et al. Treatment of torsade de pointes with magnesium sulfate. Circulation 1988;77:392.

44. Takanaka C, Singh BN. Barium-induced nondriven action potential as a model of triggered potentials from early afterdepolarizations: significance of slow channel activity and differing effects of quinidine and amiodarone. J Am Coll Cardiol 1990;15:213.

45. Vincent GM, Timothy KW, Leppert M, Keating M. The spectrum of symptoms and QT intervals in carriers of the gene for the long-QT syndrome. N Engl J Med 1992;327:846.

46. Schwartz PJ, Vanoli E, Stramba-Badiale M, et al. Autonomic mechanisms and sudden death. New insights from analysis of baroreceptor reflexes in conscious dogs with and without a myocardial infarction. Circulation 1988;78:969.

47. Patterson E, Eller BT, Abrams GD, Vasiliades J, Jucchesi BR. Ventricular fibrillation in a conscious canine preparation of sudden coronary death: prevention by short- and long-term amiodarone administration. Circulation 1983;68:857.

48. Norris RM, Brown MA, Clarke ED, et al. Prevention of ventricular fibrillation during acute myocardial infarction by intravenous propranolol. Lancet 1984;2:883.

49. Patterson E, Lucchesi BR. Electrophysiologic and antiarrhythmic actions of nadolol. Acute ischemia in the presence of previous myocardial infarction. Am Heart J 1988;116:1223.

50. Birnbaum Y, Sclarovsky S, Ben-Ami R, et al. Polymorphous ventricular tachycardia early after acute myocardial infarction. Am J Cardiol 1993; 71:745.

Sunil Nath and John P. DiMarco

Postinfarction Arrhythmias: Identification and Treatment of 7 High-Risk Patients

Sudden cardiac death is responsible for more then 300,000 deaths in the United States each year. The mechanism(s) of sudden cardiac death remain incompletely understood. However, it is generally accepted that the majority of patients have a fatal ventricular arrhythmia, either ventricular tachycardia that degenerates into ventricular fibrillation or primary ventricular fibrillation, as the initiating event. Experimental and clinical studies have indicated that the mechanisms responsible for ventricular tachycardia and fibrillation are both diverse and complex.[1] Acute ischemia, autonomic nervous system influences, circulating catecholamines, electrolyte abnormalities, and antiarrhythmic drugs are all known trigger factors in precipitating ventricular arrhythmias, particularly in patients with underlying structural heart disease. The most common underlying cardiac disorder responsible for sudden death in the United States is coronary artery disease, which accounts for approximately 80% of cases. Autopsy studies have shown that most victims of sudden death have multivessel coronary artery disease, with about 50% of the patients also having evidence of one of more prior healed myocardial infarctions.[2]

Identifying the patient at high risk for a potentially fatal ventricular arrhythmia is important because only approximately 20% of patients who experience an episode of sudden death survive to hospital discharge. Since the most common cause of sudden death in North Amer-

Clinical Cardiac Electrophysiology: Perioperative Considerations
Edited by Carl Lynch III. J.B. Lippincott Company, Philadelphia, PA ©1994

ica is associated with a prior myocardial infarction, this chapter will discuss the diagnostic tools currently available to identify patients who are at high risk for sudden death following an infarction. The management of the high-risk patient is then discussed in a subsequent section.

LEFT VENTRICULAR FUNCTION

Probably, the single most important factor in risk stratification following a myocardial infarction is assessment of left ventricular systolic function. Numerous studies have shown that patients with depressed left ventricular function have an increased long-term cardiac mortality rate, including an increased risk of sudden death following a myocardial infarction. One of the larger studies, the Multicenter Postinfarction Research Group, assessed left ventricular ejection fraction by radionuclide angiography in 866 postinfarct patients.[3] The study found an ejection fraction <40% to be a significant risk factor for subsequent cardiac mortality. Patients with an ejection fraction >40% had a 1-year cardiac mortality of <5%, compared to 10 to 15% for patients with an ejection fraction of 20 to 39% and 45 to 50% for patients with an ejection fraction <20% (Fig. 7-1). Typically, two-dimensional echocardiography, radionuclide angiocardiography, and/or contrast ventriculography are used to obtain an assessment of global and regional left ventricular systolic function in patients recovering from a myocardial infarction.

ISCHEMIA

Stress testing with or without nuclear perfusion imaging has been found to be useful for risk stratification after myocardial infarction. Several studies have shown that the occurrence of exercise-induced

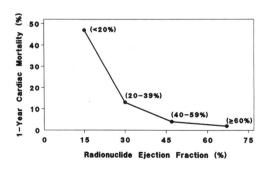

FIGURE 7-1. Cardiac mortality rate at 1 year following a myocardial infarction in patients with four categories of radionuclide left ventricular ejection fraction determined before discharge. (Reproduced with permission from The Multicenter Postinfarction Research Group. Risk stratification and survival after myocardial infarction. N Engl J Med 1983; 309:331.)

angina, exercise-induced ventricular arrhythmias, exercise-induced diagnostic ST-segment changes, or nuclear perfusion abnormalities suggestive of residual ischemia 1 to 2 weeks after a myocardial infarction to be predictive of future cardiac events including sudden death. Patients with inducible ischemia should be considered for prophylactic β-blocker therapy and/or early cardiac catheterization with a view to possible revascularization if indicated (see below).

ELECTRICAL INSTABILITY

Holter Monitoring

Holter monitoring has been extensively used in risk stratification in the past two decades. Several studies have examined the relationship between spontaneous ventricular ectopy recorded on a 24-hr Holter monitor in the first several weeks after a myocardial infarction and the occurrence of a subsequent major arrhythmic event (i.e., sudden cardiac death or sustained ventricular tachycardia).[4–8] These studies showed that patients with frequent premature ventricular complexes (≥10/hr), multiform premature ventricular complexes, couplets, or nonsustained ventricular tachycardia (three or more consecutive premature ventricular complexes) had an incidence of sudden cardiac death ranging from 8 to 15%. In contrast, patients who had none of these findings had a 3 to 6% incidence of sudden cardiac death. More recently, several large prospective studies have examined the relationship between spontaneous ventricular ectopy on Holter monitoring, left ventricular ejection fraction, and subsequent sudden cardiac death.[7–9] These studies found that patients with both high-risk ventricular ectopy and left ventricular dysfunction had a high incidence of sudden cardiac death, ranging from 18 to 31%. In comparison, patients with little or no ventricular ectopy and preserved left ventricular function had a incidence of sudden death of only 0 to 2%. However, a recent study called the Cardiac Arrhythmia Suppression Trial (CAST) reported a 2.5-fold higher total mortality and a 3.6-fold higher arrhythmic mortality among patients with a previous myocardial infarction whose ventricular ectopy had been successfully suppressed with the class Ic antiarrhythmic drugs, encainide or flecainide, as compared to patients taking placebo.[10] It is interesting to note that the arrhythmic mortality in the placebo arm of the trail was only 1.2%, suggesting that the study population was already at low risk. In light of the findings of this trial, the general recommendation now is to not treat asymptomatic ventricular ectopy following a myocardial infarction with class Ic antiarrhythmic drugs.

Signal-Averaged Electrocardiography

Studies in experimental myocardial infarction have shown that fragmented and delayed electrical activity can be recorded from the border zones surrounding the infarction. These zones, which have been shown to be composed of both scar tissue and viable myocardium, are thought to represent the critical substrate responsible for the genesis of reentrant ventricular arrhythmias late after myocardial infarction. Electrical activity from these zones is usually undetectable on the standard surface electrocardiogram (ECG) because the voltage is of too low an amplitude. The signal-averaged ECG allows detection of these low-amplitude electrical signals, referred to as late potentials, by first amplifying the incoming QRS signal. The amplified signal is then averaged over 200 to 400 beats to eliminate any incoming background noise (predominantly from skeletal muscle). After signal-averaging, the QRS complex is high-pass filtered (usually at a cutoff frequency of 25 or 40 Hz). High-pass filtering takes advantage of the fact that late potentials are composed predominantly of high-frequency signals, whereas the QRS and ST segments, which obscure the late potentials on the standard surface ECG, are composed of lower frequencies. The presence of late potentials is considered if one or more of the following three quantitative criteria are met: 1) the filtered QRS is greater than 114 to 120 msec, 2) the duration of low-amplitude signals of <40 µV (LAS) is longer than 38 msec, and 3) the root-mean-square voltage of the terminal 40 msec of the QRS complex is <20 µV (Fig. 7-2). No time domain criteria have been established for the definition of late potentials in the presence of a bundle branch block, and, therefore, the signal-averaged ECG is less helpful in patients with a preexisting bundle branch block on their standard surface ECG. Several prospective studies have assessed the prognostic significance of the signal-averaged ECG after myocardial infarction. The six studies comprised 1068 patients who were studied anywhere between 6 days to 6 weeks following the infarction and followed for a mean of 11 months.[11–16] Although the precise definition of late potentials and the recording technique varied between studies, all the studies found a significant association between the presence of late potentials and the occurrence of an arrhythmic event (either sustained ventricular tachycardia or sudden death) during long-term follow-up. Twenty percent of patients with late potentials had an arrhythmic event, compared to only 3% of patients without late potentials. However, despite the reasonable sensitivity and specificity (77 and 72%, respectively) of the signal-averaged ECG in these studies, the positive predictive value of an abnormal signal-averaged ECG for a subsequent arrhythmic event was only 20%. The negative predictive value was much better at 97%. In clinical

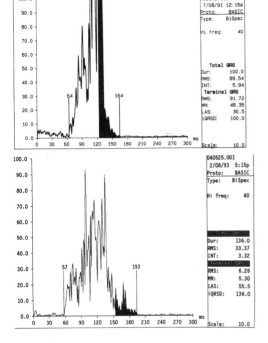

FIGURE 7-2. *Top*, normal signal-averaged electrocardiogram. *Bottom*, abnormal signal-averaged electrocardiogram showing the presence of late potentials. The last 40 msec of the QRS are shaded in *black*. *DUR*, duration; *RMS*, root mean square; *IN*, integral; *LAS*, low-amplitude signals.

decision-making, a normal signal average is useful in risk stratification; conversely, the presence of late potentials is less helpful if used alone for prognostication.

Heart Rate Variability

Measurement of heart rate variability is a relatively new technique for assessing the risk of sudden cardiac death following myocardial infarction. Normal heart rate variability depends on a balance between sympathetic and parasympathetic inputs to the sinus node. Reduced heart rate variability is thought to result from either an increased sympathetic and/or a decreased parasympathetic influence on the heart, resulting in an increased risk of ventricular arrhythmias. Heart rate variability can be analyzed in the time or frequency domain. The simplest way to measure heart rate variability is in the time domain, by determining the standard deviation of the RR interval about the mean RR interval for a specified number of beats or time.

Several investigators have demonstrated that decreased heart rate variability is associated with an increased risk of long-term mortality following myocardial infarction. In one of these studies, Kleiger et al.[17]

showed that patients with normal heart rate variability had a 3-year mortality rate of less than 10% compared to greater than 30% in patients with abnormal heart rate variability. More recently, Farrell et al.[18] found reduced heart rate variability to be more predictive of arrhythmic events (sustained ventricular tachycardia or sudden death) following myocardial infarction than an abnormal signal-averaged ECG, frequent or complex ventricular ectopy on Holter monitoring, or left ventricular ejection fraction. If these results are confirmed by additional studies, the way patients are risk stratified postinfarction may be changed.

Combined Noninvasive Tests

Given the low positive predictive value of any single diagnostic test, several studies have investigated the value of a combination of these tests to determine whether the predictive accuracy can be improved. Three recent prospective studies have assessed the combination of the left ventricular ejection fraction, the signal-averaged ECG, and 24-hour Holter monitoring with respect to risk stratification postmyocardial infarction.[11,13,19] The combination of an ejection fraction <40% and the presence of late potentials on the signal-averaged ECG had a positive predictive value of 34 to 37% for an arrhythmic event in the first year following the infarction. The combination of all three tests (i.e., an abnormal ejection fraction, the presence of late potentials, and high grades of ventricular ectopy on the Holter monitor recording) carried a 50 to 58% arrhythmic event rate, in contrast to the 0 to 2% event rate when all three tests were normal. Recently, Farrell et al.[18] found the combination of abnormal heart rate variability, the presence of late potentials, and nonsustained ventricular tachycardia on the Holter monitor to have a positive predictive value of 58% for an arrhythmic event.

Invasive Electrophsyiologic Testing

In recent years, electrophysiologic study has been shown to be useful in the diagnosis and management of patients with sustained ventricular tachycardia who had an out-of-hospital cardiac arrest. More recently, this technique has been used in identifying patients at increased risk for sudden death after a myocardial infarction. The reported incidence of inducible sustained ventricular arrhythmias in postinfarct patients has varied from 11 to 44% and seems to depend on the patient population being studied, the time at which electrophysiologic study was performed postinfarction, the stimulation protocol used, and the use of concomitant antiarrhythmic therapy.[20-24] Several studies have emphasized the significance of the type of arrhythmia induced. Although a rapid polymorphic ventricular tachycardia or ventricular fibrillation is inducible in 10 to 20% of patients, its inducibility does not correlate with the risk of a subsequent arrhythmic event. Conversely, the inducibility of sustained monomorphic ventricular tachycardia appears to be a finding of clinical significance. Several small studies have reported that these patients have a 21 to 75% incidence of any arrhythmic event during long-term follow-up, although the risk of sudden death appears to be much smaller.[21,23,24] However, whether the inducible ventricular arrhythmia provides prognostic information independent of other known clinical variables and diagnostic tests and whether electrophysiologic study-guided antiarrhythmic therapy is beneficial in reducing mortality are presently not known.

MANAGEMENT OF THE HIGH-RISK PATIENT

Previous and Ongoing Multicenter Studies

Using a combination of noninvasive tests, it is possible to risk stratify patients for future arrhythmic events. Patients with an ejection fraction <40% or abnormal heart rate variability combined with the presence of late potentials on the signal-averaged ECG and nonsustained ventricular tachycardia on Holter or ambulatory monitoring are considered to be at high risk. However, the exact management of high-risk patients is controversial. Currently, there is no evidence that specific antiarrhythmic treatment will decrease their future risk of an arrhythmic event. In fact, studies that have examined the empiric use of class I antiarrhythmic drugs in patients recovering from a myocardial infarction have shown negative results. The CAST findings[10] (see previously) and data from a meta-analysis of all the trials[25] indicate a higher mortality in patients treated with class I antiarrhythmic drugs, compared to placebo (Fig. 7-3).

Recently three small studies have suggested that the class III antiarrhythmic drug, amiodarone, may reduce mortality, including death from arrhythmia postinfarction.[26-28] Currently, two large multicenter studies, the Canadian Amiodarone Myocardial Infarction Arrhythmia Trial (CAMIAT) and the European Myocardial Infarction Arrhythmia Trial (EMIAT), are assessing the effect of amiodarone on arrhythmic events following a myocardial infarction.

In contrast to the antiarrhythmic drug therapy trials, there are data from at least five studies to suggest that long-term prophylactic β-blocker administration significantly reduces the risk of sudden

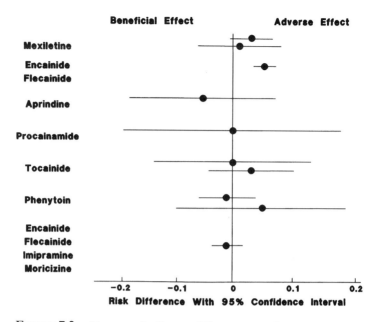

FIGURE 7-3. Mean case-fatality rate difference (rate of death in control group subtracted from the rate in the treatment group) of several randomized controlled trials of therapy with class I antiarrhythmic drugs after myocardial infarction. (Reproduced with permission from Hine LK, Laird NM, Chalmers TC. Meta-analysis of empirical long-term antiarrhythmic therapy after myocardial infarction. JAMA 1989;262:3037.)

death after acute myocardial infarction. In the multicenter Norwegian timolol study,[29] 1884 patients were randomly selected to receive placebo or timolol (10 mg twice daily) for 7 to 28 days following the infarction. The study demonstrated a 50% reduction in the sudden cardiac death rate during long-term follow-up in the treatment arm, compared to placebo (Fig. 7-4). Similarly, the Beta-Blocker Heart Attack Trial (BHAT) showed a 4.6% sudden death mortality in the placebo group versus 3.3% among patients taking propranolol (180 to 240 mg/day) during long-term follow-up.[30] In addition, these studies demonstrated that the absolute benefit of β-blocker treatment postinfarction appears to be greatest in the highest risk patients.

More recently, the Survival and Ventricular Enlargement trial (SAVE) investigated the long-term outcome of administering the angiotensin-converting enzyme inhibitor, captopril, to survivors of acute myocardial infarction with asymptomatic left ventricular dysfunction.[31] The study included 2231 patients with left ventricular ejection fractions ≤40% randomly assigned to receive placebo or captopril 3 to 16

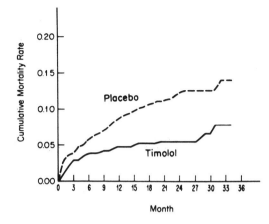

FIGURE 7-4. Life table cumulated rates of sudden cardiac death during administration of timolol versus placebo. (Reproduced with permission from The Norwegian Multicenter Study Group. Timolol-induced reduction in mortality and reinfarction in patients surviving acute myocardial infarction. N Engl J Med 1981;304:801.)

days after a myocardial infarction. Patients taking captopril had a significantly lower total mortality and cardiovascular mortality than the control group during a mean follow-up of 42 months (the risk reduction in the treatment group was 19 and 21%, respectively). Other major trials have reported a significant reduction in total mortality including sudden deaths among patients with congestive heart failure treated with angiotension-converting enzyme inhibitors.[32,33]

Data from the Coronary Artery Surgery Study (CASS) registry indicate that surgical revascularization may significantly reduce the risk of sudden death.[34-36] The 5-year sudden death rate was 6% in the medically treated patients, compared to 2% in the surgically treated patients. Among patients with two- or three-vessel disease and a history of congestive heart failure, the 5-year incidence of sudden death was reduced from 17 and 31%, respectively, in the medically treated group, to 2 and 9%, respectively, in the surgical cohort. In the European Coronary Surgery Study (ECSS), which included only patients with left ventricular ejection fractions ≥50%, surgical revascularization was also associated with a significantly lower rate of sudden death, compared to medical therapy (3 versus 9%, respectively, at 8 years follow-up) (Fig. 7-5).[37,38] Whether percutaneous revascularization similarly reduces the risk of sudden death as surgical revascularization is presently not known. Trials comparing percutaneous and surgical revascularization among patients with multivessel coronary disease are currently ongoing.

A multicenter study called the Coronary Artery Bypass Grafting-Patch Trial (CABG-Patch) is currently assessing the role of prophylactic implantation of a cardioverter-defibrillator (ICD) in reducing total mortality among high-risk patients undergoing surgical revasculariza-

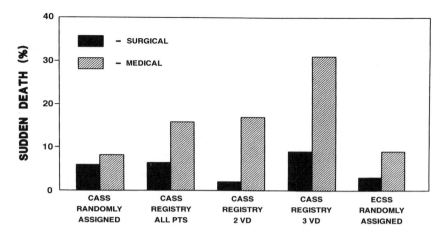

FIGURE 7-5.   Effects of surgical revascularization on sudden cardiac death. Data comparing medically and surgically treated patients (*PTS*) in the randomized Coronary Artery Surgical Study (*CASS*), all patients in the CASS registry with coronary artery disease, patients with two- and three-vessel disease (*VD*) in the CASS registry, and in the European Cooperative Surgical Study (*ECSS*) are shown. (Reproduced with permission from Nath S, Haines DE, Hobson CE, et al. Ventricular tachycardia surgery. J Cardiovasc Electrophysiol 1992;3:160.)

tion who have not had a prior arrhythmic event. The study is enrolling patients with a left ventricular ejection fraction <36% and a positive signal-averaged ECG and randomly assigning them to either ICD implantation or no ICD implantation in the operating room. A similar study known as the Multicenter Automatic Defibrillator Implantation Trial (MADIT) is examining the efficacy of ICD compared to medical therapy in preventing sudden death. Patients are eligible for random selection if they have coronary artery disease, an ejection fraction of ≤35%, electrocardiographically documented nonsustained ventricular tachycardia, and inducible sustained ventricular tachycardia or fibrillation during programmed stimulation.

Finally, the Multicenter Unsustained Tachycardia Trial (MUSTT) is a multicenter study that is investigating the outcome of prophylactic electrophysiology-guided antiarrhythmic therapy (drug or ICD treatment) versus no therapy in high-risk patients. Patients are eligible for entry into the study if they have documented coronary artery disease, a left ventricular ejection fraction of ≤40%, nonsustained ventricular tachycardia, and a positive signal-averaged ECG. Eligible patients undergo a diagnostic electrophysiology study, and those who have ventricular tachycardia induced by programmed stimulation are randomly assigned to either therapy or no therapy.

Current Recommendations

Presently, until the results of the above multicenter trials are made available, no specific antiarrhythmic therapy can be recommended for the high-risk postinfarct patient outside of entering the patient into one of the above studies. Currently, these patients should be treated with a β-blocker if no specific contraindication to its use exists. In addition, patients with a left ventricular ejection fraction ≤40% and/or symptomatic heart failure may benefit from the administration of an angiotensin-converting enzyme inhibitor. Finally, patients with multivessel coronary disease and reduced left ventricular function should be considered for surgical revascularization, particularly if they have evidence of spontaneous or inducible myocardial ischemia.

References

1. Janse M, Wit AL. Electrophysiological mechanisms of ventricular arrhythmias resulting from myocardial ischemia and infarction. Physiol Rev 1989;69:1049.
2. Barbour DJ, Warnes CA, Roberts WC. Cardiac findings associated with sudden death secondary to atherosclerotic coronary artery disease: comparison of patients with and those without previous angina pectoris and/or healed myocardial infarction. Circulation 1987;75(suppl 2):9.
3. The Multicenter Postinfarction Research Group. Risk stratification and survival after myocardial infarction. N Engl J Med 1983;309:331.
4. Moss AJ, Davis HT, DeCamilla J, Bayer LW. Ventricular ectopic beats and their relation to sudden and nonsudden cardiac death after myocardial infarction. Circulation 1979;60:998.
5. Bigger JT, Fleiss JL, Kleiger K, et al. The multicenter post-infarction research group. The relationship between ventricular arrhythmias, left ventricular dysfunction and mortality in the two years after myocardial infarction. Circulation 1984;69:250.
6. Kostis JB, Byington R, Friedman LM, et al. Prognostic significance of ventricular ectopic activity in survivors of acute myocardial infarction. J Am Coll Cardiol 1987;10:231.
7. Mukharji J, Rude RE, Poole WK, et al. The MILIS study group. Risk factors for sudden death after acute myocardial infarction: two-year follow-up. Am J Cardiol 1984;54:31.
8. Ruberman W, Weinblatt E, Goldberg JD, et al. Ventricular premature beats and mortality after acute myocardial infarction. N Engl J Med 1977;297:750.
9. Schulze RA Jr, Strauss HW, Pitt B. Sudden death in the year following myocardial infarction: relation to ventricular premature contractions in the late hospital phase and left ventricular ejection fraction. Am J Med 1977;62:192.
10. The Cardiac Arrhythmia Suppression Trial (CAST) Investigators Preliminary Report. Effect of encainide and flecainide on mortality in a randomized trial of arrhythmia suppression after myocardial infarction. N Engl J Med 1989;321:406.

11. Kuchar DL, Thorburn CW, Sammel L. Prediction of serious arrhythmic events after myocardial infarction: signal-averaged electrocardiogram, Holter monitoring and radionuclide ventriculography. J Am Coll Cardiol 1987;9:531.
12. Gomes JA, Winters SL, Martinson M, et al. The prognostic significance of quantitative signal-averaged variables relative to clinical variables, site of myocardial infarction, ejection fraction and ventricular premature beats: a prospective study. J Am Coll Cardiol 1988;13:377.
13. El-Sherif N, Ursell SN, Bekheit S, et al. Prognostic significance of the signal-averaged electrocardiogram depends on the time of recording in the postinfarction period. Am Heart J 1989;118:256.
14. Breithardt G, Borggrefe M, Haarten K. Role of programmed ventricular stimulation and non-invasive recording of ventricular late potentials for the identification of patients at risk of ventricular arrhythmias after acute myocardial infarction. In: Zipes DP, Jalife J, eds. Cardiac electrophysiology and arrhythmias. Orlando, FL: Grune & Stratton, 1984:553.
15. Denniss AR, Richards DA, Cody DV, et al. Prognostic significance of ventricular tachycardia and fibrillation induced at programmed stimulation and delayed potentials detected on the signal-averaged electrocardiograms of survivors of acute myocardial infarction. Circulation 1986;74:731.
16. Cripps T, Bennett ED, Camm AJ, Ward DE. High-gain signal-averaged electrocardiogram combined with 24-hour monitoring in patients early after myocardial infarction for bedside prediction of arrhythmic events. Br Heart J 1988;60:181.
17. Kleiger RE, Miller JP, Bigger JT, et al. Decreased heart rate variability and its association with increased mortality after acute myocardial infarction. Am J Cardiol 1987;59:256.
18. Farrell TG, Bashir Y, Cripps T, et al. Risk stratification for arrhythmic events in postinfarction patients based on heart rate variability, ambulatory electrocardiographic variables, and the signal-averaged electrocardiogram. J Am Coll Cardiol 1991;18:687.
19. Gomes JA, Winters SL, Stewart D, et al. A new noninvasive index to predict sustained ventricular tachycardia and sudden death in the first year after myocardial infarction: based on signal-averaged electrocardiogram, radionuclide ejection fraction and Holter monitoring. J Am Coll Cardiol 1987;10:349.
20. Marchlinski FE, Buxton AE, Waxman HL, Josephson ME. Identifying patients at risk of sudden death after myocardial infarction: value of the response to programmed stimulation, degree of ventricular ectopic activity and severity of left ventricular dysfunction. Am J Cardiol 1983;52:1190.
21. Bhandari AK, Hong R, Kotlewski A, et al. Prognostic significance of programmed ventricular stimulation in survivors of acute myocardial infarction. Br Heart J 1989;61:410.
22. Denniss AR, Richards DA, Cody DV, et al. Prognostic significance of ventricular tachycardia and fibrillation induced at programmed stimulation and delayed potentials detected on the signal-averaged electrocardiograms of survivors of acute myocardial infarction. Circulation 1986;74:731.
23. Cripps T, Bennett ED, Camm AJ, Ward DE. Inducibility of sustained monomorphic ventricular tachycardia as a prognostic indicator in survivors of recent myocardial infarction: a prospective evaluation in relation to other prognostic variables. J Am Coll Cardiol 1989;14:289.

24. Bhandari A, Hong R, Kotlewski A, et al. Prognostic significance of programmed stimulation in high risk patients surviving acute myocardial infarction. J Am Coll Cardiol 1988;11:6A. Abstract.
25. Hine LK, Laird NM, Chalmers TC. Meta-analysis of empirical long-term antiarrhythmic therapy after myocardial infarction. JAMA 1989;262:3037.
26. Burkart F, Pfisterer M, Kiewski W, et al. Effect of antiarrhythmic therapy on mortality in survivors of myocardial infarction with asymptomatic complex ventricular arrhythmias: Basel Antiarrhythmic Study of Infarct Survival (BASIS). J Am Coll Cardiol 1990;16:1711.
27. Cairns JA, Connolly SJ, Gent M, Roberts R. Post-myocardial infarction mortality in patients with ventricular premature depolarizations: Canadian Amiodarone Myocardial Infarction Arrhythmia Trial pilot study. Circulation 1991;84:550.
28. Ceremuzynski L, Kleczar E, Krzeminska-Pakula M, et al. Effect of amiodarone on mortality after myocardial infarction: a double-blind, placebo-controlled, pilot study. J Am Coll Cardiol 1992;20:1056.
29. The Norwegian Multicenter Study Group. Timolol-induced reduction in mortality and reinfarction in patients surviving acute myocardial infarction. N Engl J Med 1981;304:801.
30. Beta-Blocker Heart Attack Trial Research Group. A randomized trial of propranolol in patients with acute myocardial infarction. I. Mortality results. JAMA 1982;247:1707.
31. Pfeffer MA, Braunwald E, Moyé LA, et al, on behalf of the SAVE Investigators. Effect of captopril on mortality and morbidity in patients with left ventricular dysfunction after myocardial infarction. Results of the Survival and Ventricular Enlargement Trial. N Engl J Med 1992;327:669.
32. The SOLVD Investigators. Effect of enalapril on survival in patients with reduced left ventricular ejection fractions and congestive heart failure. N Engl J Med 1991;325:293.
33. Cohn JN, Johnson G, Ziesche S, et al. A comparison of enalapril with hydralazine-isosorbide dinitrate in the treatment of chronic congestive heart failure. N Engl J Med 1991;325:303.
34. Holmes DR, Davis KB, Mock MR, et al. The effect of medical and surgical treatment on subsequent sudden cardiac death in patients with coronary artery disease: a report from the Coronary Artery Surgery Study. Circulation 1986;73:1254.
35. Passamani E, Davis KB, Gillespie MJ, et al. A randomized trial of coronary artery bypass surgery. Survival of patients with a low ejection fraction. N Engl J Med 1985;312:1665.
36. Alderman EL, Bourassa MG, Cohen LS, et al. Ten-year follow-up of survival and myocardial infarction in the randomized Coronary Artery Surgery Study. Circulation 1990;82:1629.
37. European Coronary Surgery Study Group. Long-term results of prospective randomised study of coronary artery bypass surgery in stable angina pectoris. Lancet 1982;2:1173.
38. Varnauskas E and the European Coronary Surgery Study Group. Survival, myocardial infarction, and employment status in a prospective randomized study of coronary bypass surgery. Circulation 1985;72(suppl V): V-90.

Gerald V. Naccarelli
Anne H. Dougherty
Deborah Wolbrette
Sohail Jalal
Hue-Teh Shih

8 | Advances in Implantable Cardioverter/Defibrillators

A major advance in the treatment of patients with life-threatening ventricular tachyarrhythmias was the development of the automatic implantable cardioverter/defibrillator (AICD) by Michel Mirowski.[1] The first human implant was performed in 1980 and the first implant device was commercially released in 1986. Since that time, there has been a marked growth in the insertion (>30,000 implanted) of implantable cardioverter/defibrillators (ICDs). Because of the marked increase in the use of these devices for the prevention of sudden cardiac death, physicians of multiple specialties, including cardiologists, cardiac electrophysiologists, cardiac surgeons, cardiac anesthesiologists, and emergency room physicians, need to understand the prescription, preoperative, perioperative, and postoperative management of patients with such devices.

ICD FUNCTION

ICD function consists of tachycardia detection and tachycardia therapy. ICD systems monitor heart rate activity continuously through pacing lead systems that are positioned endocardially or epicardially. With ICDs implanted via a thoracotomy approach, sensing leads and defibrillator patches are connected from the epicardium of the heart

Clinical Cardiac Electrophysiology: Perioperative Considerations
Edited by Carl Lynch III. J.B. Lippincott Company, Philadelphia, PA ©1994

through leads to the pulse generator (Fig. 8-1). When a rapid heart rate is sensed by the device, it will automatically charge and deliver a direct current shock to terminate an arrhythmia (Figs. 8-2 and 8-3). The predominant detection parameter used to determine when the device will charge and shock is called the heart rate cutoff or the tachycardia detection interval. The concept of rate is a simple one. In order to fulfill the rate parameter, the patient's heart rate must exceed the rate cutoff for a period of 5 to 20 sec. The device then detects the arrhythmia and begins charging energy to the capacitors in order to deliver a countershock. Optimally, the patient's rate cutoff should be set at a rate faster than his or her maximal sinus rate, but slower than the ventricular tachyarrhythmia rate. With Cardiac Pacemaker Incorporated (CPI) devices, a separate parameter of probability density function can also be used to differentiate sinus tachycardia, atrial fibrillation, and ventricular tachycardia. With newer devices, more sophisticated tachyarrhythmia algorithms exist in order to more accurately distinguish ventricular tachycardia from sinus tachycardia and atrial fibrillation. Some of these parameters include onset and rate stability criteria.

ADVANCES IN ICD THERAPY

The initial AICD was a simple shock box that had a factory preset tachycardia detection interval, and one could only program the device on or off. With the commercially available CPI Ventak series, newer

FIGURE 8-1. Photograph of a Ventak series ICD attached to two epicardial screw-in sense/pace leads and two epicardial shock patches. An ICD programmer and an electrocardiographic strip of an ICD shock termination of rapid ventricular tachycardia/ventricular fibrillation are also depicted.

AICD Function

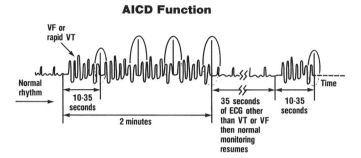

FIGURE 8-2. Diagrammatic representation of a normal ICD recognition and function. Multiple attempts (if necessary) of ICD shocks are initiated once rapid ventricular tachycardia/ventricular fibrillation is sensed by the ICD device.

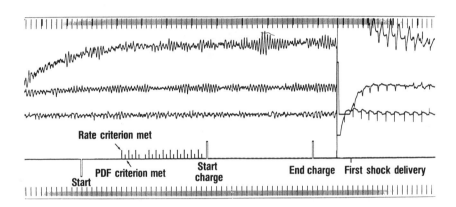

FIGURE 8-3. Electrocardiographic strips from the operating room depicting induced ventricular fibrillation, signals demonstrating rate and probability density function criteria being met, start and end charge artifact, and shock delivery and termination of ventricular fibrillation from a CPI Ventak P series AICD.

software added the ability to program the tachycardia detection interval and the shock levels down to very low energies (<1.0 J). More recently, the Medtronic PCD and Ventritex Cadence® devices that are capable of tiered therapy (bradycardia pacing, antitachycardia pacing, low energy cardioversion, and higher energy defibrillation) have been released commercially. Many other manufacturers, including CPI, Telectronics, and Intermedics have similar devices under investigational study at this time. The capabilities of these newer tiered therapy devices include programmable rate and shock energy, VVI bradycardia and antitachycardia pacing and algorithms to terminate arrhythmia without the need for shock (Figs. 8-4). With tiered therapy devices, the

**RECURRENT EPISODES OF VT APPROPRIATELY
TERMINATED BY ATP**

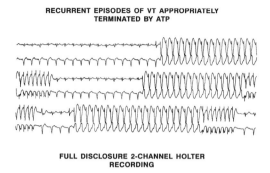

**FULL DISCLOSURE 2-CHANNEL HOLTER
RECORDING**

FIGURE 8-4. Two-channel Holter recording of recurrent VT successfully terminated by burst pacing by a Ventritex Cadence model V-1000 device. (Reprduced with permission from Luceri RM. Implantable defibrillators: physician's role after hospital discharge. Naccarelli GV, Veltri EP, eds. Implantable cardioverter-defibrillators. Boston, Blackwell Scientific Publications, 1993;150.)

patient can have his or her rhythm monitored, antibradycardia pacing is initiated if the rate goes too slowly,[3] antitachycardia pacing occurs if the rate increases to a predefined rate, and shock therapy at low or high energy is delivered at faster rates. Newer devices have various forms of event logs and electrogram storage capabilities. These logs can be used to assess whether the shocks are appropriate.

Third generation devices have noninvasive programmed stimulation, so that by using the pacing system programmer one can pace the heart and induce tachycardia without the need for inserting catheters. Another advance in ICD therapy includes the development of smaller ICD generators. The initial AICD weighed 260 g and the Medtronic PCD is currently the smallest ICD at 196 g. Many companies are working on developing generators that are smaller so that ICDs may be implanted pectorally, as is done with pacemakers. Some of the newer devices have noncommitted therapy. That is, once the arrhythmia is detected, they will charge the batteries and, before delivering the shock, will take a second look to make sure the patient is still in an abnormal rhythm before delivering a shock.

The biggest advance in ICD technology has been the development of nonthoracotomy approaches for implanting ICDs.[2] With this technique, the sense/pace lead is implanted endocardially as is done with a pacemaker. Shock coils and pacing electrodes are built into the endocardial leads and a separate patch can be implanted subcutaneously over the pectoralis muscles instead of epicardially (Fig. 8-5). The development of generators that have biphasic wave forms[4] which compensate for smaller lead surface area and still achieve acceptable defibrillation thresholds (DFTs) has been a major breakthrough. With the CPI Ventak P-2 device and a lead system alone with biphasic wave forms, 80% of patients can have implantation of the device without a subcutaneous patch or a thoracotomy approach.[5]

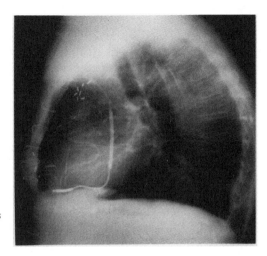

FIGURE 8-5. Lateral chest x-ray depicting an endocardial lead with a distal right ventricular apex and proximal high right atrium shock electrode along with a left pectoralis muscle subcutaneous patch from a CPI Endotak system.

RESULTS OF ICD IMPLANTATION

No prospective, randomized trial has been performed to determine the efficacy of ICDs. However, once implanted, the device appears to reduce sudden death mortality to less than 3% per year.[6-8] Historical comparisons of sudden death mortality in similar patient groups range as high as 40 to 60% in 3 years (Fig. 8-6). Therefore, it appears that ICDs in patients with life-threatening ventricular tachyarrhythmias reduce sudden cardiac death by a significant fraction.

INDICATIONS FOR DEVICE THERAPY

Class I indications[9] for ICD implantation include: 1) patients with ventricular tachycardia (VT) or ventricular fibrillation (VF) in whom electrophysiologic (EP) testing or Holter monitoring cannot be used

FIGURE 8-6. Life survival curves depicting decreased sudden and total death rates of patients with AICDs compared to historical (expected) control subjects. (Reproduced with permission from Mirowski M, Reid PR, Winkle RA, et al. Mortality in patients with implanted automatic defibrillators. Ann Intern Med 1983;98:585.)

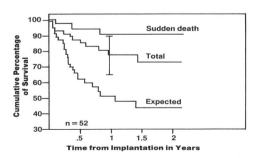

to predict the efficacy of therapy (includes patients whose VT/VF cannot be induced by programmed stimulation); 2) patients with recurrent VT or VF despite drug therapy that is guided by EP testing or Holter monitoring (includes patients whose clinical arrhythmia is not suppressed by antiarrhythmic drugs); 3) patients with spontaneous VT or VF in whom drugs are not tolerated or not acceptable; 4) patients whose VT or VF remains persistently inducible despite the best drug therapy, surgical therapy, or catheter ablation. As we learn more about these devices and their use, these indications will expand in the future.

SURGICAL PROCEDURE

Several surgical options exist for the placement of epicardial patches and leads.[10] These include: median sternotomy, left lateral thoracotomy, subxiphoid, and subcostal approaches. In patients who require concomitant cardiac surgery, the median sternotomy approach is used most often. Subxiphoid and subcostal approaches are reserved for patients who undergo ICD implantation without concomitant cardiac surgery. More recently, nonthoracotomy approaches for insertion of ICDs have been developed.[11] Because of the size of the ICD generator and weights ranging from 196 to 270 g, the generator is implanted in the left upper quadrant of the abdomen. Leads are then tunneled from the heart to this pocket site.

ICD IMPLANTATION INTRAOPERATIVE REQUIREMENTS

Personnel requirements for an ICD procedure include a cardiothoracic surgeon, a cardiac electrophysiologist, an anesthesiologist, a cardiac electrophysiology nurse or technician, and a biomedical technician or engineer. Equipment requirements include an external cardioverter/defibrillator, an external defibrillator, sterile hand-held extra-cardiac defibrillating paddles, fluoroscopy for implantation of the nonthoracotomy units, a manufacturer's specific pacing systems analyzer and programmer, a physiologic recorder, a programmable stimulator, and alternating current (AC) fibrillator.[12] Table 8-1 depicts our typical ICD implant procedure protocol. Table 8-2 lists acceptable pacing system analyzer measurements for ICD implants. It is important that the entire medical team understands that proper testing needs to be performed as part of the operation. Acceptable pacing thresholds, sensing of R waves, and DFTs are required for the device to function as prescribed after implantation.

TABLE 8-1. ICD IMPLANT PROCEDURE

1. Set up and check all equipment to be used during the ICD implant.
2. Implant the sense/pace and defibrillation lead systems.
3. Test the performance of all the leads.
4. Induce VT/VF and perform DFT testing.
5. Program the ICD.
6. Form the ICD generator pocket.
7. Tunnel the leads to the generator pocket.
8. Test the performance of all the leads after tunneling.
9. Connect the leads to the ICD.
10. Implant the ICD.
11. AC fibrillate and test sensing and shock function of the ICD.
12. Perform a final interrogation of the ICD postshock.
13. Inactivate the ICD during closure.
14. Reactivate the ICD soon after completion of the ICD implant.

TABLE 8-2. ACCEPTABLE PACING SYSTEM ANALYZER MEASUREMENTS AT ICD IMPLANT

Pacing/sense lead R wave	>5 mV
Pacing/sense lead slew rate	>0.75 V/sec
Rate sensing duration	<100 msec
Pacing threshold	<1.5 V (endocardial)
	<2.0 V (epicardial)
Pacing lead impedance	200–800 ohms
Patch impedance	20–100 ohms
Morphology lead R wave	>1 mV
Morphology lead duration	<150 msec

DEFIBRILLATION THRESHOLD DETERMINATION

The DFT is the least amount of energy in joules or volts delivered to the heart that reproducibly reverts ventricular fibrillation to normal rhythm. It differs from the cardioversion energy requirements, which is defined as the least amount of energy necessary to revert VT to sinus rhythm.[13] In order to determine the defibrillation threshold, ventricular fibrillation needs to be induced by programmed stimulation or an AC fibrillator. Once VF is induced, it is customary to wait 10 to 15 sec before delivering the precharged energy through an external cardioverter/defibrillator. This amount of time mimics the time the device requires to recognize the arrhythmia, charge the battery, and synchronize the shock. In general, 100% safety factors are desirable for measuring the defibrillation threshold; however, a 10-J safety factor is considered both practical and acceptable.[14] Since DFTs may be elevated by antiarrhythmic drugs,[15–17] it is desirable to test patients on the antiarrhythmic drugs that the physician plans to prescribe upon discharge. When DFTs are elevated, other physiologic or mechanical factors also need to be considered.[18] Anesthetic agents or intravenous antiarrhythmic drugs that may adversely affect the DFT should be avoided.

Currently available ICDs deliver 30 to 40 J of energy; thus, it is quite rare not to obtain an acceptable safety margin. In order to adequately determine the DFT, several inductions of ventricular fibrillation followed by the delivery of direct current shocks with predetermined energies need to be performed. Although this testing prolongs operative time, this gives the most accurate DFT measurement and ensures patient safety. If a shock is unsuccessful in reverting VF to normal sinus rhythm, a rescue shock of 40 J should quickly be delivered through the ECD box, and the external defibrillator should be charged and ready. If multiple shocks fail to revert the VF, a 360- to 400-J external shock or a 40- to 50-J internal paddle shock on the epicardium should be initiated quickly. If VF continues, oxygenation and appropriate acid-base balance should be maintained while advanced cardiac life support is instituted: epinephrine, sodium bicarbonate, and calcium chloride should all be considered for resuscitation. When the DFT is found to be satisfactory, the ICD generator is connected to the leads and patches and once more ventricular fibrillation is induced and converted. This is done to make sure that the entire implanted system senses the arrhythmia properly and functions as programmed (Fig. 8-2).

POSTOPERATIVE AND IN-HOSPITAL MANAGEMENT OF PATIENTS WITH ICDS

Following ICD implantation via thoracotomy, an intensive care unit stay of 24 to 72 hr is typical. Nonthoracotomy implantations usually require only intermediate care follow-up, since patients are usually extubated before they leave the recovery room. Perioperative mortality from thoracotomy insertion of ICDs averages 3%, and the most common causes are intractable ventricular arrhythmias, heart failure, acute myocardial infarction, and respiratory failure.[19] The usual care of arrhythmias, fluid management, respiratory care, and meticulous attention to wound care to avoid infection cannot be overemphasized. We usually activate the ICD device as soon as the patient leaves the operating room. As occurs with other thoracotomy procedures, paroxysms of rapid atrial fibrillation are common in the first few postoperative days and a ventricular response may be rapid enough to trigger the ICD device. Although early activation of the device may cause a few false discharges in patients with rapid atrial tachyarrhythmias, it terminates a life-threatening ventricular tachyarrhythmia more promptly than can paramedical personnel. Prior to discharge, a post-operative electrophysiology study is routinely performed to adequately test the device and make final changes in the prescription of the tachycardia detection interval and shock levels.

ANTIARRHYTHMIC DRUG-ICD INTERACTIONS

Forty to 70% of ICD patients receive concomitant therapy with antiarrhythmic drugs.[7,8,16,17,18] As stated earlier, antiarrhythmic drugs may alter the defibrillation threshold (Table 8-3).[11-13] In addition, antiarrhythmic drugs may slow the rate of ventricular tachycardia (Table 8-4).[16,17,19,20] Although slower tachycardia is easier to terminate with antitachycardia pacing techniques,[8] the rate of VT recurrences on drug treatment may be so slow as to fall below the tachycardia detection interval. Thus, repeat testing of an ICD is useful if an antiarrhythmic drug has been added to the treatment regimen.

EXPANSION OF ICD INDICATIONS

ICD implantation has been restricted to patients with preexisting sustained ventricular tachyarrhythmias. Since >250,000 people per year die suddenly out-of-hospital without successful resuscitation, there has been an interest in expanding these indications to patients with potentially lethal ventricular arrhythmias, such as patients with non-sustained VT associated with coronary artery disease and patients with significantly depressed ejection fraction undergoing bypass surgery. Several clinical trials are underway to determine if expansion of the indications for ICDs is beneficial.

TABLE 8-3. ANTIARRHYTHMIC DRUG EFFECTS ON DFTs

Increase	Mixed	Decrease
Lidocaine	Procainamide	Sotalol
Phenytoin	Bretylium	Clofilium
Mexiletine		N-Acetylprocainamide
Encainide		
Flecainide		
Propafenone		
Recainam		
Propranolol		
Verapamil		
Amiodarone		
Quinidine		
Disopyramide		
Moricizine		

TABLE 8-4. ANTIARRHYTHMIC DRUG EFFECTS ON PROLONGING VENTRICULAR TACHYCARDIA CYCLE LENGTH

Minimal	Moderate	Marked
Propranolol	Quinidine	Encainide
Mexiletine	Procainamide	Flecainide
Tocainide	Disopyramide	Propafenone
Sotalol	Moricizine	Amiodarone
		Classes IA + IB

The CABG PATCH trial[21] is a trial to test whether patients undergoing elective bypass surgery who have an ejection fraction of 36% or less and a positive signal-averaged electrocardiogram will benefit from ICD implantation. All patients will have bypass surgery, but half of the patients will also have an ICD implanted. Since the 2-year mortality in this patient group is 30%, and 40% of that subset die suddenly, the hypothesis is that the ICD may prolong survival in this patient group.

In patients with nonsustained ventricular tachycardia and coronary disease and low ejection fractions, the Multicenter Unsustained Tachycardia Trial and the Multicenter Automatic Defibrillator Trial are ongoing to see if aggressive therapy of patients with inducible sustained VT but a clinical history of nonsustained VT would benefit from therapies such as the ICD.[21] If these trials are successful in reducing mortality, the ICD indications will expand.

ICDS: WHAT IS THE FUTURE?

In the future, most ICDs will be implanted via a nonthoracotomy system. This will minimize mortality and morbidity and be associated with significant cost savings. Smaller devices that can be implanted through the pectoral approach and smarter ICDs with better arrhythmia algorithms and biosensors to assess hemodynamic stability will be developed. More sophisticated pacing systems, including AV sequential pacing systems with rate responsiveness, will eventually replace simple ventricular demand pacing systems currently available in present day tiered therapy devices. The development of transtelephonic interrogation and programming devices to minimize patient follow-up requirements are currently under development. In summary, the future appears bright. The above enhancements in ICD developments should make the prescription, implantation, and maintenance of such devices more attractive to physicians who feel their patients would benefit from such a therapy.

References

1. Mirowski M, Reid PR, Mower MM, et al. Termination of malignant ventricular arrhythmias with an implanted automatic defibrillator. N Engl J Med 1980:303:322.

2. Klein LS, Miles WM, Zipes DP. Antitachycardia devices: realities and promises. J Am Coll Cardiol 1991;18:1349.

3. Luceri RM. Implantable defibrillators: physician's role after hospital discharge. In: Naccarelli GV, Veltri EP, eds. Implantable cardioverter-defibrillators. Boston, Blackwell Scientific Publications, 1993:150.

4. Winkle RA, Mead RH, Ruder MA, et al. Improved low energy defibrillation efficacy in man with the use of a biphasic truncated exponential waveform. Am Heart J 1989;117:122.

5. Naccarelli GV, the P-2/Endotak Investigators. Superiority of biphasic waveforms wih the Ventak P2/Endotak ICD System [Abstract]. Circulation 1993;88:I-215.

6. Mirowski M, Reid PR, Winkle RA, et al. Mortality in patients with implanted automatic defibrillators. Ann Intern Med 1983;98:585.

7. Winkle RA, Mead RH, Ruder MA, et al. Long-term outcome with the automatic implantable cardioverter-defibrillator. J Am Coll Cardiol 1989; 13:1353.

8. Kelly PA, Cannom DS, Garan H, et al. The automatic implantable cardioverter-defibrillator: efficacy, complications and survival in patients with malignant ventricular arrhythmias. J Am Coll Cardiol 1988;11:1278.

9. Cannom DS. Current indications an contraindications for ICD implantation. In: Naccarelli GV, Veltri EP, eds. Implantable cardioverter-defibrillators. Boston, Blackwell Scientific Publications, 1993:264.

10. Watkins L, Taylor C. Surgical approaches of automatic implantable cardioverter-defibrillators implantation. PACE Pacing Clin Electrophysiol 1991;14:953.

11. Bardy GH, Hofer B, Johnson G, et al. Implantable transvenous cardioverter-defibrillators. Circulation 1993;87:1152.

12. Naccarelli GV, Veltri EP. ICD implantation: intraoperative requirements. In: Naccarelli GV, Veltri EP, eds. Implantable cardioverter-defibrillators. Boston, Blackwell Scientific Publications, 1993:536.

13. Rattes MF, Jones DL, Sharma AD, Klein GJ. Defibrillation threshold: a simple and quantitative estimate of the ability to defibrillate. PACE Pacing Clin Electrophysiol 1987;10:70.

14. Marchlinski FE, Flores B, Miller JM, Gottlieb CD, Hargrove WC. Relation of the intraoperative defibrillation threshold to successful postoperative defibrillation with an automatic implantable cardioverter-defibrillator. Am J Cardiol 1988;62:393.

15. Reiffel JA, Coromilas JM, Zimmerman JM. Drug-device interactions: clinical considerations. PACE Pacing Clin Electrophysiol 1985;8:369.
16. Echt DS, Black JN, Barbey JT, Coxe DR, Cato EL. Evaluation of antiarrhythmic drugs on defibrillation energy requirements in dogs: sodium channel block and action potential prolongation. Circulation 1989;79:1106.
17. Naccarelli GV. Serial electrophysiologic guides drug testing vs ICD implantation in the survival of sudden cardiac death. In: Luceri R, ed. Sudden cardiac death: strategies for the 1990s. Miami: Peritus Publishing Corp, 1992:57.
18. Epstein AE, Ellenbogen KA, Kirk K, et al. Clinical characteristics and outcome of patients with high defibrillation thresholds: A multicenter study. Circulation 1992;86:1206.
19. Lehmann MH, Steinman RT, Meissner MD. Operative mortality and morbidity with ICD therapy. In: Naccarelli GV, Veltri EP, eds. Implantable cardioverter-defibrillators. Boston, Blackwell Scientific Publications, 1993:102.
20. Naccarelli GV, Zipes DP, Rahilly GT, Heger JJ, Prystowsky EN. Influence of tachycardia cycle length and antiarrhythmic drugs on pacing termination and acceleration of ventricular tachycardia. Am Heart J 1993;105:1-5.
21. Bigger JT. Should defibrillators be implanted in high-risk patients without a previous sustained ventricular tachyarrhythmia? In: Naccarelli GV, Veltri EP, eds. Implantable cardioverter-defibrillators. Boston, Blackwell Scientific Publications, 1993:284.

Ld R. Herzog
Carl Lynch III

9

Arrhythmias Accompanying Cardiac Surgery

Arrhythmias are common in the perioperative period with both cardiac and noncardiac surgery. Those seen during noncardiac surgery and in those patients without cardiac disease are often relatively benign. They become significant when they compromise hemodynamics, alter a favorable myocardial oxygen balance, or predispose a patient to other life-threatening arrhythmias. For example, with an atrioventricular junctional rhythm, myocardial performance may be compromised by asynchronous atrial and ventricular contractions, particularly in conjunction with preexisting valvular or ventricular dysfunction. Ventricular diastolic filling may be impaired, resulting in reduced cardiac output and tissue perfusion. Arrhythmias, such as supraventricular tachyarrhythmia, which degrade myocardial oxygen balance, may have serious implications for patients with reduced myocardial oxygen supply, e.g., coronary artery disease. Additionally, sustained arrhythmias and their resultant hemodynamic and acid-base alterations may predispose the patient to other possibly life-threatening arrhythmias. It is also important to note that in some clinical settings, certain arrhythmias, are likely to deteriorate into more life-threatening arrhythmias. In the perioperative setting, new arrhythmias in a patient with or without a known history of arrhythmia can also be viewed as a valuable indicator signifying some anatomic,

Clinical Cardiac Electrophysiology: Perioperative Considerations
Edited by Carl Lynch III J.B. Lippincott Company, Philadelphia, PA ©1994

physiologic, or pharmacologic abnormality, which should prompt a reassessment of the patient's clinical state.

Arrhythmias that are benign in healthy patients may not be well tolerated in patients with cardiac disease or undergoing open heart surgery. Furthermore, during or after open heart surgery, alterations in cardiac rhythm may often have more profound clinical significance, possibly indicating ischemia or surgically induced deterioration in myocardial conduction. Attempts should be made to identify the cause of the rhythm disturbance whether anatomic, pharmacologic, or physiologic, in order to guide appropriate therapeutic interventions. While there has been a significant reduction in the risk of cardiac surgery in the past 20 years, arrhythmias after cardiac surgery, however, continue to be a major cause of postoperative mortality and morbidity, possibly delaying recovery and prolonging hospital stay.[1,2] Until their etiology and risk factors are more clearly delineated and predictive monitors become available (e.g., heart rate variability monitoring), electrocardiogram (ECG) monitoring remains the most important tool in allowing immediate diagnosis and treatment of arrhythmias encountered in the perioperative period.

PRE- AND INTRAOPERATIVE ARRHYTHMIAS

Studies evaluating arrhythmias associated with anesthesia and noncardiac surgery have shown an incidence between 60 and 80%.[3,4] Most arrhythmias seen in the noncardiac surgical population have a very low incidence of clinical significance. While many studies have evaluated the occurrence of arrhythmias following (postoperative) open heart surgery, fewer by comparison have systematically examined the occurrence of arrhythmias early in the perioperative period of cardiac surgery/anesthesia.[5,6] Two recent "complete" studies, in which the ECG was continuously monitored and recorded, determined the incidence of arrhythmias without regard to the particular anesthetic approach. Casey et al.[5] studied the prebypass incidence of arrhythmias in patients undergoing coronary artery bypass grafting (CABG). Arrhythmias, as defined in their study, were seen in all patients (100% incidence) during the prebypass period. Premature ventricular beats were seen in 92% of patients, nonsustained ventricular tachycardia in 76%, supraventricular arrhythmias in 75%, sinus bradycardia (<60 beats/min) in 39%, and other conduction abnormalities in 20% of patients. Such arrhythmias were most frequently associated with pulmonary artery catheter insertion and intraoperative aortic dissection. Also of interest, this study found that the preoperative use of calcium channel blockers decreased the incidence of supraventricular arrhythmias associated with pulmonary artery catheter placement.

Dewar et al.[6] studied the pre-, intra-, and early postoperative incidence of arrhythmias associated with cardiac surgery. Holter monitoring was used to continuously evaluate patients undergoing CABG, valve replacement, or a combined procedure. This study looked at the occurrence of a variety of arrhythmias including premature ventricular beats, second and third degree atrioventricular block, ventricular tachycardia, ventricular fibrillation, supraventricular tachycardia, and sinus tachycardia (>130 beats/min). Arrhythmias were studied in terms of their "temporal occurrence": phase I was the 1-hour period prior to surgery and lasted until the aortic cross-clamp was placed; phase II was the "reperfusion" period; and phase III was the period from the discontinuation of bypass into the early postoperative period. Overall, patients undergoing valve replacement had the highest incidence of arrhythmias with the most dramatic difference noted during phase I when arrhythmias were noted in 100% of patients compared to a 38% incidence in patients undergoing CABG.

Frequently patients with cardiac disease have preexisting cardiac conduction or rhythm abnormalities, which may be influenced by the placement of pulmonary artery (PA) catheters commonly used for intra- and postoperative monitoring of cardiac function. A PA catheter with pacing capabilities is sometimes used prophylactically, especially in patients with left bundle branch block (BBB), who are felt to be at risk for complete heart block if right BBB were to develop during passage of the PA catheter.[7,8] In a study of 600 patients, Risk et al.[9] found that only 34 of 180 patients who received pacing PA catheters required pacing, while 4 of the 420 other patients (nonpacing PA catheters) required pacing prior to cardiopulmonary bypass (CPB). Of note, none of the 41 patients with left BBB developed complete heart block. The authors suggest that the increased cost for routine placement of a pacing PA catheter is not justified. Nevertheless, the overall incidence of 6.3% of patients requiring pre-CPB pacing suggests that pacing capability should be available. In such settings, noninvasive pacing may be quickly established using transcutaneous pacing[10] or possibly esophageal pacing.[11] The former may be used even in the presence of the patch electrodes used for automatic implantable cardiac defibrillators.[12]

A variety of anesthetic techniques have been used successfully for all types of cardiac surgery. Few studies have evaluated anesthetic technique as it relates to outcome after cardiac surgery. Tuman et al.[13,14] recently found that anesthetic technique/choice seemed to have little effect on outcome after coronary bypass or valvular cardiac surgery. Many of the agents/techniques used in the delivery of cardiac anesthesia and its adjunct care may be factors associated with the development of perioperative arrhythmias. Volatile inhalation agents have been classically described as dysrhythmogenic with regard to their cardiac-sensitizing properties, i.e., multiple studies have shown that inhalation agents reduce the amount of epinephrine needed to produce ventricular arrhythmias.[15] Inhalational agents, specifically halothane and enflurane, have also been shown to depress sinoatrial (SA) node automaticity and prolong atrioventricular (AV) nodal conduction.[16,17] When using these inhalation agents, especially halothane, it is not uncommon for a patient to develop AV dissociation, with a junctional pacemaker controlling heart rate. Although the exact mechanism of this rhythm disturbance remains controversial, it may be secondary to the electrophysiologic effects of the inhalation agent in combination with the underlying autonomic tone associated with surgery.[18] Enflurane, in combination with calcium channel-blocking agents, seems particularly prone to induce complete heart block, both clinically and experimentally.[19,20] Isoflurane has been shown to be less likely than halothane to potentiate catecholamine-induced ventricular arrhythmias.

Despite its widespread clinical use, the electrophysiologic effects of nitrous oxide have not been studied extensively. Nitrous oxide has been shown to increase sympathetic tone and therefore may play a role in the development of intraoperative arrhythmias, especially when used in combination with other anesthetic agents. Nitrous oxide has had limited use in cardiac anesthesia, given its solubility properties, which can cause enlargement of intravascular air emboli.

Narcotics are used by many as the foundation of their cardiac anesthetic technique and may have significant implications with regard to cardiac arrhythmias. Opiates have been shown to slow sinus node discharge and therefore decrease heart rate, at times leading to bradycardia.[21,22] Some reports have arguably indicated that opiates (fentanyl) depresses AV nodal conduction as well as AV nodal and ventricular refractoriness.[23] These actions probably arise from alterations in central nervous system pathways, leading to a decrease in sympathetic outflow. Additionally, narcotics have been shown to increase the ventricular fibrillation threshold[24] and may therefore be advantageous in patients with underlying cardiac arrhythmias undergoing cardiac surgery.

Certain muscle relaxants in common use have also been implicated in the occurrence of cardiac arrhythmias. Succinylcholine administration has been implicated in the development of a variety of rhythm disturbances including bradycardia and tachyarrhythmias. Succinylcholine stimulates all cholinergic receptors (including nicotinic and muscarinic receptors), and for this reason its effects on cardiac rhythm may be quite variable. A common clinical occurrence is sinus bradycardia or transient asystole following a repeated dose of succinylcholine in adults.[25] Given the relative imbalance between parasympathetic and sympathetic tone in children/infants, these rhythm disturbances may occur after a single dose of succinylcholine when used in this patient population.[26] In addition, it is well known that succinylcholine may affect cardiac rhythm secondarily to exaggerated release of potassium in certain clinical situations (e.g., burns, neuromuscular disorders, or multiple trauma).

Most of the commonly used nondepolarizing muscle relaxants are not directly dysrhythmogenic, although their effects on the autonomic nervous system and activation of histamine release may predispose the patient to rhythm disturbances. Histamine release, like β-adrenergic agents, stimulates cyclic AMP production, Ca^{2+} entry, and the inotropic cascade.[27,28] Pancuronium is well known to increase heart rate and shorten AV nodal conduction time,[29] probably by the combined effect of blocking muscarinic receptors,[30] as well as increasing sympathetic tone by permitting a greater release of catecholamines. Its sympathomimetic effects may lead to increased myocardial oxygen consumption and ischemia. Pancuronium has also been implicated in the development of ventricular arrhythmias when used in combination with halothane and tricyclic antidepressants.[31] Vecuronium, which has few if any hemodynamic side effects, has been reported to cause severe bradycardia and even asystole when used in combination with high-dose narcotics, which frequently is the case during cardiac anesthesia.[32,33] The mechanism for this finding remains unclear although it appears to be associated with clinical situations in which there is increased parasympathetic (vagal) tone.[34] Obviously, any anesthetic agent or adjunct that modulates sympathoadrenal and/or vagal tone, such as a muscle relaxant reversal agent[35] or naloxone,[36] may also have the potential to induce or aggravate arrhythmias.

ATRIAL ARRHYTHMIAS FOLLOWING CARDIAC SURGERY

Atrial tachyarrhythmias (atrial fibrillation, atrial flutter, and atrial tachycardia) are the most common arrhythmias seen following cardiac surgery. Of these atrial tachyarrhythmias atrial fibrillation is most common.[37,38] The clinical implications of atrial arrhythmias vary tremendously. Atrial arrhythmias may have significant hemodynamic consequences, when they occur during the vulnerable early postoperative period.

Factors that affect how well atrial arrhythmias are tolerated include: the associated ventricular rate, the degree of underlying cardiac dysfunction, and the duration of the arrhythmia. A reduced ventricular filling time, in combination with an increased myocardial oxygen consumption, may have significant clinical consequences in patients with compromised cardiac function. For example, during or after coronary artery bypass with partial or incomplete revascularization, rapid atrial fibrillation may be poorly tolerated. The hemodynamic consequences of atrial arrhythmias may be exaggerated in patients with valvular disease, hypertension, or cardiomyopathy, and early treatment of the arrhythmia and prevention of its recurrence are of upmost importance.

Although atrial tachyarrhythmias may occur at any time after surgery, most occur between postoperative days 3 and 5 following CABG.[39] Besides their hemodynamic consequences, atrial arrhythmias may cause additional morbidity through associated embolic phenomenon and the possibility of sustained or recurrent atrial arrhythmias.[40]

Incidence

The incidence of atrial arrhythmias following cardiac surgery varies tremendously with reported incidences ranging from 11 to 100%.[37,41–43] Variability in reported incidence is associated with variations in the definition of arrhythmia and type of cardiac surgery performed, as well as various modes and time intervals of monitoring. Vecht et al.[43] reviewed 16 published studies of atrial tachyarrhythmias after coronary bypass surgery and reported a mean incidence of 33.4%. White et al.,[42] using 24-hour Holter monitoring for the first 7 postoperative days following coronary bypass, reported that 100% of patients had at least one episode of atrial tachyarrhythmia although most did not have detrimental clinical consequences and most were not sustained.

The incidence of atrial arrhythmias is higher in patients undergoing valvular surgery. Approximately 60% of patients undergoing cardiac valvular surgery experience postoperative atrial arrhythmias,[44–46] with increased age increasing the incidence.[43,45] The incidence may exceed 60% in patients undergoing valvular surgery in combination with revascularization.

Risk Factors

The etiology and risk factors associated with atrial arrhythmias following cardiac surgery are wide ranging (Table 9-1) and are still being debated.[38,41,42,46–51] It is known that the stress response associated with cardiac surgery leads to elevated levels of catecholamines in the post-

TABLE 9-1. FACTORS IMPLICATED IN THE DEVELOPMENT OF ATRIAL ARRHYTHMIAS AFTER CARDIAC SURGERY

History of hypertension[38]
Previous cardiac surgery[46]
Increasing age[47]
Type of surgery[41]
History of myocardial infarction[41]
Number of grafts performed during CABG[38]
Withdrawal of β-blockers[42]
Operative ischemic time[48]
Inadequate atrial preservation/intraoperative activity[49–51]

operative period.[52] These circulating catecholamines are felt to play a significant role in the development of arrhythmias following cardiac surgery. Evidence supporting this includes multiple studies, which have shown that the incidence of postoperative arrhythmias can be significantly reduced with the administration of β-blockers in the postoperative period.[38,43,53,54]

It is generally accepted that local trauma to myocardial tissue during cardiac surgery is an important factor associated with the development of intraoperative as well as postoperative arrhythmias. Surgical requirements such as ventriculotomy and atriotomy may lead to disruption of normal function in the cardiac conduction system, including the SA and AV nodes. The increased surgical trauma associated with valvular surgery in combination with frequent preexisting atrial distention is felt to be the reason for the increased incidence of atrial arrhythmias following valve repair.[41] During CPB, electrical activity frequently persists in the atria[50] in spite of generalized cardioplegia and evidence of electrical quiescence of the ventricle. Failure of atrial preservation has also been demonstrated with regard to atrial mechanical activity[55] and microscopically with regard to mitochondrial integrity.[56] Ongoing ischemic electrical activity in the atria during CPB appears to correlate with the postoperative appearance of supraventricular tachyarrhythmias.[49] In addition, posttraumatic (postoperative) sequelae such as pericardial or mediastinal inflammation may play a role in the development of arrhythmias.[44]

Systemic abnormalities, which are frequently seen in the postoperative period following cardiac surgery, may also play a role in the development of atrial arrhythmias. Factors such as electrolyte or acid-base disturbances, hemodynamic abnormalities, hypoxia, or hypercarbia alone or in combination may play a role in the development and persistence of atrial arrhythmias. Given the frequency with which these abnormalities occur following cardiac surgery, preventive measures should be taken where possible, and correction of the abnormality should be apart of the initial treatment regimen.

Prophylaxis

Given the potential morbidity and mortality associated with postoperative atrial arrhythmias following cardiac surgery, multiple studies dies in the recent past have attempted to identify an agent that might be effective in decreasing their occurrence. The majority of these studies have used coronary bypass surgery as the model with few looking separately at prophylaxis for other types of cardiac surgery. It is likely that various types of patients after cardiac surgery share similarities in etiology of postoperative atrial arrhythmias. Given these similarities, those studies looking at various prophylactic regimens may provide insight into postoperative atrial arrhythmias in general and will be discussed as such.

Digoxin and β-blockers as prophylactic agents have been studied most extensively. Given the various investigational designs, these studies have at times been difficult to compare and evaluate. Tables 9-2 and 9-3 present recent studies evaluating β-blockers and digoxin, respectively, as prophylactic agents following cardiac surgery. The use of β-blockers as prophylaxis for supraventricular arrhythmias has been shown to be both safe and effective. It is well known that withdrawal of β-blockers leads to exaggerated sympathetic responses (including arrhythmias), felt to be secondary to up-regulation or increased sensitivity of the β-receptors. For this reason, most clinicians recommend continuation of treatment with β-blockers throughout the perioperative period. β-Blockers are felt to attenuate the cardiac arrhythmic effects of increased catecholamines known to be present in the postoperative period without significantly compromising ventricular function. β-Blockers have therefore become the most accepted choice for prophylaxis of supraventricular arrhythmias following cardiac surgery, although most clinicians do not advocate their routine use. Obviously, when adrenergic agents are used immediately following cardiopulmonary bypass to provide inotropic stimulation, the benefit of β-adrenergic blockade may be lost.

The use of digoxin as a prophylactic agent remains controversial. As shown in Table 9-3, conflicting results have been found in multiple studies with some studies demonstrating an increased incidence of supraventricular arrhythmias with the use of digoxin.[61] Additionally, some authors have cautioned against the routine use of digoxin as a prophylactic agent given its potential for toxicity and the likelihood of increased sensitivity to digoxin in the postoperative period.[62] Other

TABLE 9-2. EFFECTIVENESS OF β-ADRENERGIC BLOCKADE IN
PREVENTING ATRIAL ARRHYTHMIAS AFTER CARDIAC SURGERY

Study	Agent	n	Atrial Arrhythmia Incidence (%)		P
			Placebo	Drug	
Stephenson et al., 1980[54]	Propranolol, 10 mg qid 18 hr postoperatively	223	18	8	P < 0.05
Ivey et al., 1983[57]	Propranolol, 80 mg/day 24 hr postoperatively	116	16	13	NS
White et al., 1984[42]	Timolol, 0.5 mg iv bid then 10 mg po bid postoperatively	41	581 events	84 events	P < 0.05
Matangi et al., 1985[58]	Propranolol, 5 mg po qid postoperatively	164	23	9.8	P < 0.02
Daudon et al., 1986[59]	Acebutolol, 200–400 mg 36 hr postoperatively	100	40	0	P < 0.001
Vecht et al., 1986[43]	Timolol, 5 mg bid 24 hr, 10 mg bid postoperatively	132	20	8	P < 0.05
Lamb et al., 1988[53]	Atenolol, 50 mg/day 72 hr postoperatively	60	37	3	P < 0.001
Matangi et al., 1989[60]	Atenolol, 5 mg iv for 2 doses then 50 mg po qid postoperatively	70	938 mild 142 moderate 21 severe events	982 mild 30 moderate 5 severe events	NS P < 0.0005 P < 0.0005

authors also caution against its use given the multiple additional abnormalities such as electrolyte disturbances, hypoxia, and compromised renal function that may be present in the postoperative/postbypass period predisposing the patient to digoxin toxicity.[63] Given these continued controversies most authorities do not advocate the use of digoxin as a prophylactic agent for supraventricular arrhythmias following cardiac surgery.

Prophylactic regimens combining β-blockers and digoxin have also been studied and seem to demonstrate some success.[47,64] Given the study methods it remains unclear as to whether the decreased incidence of postoperative supraventricular arrhythmias is secondary to the combination of drugs or merely because of the known benefits of continued β-blocker administration in the perioperative period. Verapamil has also been studied as a prophylactic agent and has demonstrated no significant benefit.[39,65,66]

TABLE 9-3. EFFECTIVENESS OF DIGOXIN IN PREVENTING ATRIAL
ARRHYTHMIAS AFTER CARDIAC SURGERY

Study	Dosing Regimen	Atrial Arrhythmia Incidence (%)		P
		Placebo	Digoxin	
Johnson et al., 1976[108]	1 mg po 2–3 days preoperatively, continued postoperatively	26	6	Decrease: P < 0.01
Tyras et al., 1979[61]	1–1.5 mg po 1 day preoperatively	11	31 (1 mg) 21 (1.5 mg)	Increase: P < 0.05
Csicsko et al., 1981[109]	1 mg iv 4–12 hr postoperatively	15	2	Decrease: P < 0.01
Chee et al., 1982[41]	0.75 mg po 1 day preoperatively	72	5	Decrease: P<0.01
Weiner et al. 1986[110]	1 mg iv 6–12 hrs postoperatively	16	15	NS

VENTRICULAR ARRHYTHMIAS FOLLOWING CARDIAC SURGERY

Ventricular rhythm disturbances can have clinical manifestations as well as electrocardiographic findings similar to atrial arrhythmias. Given these similarities the diagnosis of ventricular arrhythmias after cardiac surgery may be difficult. Temporary epicardial wires, which many clinicians routinely use, can be extremely helpful in the diagnosis as well as treatment of these arrhythmias. Ventricular arrhythmias, as with atrial arrhythmias, have a high incidence of occurrence preoperatively as well as intraoperatively in cardiac surgical patients.[5,6] Placement of invasive monitors, induction of anesthesia, aortic dissection, cannulation for cardiopulmonary bypass, rewarming, and weaning from bypass have all been associated with increased pre- and intraoperative ventricular arrhythmias.

Although all types of cardiac surgery and the associated occurrence of postoperative arrhythmias have not been systematically studied, some general information can be obtained by evaluating various general types of cardiac surgery (revascularization, valve surgery, etc.) and their potential associated ventricular rhythm disturbances. Many authors classify ventricular arrhythmias into three categories: benign ((unifocal and multifocal premature ventricular complexes (PVCs)); potentially lethal (nonsustained ventricular tachycardia); and lethal (sustained ventricular tachycardia, ventricular fibrillation, and torsade de pointes). Although this classification may be useful, a particular ventricular arrhythmia may affect individual patients differently. Ventricular arrhythmias following cardiac surgery may therefore cause increased morbidity and mortality, which will be dependent on the type of cardiac surgery performed as well as on the patient's underlying cardiac disease state.

Incidence

Studies investigating ventricular rhythm disturbances have examined a wide variety of parameters defining the arrhythmia and used a wide variety of methodologies. As one would expect, the incidence of these ventricular arrhythmias is variable, depending on the type of cardiac surgery performed. Few investigators have systematically evaluated patients throughout the perioperative period to determine the incidence of ventricular arrhythmias following cardiac surgery. Many authors/investigators have empirically classified ventricular arrhythmias as "early" (occurring within a few days of the operation) and "late" (occurring weeks to months following surgery). PVCs frequently occur early after cardiac surgery. The reported incidence of PVCs following coronary bypass is higher than that seen after valvular surgery. Michelson et al.[44] found that 36% of patients undergoing coronary bypass and 20% of those undergoing valvular surgery developed PVCs on Holter monitoring before hospital discharge. PVCs were new in 28% of the bypass patients and 13% of the valve surgery patients. Rubin et al.[67] found a 57% incidence of PVCs documented by Holter monitor following coronary bypass. Hoie et al.[68] found only a 4% incidence of PVCs in patients after aortic valve replacement reinforcing the point that, in general, ventricular arrhythmias (PVCs) appear to be less frequent following valve surgery compared to coronary bypass. In contrast to early ventricular ectopy, valve surgery has, however, been shown to be associated with a high incidence ventricular ectopy late after surgery. Gradman et al.[69] reported an 89% incidence of frequent or complex ventricular ectopy in 45 patients who underwent aortic valve replacement 6 weeks to 10 years prior to Holter monitoring.

Many authors/investigators divide ventricular tachycardia (VT) into nonsustained and sustained, sustained VT being defined as VT lasting >30 seconds or resulting in hemodynamic collapse. Even more ominous is the frequent deterioration of VT into ventricular fibrillation (VF). The incidence of sustained VT early after CABG is low, occurring in about 1% of patients[63,70] with a somewhat higher incidence of nonsustained VT after CABG, ranging from 5 to 36%.[43,44] The reported incidence of VT (sustained or nonsustained) early after valve surgery is lower than that seen following CABG. Hoie et al.[68] studied

44 patients undergoing valve surgery and found a 0% incidence of VT early after surgery, although they noted one patient who experienced VF that was felt to be secondary to an extracardiac complication. Other investigators have found a similarly low incidence of VT early after valve surgery.[44]

Ventricular arrhythmias (VT or VF) occurring late after cardiac surgery are important in the reported long-term morbidity and mortality, with ventricular arrhythmias presumed to account for the majority of sudden deaths following valve surgery. Nonsustained VT has a reported incidence of 35% late following aortic valve replacement.[69] The presence of sustained VT or VF late (within the first year) after valve surgery accounts for 24% of the annual deaths seen in this group.[71] Foppl et al.[71] found a mortality rate of 6.9% within the first 4 weeks of aortic valve replacement and an annual mortality of 3.6%.[71] VT or VF/sudden death has also been reported to have a significantly high incidence following left ventricular aneurysmectomy. Cooperman et al.[72] found an 11% incidence of sudden death within 20 weeks following aneurysmectomy. In contrast, new sustained VT occurred on average 27 days following CABG in only 0.7% of patients.[73]

Torsade de pointes can be seen following cardiac surgery, although given its low incidence of occurrence, it has not been extensively studied. Its occurrence is usually associated with congenital or acquired long QT syndrome. The specific causes of acquired long QT syndrome are not well understood although implicated factors include antiarrhythmic agents as well as other medications and electrolyte abnormalities. The important clinical significance of this arrhythmia is its frequent deterioration to VF.

Risk Factors

As with supraventricular arrhythmias, similar factors have been implicated in the development of postoperative ventricular arrhythmias. Factors relating to cardiopulmonary bypass and myocardial preservation, surgical trauma to the myocardium or a portion of its conduction system, as well as alterations hemodynamics, may all contribute to the development of postoperative ventricular arrhythmias. Although patients with poor myocardial preservation are likely to experience at least temporary contractile dysfunction postoperatively, it is not clear whether any concomitant ventricular arrhythmias arise solely from the transiently ischemic ("stunned") myocardium or are related to the accompanying inotropic agents frequently used. Arrhythmias may also develop secondary to inadequate coronary perfusion and for that reason may have major clinical significance following myocardial revascularization surgery. The endocardium and epicardium are subject to foreign body mechanical irritation from a variety of sources including mediastinal drainage tubes or prosthetic valves, which may also contribute to perioperative arrhythmias. In addition, perioperative changes in oxygenation, temperature, ventilation, electrolyte status, or addition of a dysrhythmogenic medication may cause or contribute to the development of these rhythm disturbances.

In addition to the above, there are other risk factors that have been shown to be associated with an increased incidence of ventricular arrhythmias following cardiac surgery. Some investigators have reported that ventricular ectopy, as one might expect, is seen more frequently in patients who exhibit the disturbance preoperatively.[44] Other investigators, however, have found no association between pre- and postoperative frequency or complexity of ventricular ectopy following some types of cardiac surgery.[74] Gradman et al.[69] found that preoperative ventricular function (ejection fraction) was an important factor associated with the development of ventricular arrhythmias following aortic valve replacement, noting that patients with an ejection fraction less than 50% had a significantly higher incidence of ventricular arrhythmias including PVC couplets, and VT. Ventricular function is most likely also a risk factor following CABG, although at this time it has not been studied. It has also been shown that postoperative evidence of ischemia, as one might expect, is associated with an increased incidence of ventricular arrhythmias.[67]

Additional systematic studies are still necessary to better define risk factors for the development of ventricular arrhythmias following various types of cardiac surgery. A variety of diagnostic tests have been used following cardiac surgery in hopes of establishing a method to determine the risk of ventricular arrhythmias. Twenty-four-hour Holter monitoring, exercise testing with or without thalium, signal-averaged ECG, and ventricular programmed electrical stimulation have been studied; each method seems to show some promise given a particular clinical situation.

CONDUCTION DISTURBANCES FOLLOWING CARDIAC SURGERY

SA, AV, and intraventricular conduction disturbances are also common following all types of cardiac surgery, with an overall incidence of approximately 22%.[75] The incidence of BBB early in the postoperative (CABG) period has been reported to be as high as 45%, with complete AV block occurring in 4% of cases.[76] Chu et al.[77] reported on the incidence of commonly seen conduction abnormalities following coronary bypass surgery, including right BBB (seen transiently in 60% of cases and persistently in 29%), left anterior hemiblock (26% transient, 33% persistent), and incomplete right BBB (9% transient, 26% persistent). The majority of these conduction disturbances seen following coronary bypass have been found to spontaneously resolve within the first 2 months of surgery. Transient BBBs are frequently seen with washout of hypothermic cardioplegia following cardiopulmonary bypass.[78] This phenomenon has been shown, at least in part, to be related to the concentration of potassium in the cardioplegia solution.[79,80]

Atrioventricular nodal as well as intraventricular conduction disturbances are also commonly seen following valvular surgery, with a reported incidence of 29% following aortic valve replacement.[81] Gannon et al.[82] reported a 13% incidence of complete heart block following aortic valve replacement. Interestingly, other studies have documented improvement or eventual resolution of AV conduction disturbances following aortic valve replacement.[81] Few studies have been preformed looking at these types of conduction disturbances following other types of cardiac valve surgery.

The development of these conduction abnormalities following cardiac surgery are associated with factors similar to those causing atrial or ventricular arrhythmias. Systemic factors such as electrolyte disturbances, hypoxia, or hemodynamic abnormalities have been implicated. In addition, factors associated with cardiopulmonary bypass, such as aortic cross-clamp time and duration of hypothermia, are also felt to be important with regard to the postoperative development of conduction abnormalities.[76,79] An important factor leading to the development of conduction abnormalities following valve surgery is the proximity of the surgical field to the conduction system. This "danger zone," which frequently suffers surgical trauma or develops edema following valve surgery may lead to altered conduction. Calcium deposits requiring surgical debridement in this danger zone or suture placement in this zone have been shown to result in an increased incidence of heartblock postoperatively.[81]

ARRHYTHMIAS FOLLOWING SURGICAL REPAIR OF CONGENITAL HEART DEFECTS

Patients, usually infants or children, who undergo surgical repair of congenital heart defects frequently experience postoperative arrhythmias. Given the physiologic and anatomic complexities surrounding the various types of congenital heart defects, as well as the number of surgical techniques that are used in their repair, it is not surprising that such surgery is associated with a wide variety of arrhythmias. Surgical repair of specific congenital heart defects appears to place patients at risk for the development of specific arrhythmias. Postoperative arrhythmias are many times the direct result of surgical trauma to the conduction system, although they may also be associated with sequelae such as myocardial scarring, hypertrophy, ischemia, or cyanosis. Table 9-4 lists the types of arrhythmias most frequently seen following specific types of surgery.

Sinus node dysfunction occurs frequently following surgical procedures involving the atria. Rhythms such as sinus bradycardia, sinoatrial block, ectopic atrial rhythms, AV junctional escape rhythm, and sinus arrest may occur following injuries to the sinus node.[86] These rhythm disturbances are frequently seen following surgical correction of transposition of the great vessels, tetralogy of Fallot, and atrial septal defects, as well as in other types of congenital heart defects.[87,88] These rhythms may be transient, as with cannulation of the superior vena cava, or they may be persistent and even progressive if the injury to the sinus node is secondary to direct surgical trauma, interruption of its blood supply, or postoperative scarring/fibrosis.

The AV node and the His bundle conducting tract may be injured during surgical procedures, which are in the proximity of these structures. Injuries to these specialized structures of the cardiac conduction system may lead to a variety of rhythm disturbances including AV junctional tachycardia and varying degrees of AV block, as well as intraventricular conduction defects. As with other injuries to the cardiac conduction system, surgical trauma leading to direct tissue damage, ischemia, or postoperative fibrosis/scarring can all contribute to the development of these rhythm disturbances. As noted in Table 9-4, AV conduction defects and specifically complete AV block are seen following a variety of surgical procedures for congenital heart defects. Evidence of AV nodal injury may not be clinically noted until years after the surgical repair.[89] Patients who have undergone surgical repair of a congenital heart defect and have a high-degree AV block have an increased mortality and therefore permanent pacing is a frequently used treatment.[90] Intraventricular conduction defects, such as bundle branch blocks and fascicular blocks, are also common sequelae of surgical repair of congenital heart defects. Right bundle branch block occurs in most patients following surgical repair of tetralogy of Fallot[91] and is a common finding following repair of a atrioventricular canal. A right bundle branch block in combination with a left anterior hemiblock is also a frequent finding following repair of tetralogy of Fallot.[92] These intraventricular conduction defects (especially right BBB with a left anterior hemiblock) may place the patient at increased risk of late development of complete heart block.

246

TABLE 9-4. ARRHYTHMIAS FOLLOWING SURGICAL REPAIR OF CONGENITAL HEART DEFECTS

Congenital Defect	Overall (%)	Conduction Abnormalities	%	Arrhythmias	%	Sudden Death (%)
Tetralogy of Fallot	30–60	RBBB RBBB and LAH CHB	59–100 7–25 1–2	PVC's VT	67 10–15	2–5
Transposition of great arteries (Mustard or Senning repair)*	50–85	AV junctional rhythm AV blocks	42 1	Sinus bradycardia Atrial flutter SVT VT	50 25 7 1	2–8
Functional single ventricle (Fontan repair)	25–40	CHB AV junctional rhythm	9.5 8	SVT Ventricular arrhythmias	23–42 7.5	3
Aortic and subaortic stenosis repair	10			Ventricular arrhythmias commonly seen		
Ventricular septal defect	10	CHB RBBB frequently seen	1–2			
AV canal defect	10	CHB RBBB and AV conduction defects also frequently seen	7			
Atrial septal defect	9			Sinus bradycardia, ectopic atrial rhythm, AF/F, frequently seen		
Anomalous pulmonary venous return	9			Sick sinus syndrome and SVT frequent		
Ebstein's anomaly				Ventricular	13	

Data from Krongrad,[83] Vetter,[84] and Oh et al.[85] AV, atrioventricular; AF/F, atrial fibrillation/flutter; CHB, complete heart block; SVT, supraventricular tachycardia; VT, ventricular tachycardia; RBBB, right bundle branch block; LAH, left anterior hemiblock; PVCs, premature ventricular complexes.

*The long-term effects seen with the arterial switch repair of the great arteries is not yet clear. Preliminary results at the University of Virginia (Summer et al. University of Virginia Division of Pediatric Cardiology, unpublished results) have found no arrhythmias following arterial switch (n = 26) compared to a 19% incidence following a Senning procedure (n = 16).

While repair of simple congenital cardiac abnormalities in children (e.g., ventricular or atrial septal defect) seems to have a low incidence of ventricular rhythm disturbances, ventricular arrhythmias are commonly seen following surgical repair requiring ventriculotomy or any direct surgical trauma to the ventricles (Table 9-4). This myocardial damage and subsequent scarring are felt to serve as a focus for the development of reentrant arrhythmias. The incidence of ventricular arrhythmias in this patient population appears to be associated with the extent of ventriculotomy and damage to the ventricular conduction system. Additionally, some clinicians feel that a preoperative history of ventricular arrhythmias or elevated ventricular pressures may predispose the patient to the development of postoperative ventricular rhythm disturbances. Following repair of tetralogy of Fallot, ventricular ectopy, trifascicular heart block, and an elevated right ventricular pressure were found to be risk factors for the development of ventricular arrhythmias.[93,94] Ventricular arrhythmias following surgical repair of tetralogy of Fallot may have significant implications with regard to a patient's long-term mortality, given that late sudden death thought to be secondary to ventricular arrhythmias was reported in 5 to 30% of patients who underwent repair of tetrology of Fallot.[93,95] Blake et al.[96] reported a 9% incidence of sudden death, presumed to be secondary to ventricular arrhythmias, up to 21 years following surgical repair of a ventricular septal defect.

EPICARDIAL ELECTRODES: DIAGNOSTIC AND THERAPEUTIC APPLICATIONS

Following completion of the surgical repair, placement of ventricular and/or atrial epicardial pacing wires is performed by many surgeons. Such intraoperatively placed epicardial electrodes have been shown to be beneficial in the diagnosis and/or treatment of arrhythmias following cardiac surgery in 81% of patients.[40] For this reason many clinicians consider their placement to be an important part of the postoperative care of the cardiac surgical patient, although their use is only one component in the overall approach to arrhythmias following cardiac surgery. Two bipolar electrodes are usually placed, one on the right atrium and one on the right ventricle.[97] These electrodes may be helpful in the diagnosis of arrhythmias by allowing direct recordings (electrograms), which may further define or clarify atrial and ventricular depolarizations and their relationship to each other.[97-99] In addition, these electrodes may be used for pacing, which may further assist in diagnosis, or be used as a treatment modality for certain types of arrhythmias.

Electrograms are usually obtained by connecting the temporary atrial electrodes to the right and left arm leads on a standard ECG. When lead I (bipolar, between right and left arm) is recorded, a bipolar atrial electrogram is obtained, showing little, if any, ventricular electrical activity. Recordings made using leads II and III will provide a unipolar atrial electrogram recording both atrial and ventricular electrical activity. These atrial electrograms (bipolar and unipolar) as well as a standard surface ECG will provide valuable information in the diagnosis of certain arrhythmias following cardiac surgery (Figs. 9-1 and 9-2). An alternative to such epicardial electrodes may be the use of atrial electrical activity recorded through the fluid column of a central venous catheter (or the right atrial division of a PA catheter). Donovan et al.[100] reported this technique to be helpful in accurately diagnosing the etiology of arrhythmias in 84% of 57 patients studied.

Pacing with temporary epicardial electrodes following cardiac surgery has been shown to be an effective treatment for certain types of atrial as well as ventricular arrhythmias. In addition, epicardial pacing may have certain advantages over other modes of treatment in that it is immediate, painless to the patient, and avoids potential complications associated with pharmacologic therapy. Overdrive atrial pacing has been effectively used to treat various types of atrial arrhythmias including atrial flutter and some forms of supraventricular tachycardia (Fig. 9-3).[101,102] Atrial fibrillation and sinus tachycardia are not effectively treated by pacing.

Many ventricular arrhythmias occurring postoperatively may also be effectively treated with ventricular as well as atrial pacing

Atrial Electrograms

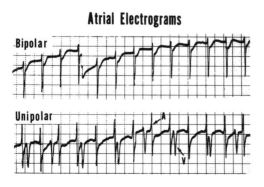

FIGURE 9-1. Bipolar and unipolar atrial electrograms recorded sequentially from the same patient. The bipolar atrial electrogram clearly defines the presence of regular atrial activity at 148 beats/min, and the unipolar atrial electrogram demonstrates the A-V relationship. Thus, this rhythm is an ectopic atrial tachycardia with variable A-V conduction. A, atrial complex; V, ventricular complex. (Reproduced with permission from Waldo AL, MacLean WAH: The diagnosis and treatment of cardiac arrhythmias following open heart surgery: emphasis on the use of epicardial wire electrodes. Mt. Kisco, NY: Futura Publishing Company, Inc., 1980, p. 69.)

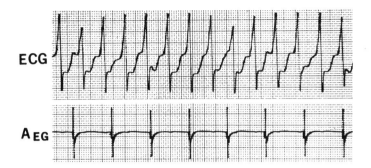

FIGURE 9-2. ECG lead recorded simultaneously with bipolar atrial electrogram (A_{EG}) during wide QRS complex tachycardia at 155 beats/min. The atrial electrogram demonstrates the presence of sinus rhythm at 90 beats/min. The demonstration of A-V dissociation with fusion beats (second and ninth QRS complex) establishes that the wide QRS complex tachycardia is ventricular in origin. (Reproduced with permission from Waldo AL, MacLean WAH: The diagnosis and treatment of cardiac arrhythmias following open heart surgery: emphasis on the use of epicardial wire electrodes. Mt. Kisco, NY: Futura Publishing Company, Inc., 1980, p. 90.)

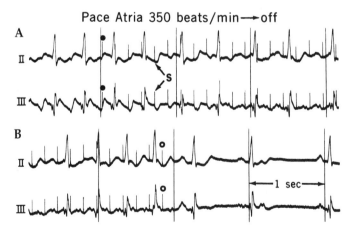

FIGURE 9-3. ECG leads II and III. A and B are not continuous traces. The dot in A marks the onset of atrial pacing at 350 beats/min. Note how rapidly the morphology of the atrial complexes changes dramatically, so that by the end of the trace in A, the atrial complexes are positive in both leads II and III. B shows the termination of 30 seconds of atrial pacing at 350 beats/min. The *circle* represents the last paced atrial beat. Note that with abrupt termination of rapid atrial pacing, sinus rhythm appears. S, stimulus artifact. Time lines are at 1-second intervals. (Reproduced with permission from Waldo AL, MacLean WAH: The diagnosis and treatment of cardiac arrhythmias following open heart surgery: emphasis on the use of epicardial wire electrodes. Mt. Kisco, NY: Futura Publishing Company, Inc., 1980, p. 129.)

techniques. Persistent ventricular ectopy that arouses clinical concern has been shown to be effectively suppressed with atrial or ventricular epicardial pacing.[97] Ventricular tachycardia (recurrent or sustained) has also been effectively treated with epicardial pacing (Fig. 9-4). Waldo et al.[99] described a variety of ventricular pacing methods that may be effective, including overdrive pacing (Fig. 9-4), ventricular "paired-pacing," and programmed interval or random PVCs.

Overdrive pacing is felt by some clinicians to be the simplest and most effective initial treatment for hemodynamically stable ventricular tachycardia. Pacing usually starts at 5 to 10 beats/min faster than the intrinsic ventricular rate.[99] When ventricular capture occurs, pacing may be terminated abruptly or gradually slowed, depending on the clinical situation. If this approach does not terminate the tachycardia, the pacing rate should be increased by 5 to 10 beats/min and the pacing repeated. When this approach is used, a critical pacing rate and duration must be achieved in order to terminate most ventricular tachycardias.[99,103] Critical ventricular pacing rates most commonly range between 120 to 125% of the intrinsic rate with higher rates (as high as 140%) occasionally being required.[78,99,104] If the critical pacing rate is not achieved while overdrive pacing is used for the treatment of ventricular tachycardia, "transient entrainment" may occur, resulting in resetting of the ventricular rate to that of the paced rate.[99] With

Vent. Tachy. Rate · 160 beats/min   Vent. Pace Rate · 165→200 beats/min

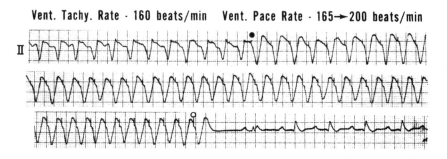

FIGURE 9-4.   These three frames are continuous recordings of ECG lead II in a patient with ventricular tachycardia at a rate of 160 beats/min. Beginning with the 13th beat in the top trace (*dot*), ventricular pacing at a rate of 165 beats/min is initiated, and the ventricular pacing rate is gradually increased to 200 beats/min. With the abrupt termination of pacing (*circle* in the bottom trace), a sinus rhythm is present. Note that as the ventricular pacing rate is increased, a progressive fusion morphology of the QRS complex is not seen, although the duration of the QRS complexes does prolong as the ventricular pacing is increased. In this patient, ventricular pacing at rates below 200 beats/min entrained but did not interrupt the ventricular tachycardia. (Reproduced with permission from Waldo AL, MacLean WAH: The diagnosis and treatment of cardiac arrhythmias following open heart surgery: emphasis on the use of epicardial wire electrodes. Mt. Kisco, NY: Futura Publishing Company, Inc., 1980, p. 151.)

higher rates of ventricular tachycardia the risk of degeneration to ventricular fibrillation increases and for this reason it is important to be prepared for electrical cardioversion if indicated.

Ventricular "paired-pacing" has also been used successfully in the treatment of ventricular tachycardias that are unresponsive to rapid ventricular pacing at clinically acceptable rates. The ventricles are paced asynchronously at a fixed rate with a premature stimulus delivered at a fixed interval following each paced beat.[99] In addition to treating automatic or reentrant ventricular tachycardia, this technique has been used successfully for refractory AV junctional tachycardia. In cases of persistent ventricular tachycardia unresponsive to other treatment, paired-pacing can in effect create ventricular bigeminy and therefore make every second beat become mechanically effective which may improve the patient's hemodynamic state.

Other, less reliable methods have also been used in an attempt to treat/terminate persistent postoperative ventricular tachycardias. Introduction of premature beats (PVCs), programmed or random, have been used with somewhat unpredictable success.[99] This technique appears to be somewhat more effective when used with relatively slow ventricular tachycardias (<150 beats/min). Intermittent "burst" ventricular pacing has also shown some promise with ventricular tachycardias resistant to conventional therapy or other pacing techniques.[105]

As is well known, ventricular tachycardia may degenerate to ventricular fibrillation and for this reason extreme caution must be taken with any treatment regimen. As many as 15% of patients with ventricular tachycardia treated with epicardial pacing will convert to a more rapid ventricular tachycardia or ventricular fibrillation.[99] Ventricular fibrillation cannot be treated with pacing. When pacing techniques are being used, it is crucial for the clinician to be prepared to administer electrical cardioversion if clinically indicated. Patients with cardiac rhythms prone to degeneration (e.g., prolonged QT) may have a decreased incidence of this degeneration with pacing at rates faster than their intrinsic rate.[106]

Other rhythm disturbances causing hemodynamic problems following cardiac surgery (AV conduction abnormalities, intraventricular conduction blocks) have also been effectively treated with epicardial pacing. Pacing electrodes are many times valuable shortly after a patient is weaned from cardiopulmonary bypass when these conduction disturbances are frequently seen. AV sequential, fixed rate (atrial or ventricular), or demand pacing may be helpful in optimizing a patient's hemodynamics until a more stable rhythm is reestablished. Temporary atrial pacing used directly after cardiac surgery not only improves hemodynamics but has also been shown to decrease the frequency of premature atrial and ventricular contractions.[107]

References

1. Angelini P, Feldman MJ, Lutschanowski R, Leachman LD. Cardiac arrhythmias during and after heart surgery: diagnosis and management. Prog Cardiovasc Dis 1974;16:469.
2. Kouchoukos NT, Oberman A, Kirklin JW, et al. Coronary bypass surgery: analysis of factors affecting hospital mortality. Circulation 1980;62(suppl 1):I84.
3. Kuner J, Enescu V, Utsu F, Boszormenyi E, Bernstein H, Corday E. Cardiac arrhythmias during anesthesia. Dis Chest 1967;52:580.
4. Bertrand CA, Steiner NV, Jameson AG, Lopez M. Disturbances of cardiac rhythm during anesthesia and surgery. JAMA 1971;216:1615.
5. Casey WF, Wynands JE, Ramsey JG, et al. The incidence of prebypass dysrhythmias in patients undergoing coronary artery surgery. J Cardiothorac Anesth 1988;2:123.
6. Dewar ML, Rosengarten MD, Blundell PE, Chui RC. Perioperative Holter monitoring and computer analysis of dysrhythmias in cardiac surgery. Chest 1985;87:593.
7. Thomson IR, Dalton BC, Lappas DG, Lowenstein E. Right bundle-branch block and complete heart block caused by the Swan-Ganz catheter. Anesthesiology 1979;51:359.
8. Morris D, Mulvihill D, Lew WYW. Risk of developing complete heart block during pulmonary artery catheterization in patients with left bundle-branch block. Arch Intern Med 1987;147:2005.
9. Risk SC, Brandon D, D'Ambra MN, Koski EG, Hoffman WJ, Philbin DM. Indications for the use of pacing pulmonary artery catheters in cardiac surgery. J Cardiothorac Vasc Anesth 1992;6:275.
10. Madsen JK, Meibom J, Videbak R, Pederson F, Grande P. Transcutaneous pacing: experience with the Zoll non-invasive temporary pacemaker. Am Heart J 1988;116:7.
11. Pattison CZ, Atlee JLI, Matthews EL, Buljubasic N, Entress JJ. Atrial pacing thresholds measured in anesthetized patients with the use of an esophageal stethoscope modified for pacing. Anesthesiology 1991;74:854.
12. Kemnitz J, Winter J, Vester EG, Peters J. Transcutaneous cardiac pacing in patients with automatic implantable defibrillators and epicardial patch electrodes. Anesthesiology 1992;77:258.
13. Tuman KJ, McCarthy RJ, Spiess BD, DaValle M, Dabir R, Ivankovich AD. Does choice of anesthetic agent significantly alter outcome after coronary bypass surgery? Anesthesiology 1989;70:189.
14. Tuman KJ, McCarthy RJ, Spiess BD, Ivankovich AD. Comparison of anesthetic techniques in patients undergoing heart valve replacement. J Cardiothorac Anesth 1990;4:159.
15. Johnston RR, Eger II EI, Wilson C. A comparative interaction of epinephrine with enflurane, isoflurane and halothane in man. Anesth Analg 1976; 55:709.
16. Atlee JL III, Brownlee SW, Burstrom RE. Conscious-state comparisons of the effects of inhalation anesthetics on specialized atrioventricular conduction times in dogs. Anesthesiology 1986;64:703.
17. Bosnjak ZJ, Kampine JP. Effects of halothane, enflurane and isoflurane on the SA node. Anesthesiology 1983;58:314.

18. Atlee JL III. Perioperative cardiac dysrhythmias: mechanisms, recognition, management. 2nd ed. Chicago: Yearbook Medical Publishers, Inc., 1990.
19. Hantler CB, Wilton N, Learned DM, Hill AEG,Knight PR. Impaired myocardial conduction in patients receiving diltiazem therapy during enflurane anesthesia. Anesthesiology 1987;67:94.
20. Gallenberg LA, Stowe DF, Kampine JP, Bosnjak ZJ. Effects of nifedipine with isoflurane, halothane, or enflurane on automaticity, conduction, and contractility in isolated guinea pig hearts. Anesthesiology 1993;78:1112.
21. Starr NJ, Sethna DH, Estafanous F, Bowman FO Jr, Gersony WM. Bradycardia and asystole following rapid administration of sufentanil and vecuronium. Anesthesiology 1986;64:521.
22. Savarese JJ, Lowenstein E. The name of the game: no anesthesia by cook book. Anesthesiology 1985;62:703.
23. Royster RL, Keeler KD, Haisty WK, Johnston WE, Prough DS. Cardiac electrophysiologic effects of fentanyl and combinations of fentanyl and neuromuscular relaxants in phenobarbital-anesthetized dogs. Anesth Analg 1988;67:15.
24. Puerto BA, Wong KC, Puerto AX, Tseng CK, Blatnick RA. Epinephrine-induced dysrhythmias: comparison during anesthesia with narcotics and with halogenated inhalation agents in dogs. Can Anaesth Soc J 1979; 26:263.
25. Abdul-Rasool IH, Sears DH, Katz RL. The effect of a second dose of succinylcholine on cardiac rate and rhythm following induction of anesthesia with etomidate or midazolam. Anesthesiology 1987;67:795.
26. Hannallah RS, Oh TH, McGill WA, Epstein BS. Changes in heart rate and rhythm after intramuscular succinylcholine with or without atropine in anesthetized children. Anesth Analg 1986;65:1329.
27. Hescheler J, Tang M, Jastorff B, Trautwein W. On the mechanism of histamine induced enhancement of the cardiac Ca^{2+} current. Pflügers Arch 1987;410:23.
28. Vigorito C, Russo P, Picotti GB, Chiariello M, Poto S, Marone G. Cardiovascular effects of histamine infusion in man. J Cardiovasc Pharmacol 1983;5:531.
29. Geha DG, Rozelle BC, Raessler KL, Groves BM, Wightman MA, Blitt CD. Pancuronium bromide enhances atrioventricular conduction in halothane anesthetized dogs. Anesthesiology 1977;46:342.
30. Son SL, Waud BE. Potency of neuromuscular blocking agents at the receptors of the atrial pacemaker and the motor endplate of the guinea pig. Anesthesiology 1977;47:34.
31. Edwards RP, Miller RD, Roizen MF, et al. Cardiac responses to imipramine and pancuronium during anesthesia with halothane and en flurane. Anesthesiology 1979;50:421.
32. Kirkwood I, Duckworth RA. An unusual case of sinus arrest. Br J Anaesth 1983;55:1273.
33. Pollok AJP. Cardiac arrest immediately after vecuronium. Br J Anaesth 1986;58:936.
34. Engbaek J, [slash]Ording H, Sorensen B, Viby-Morgensen J. Cardiac effects of vecuronium and pancuronium during halothane anest hesia. Br J Anaesth 1983;55:501.

35. Urquhart ML, Ramsey FM, Royster RL, Morell RC, Gerr P. Heart rate and rhythm following an edrophonium/atropine mixture for antagonism of neuromuscular blockade during fentanyl/N_2O/O_2 or isoflurane/N_2O/O_2 anesthes ia. Anesthesiology 1987;67:561.
36. Andree RA. Sudden death following naloxone administration. Anesth Analg 1980;59:782.
37. Kleinpeter UM, Iversen S, Tesch A, Schmeidt W, Mayer E, Oelert H. Prevention of supraventricular tachyarrhythmias post coronary artery bypass surgery. Eur Heart J 1987;8(suppl L):137.
38. Mohr R, Simolinsky A, Goor DA. Prevention of supraventricular tachyarrhythmia with low dose propranolol after coronary artery by pass surgery. J Thorac Cardiovasc Surg 1981;81:840.
39. Smith EE, Shore DF, Monro JL, Ross JK. Oral verapamil fails to prevent supraventricular tachycardia following coronary artery surgery. Int J Cardiol 1985;9:37.
40. Waldo AL, Henthorn RW, Epstein AE, Plumb VS. Diagnosis and treatment of arrhythmias during and following open heart surgery. Med Clin North Am 1984;68:1153.
41. Chee TP, Sri PS, Desser KB, Benchimo A. Postoperative supraventricular arrhythmias and the role of prophylactic digoxin in cardiac surgery. Am Heart J 1982;104:974.
42. White HD, Antmann EM, Glynn MA. Efficacy and safety of timolol for prevention of supraventicular tachycardia after coronary artery bypass surgery. Circulation 1984;70:479.
43. Vecht RJ, Nicolaiden EP, Ikwenke JK, Liassides CH, Cleary J, Cooper WB. Incidence and prevention of supraventricular tachyarrhythmias after coronary artery bypass surgery. Int J Cardiol 1986;13:125.
44. Michelson EL, Morgonroth J, MacVaugh H. Postoperative arrhythmias after coronary artery and cardiac valvular surgery detected by long-term electrocardiographic monitoring. Am Heart J 1979;97:442.
45. Douglas P, Hirshfeld JW, Edmunds LH. Clinical correlates of postoperative atrial fibrillation (abstract). Circulation 1984;70(su ppl II):165.
46. Smith R, Grossman W, Johnson L, Segal H, Collins J, Dalen J. Arrhythmias following cardiac valve replacement. Circulation 1972;4 5:1018.
47. Roffman JA, Fieldman A. Digoxin and propranolol in the prophylaxis of supraventricular tachyarrhythmias after coronary artery by pass surgery. Ann Thorac Surg 1981;31:496.
48. Ormerod OJM, McGregor CGA, Stone DL, Wispey C, Petch MC. Arrhythmias after coronary bypass surgery. Br Heart J 1984;51:618.
49. Tchervenkov CI, Wynands JE, Symes JF, Malcolm ID, Dobell ARC, Morin JE. Persistent atrial activity during cardioplegic arrest: a possible factor in the etiology of post operative supraventricular tachyarrhythmias. Ann Thorac Surg 1983;36:437.
50. Tchervenkov CI, Wynands JE, Symes JF, Malcolm ID, Dobell ARC, Morin JE. Electrical behavior of the heart following high-potassium cardioplegia. Ann Thorac Surg 1983;36:314.
51. Menasche P, Maisonblanche P, Bousseau D, Lorente P, Punica A. Decreased incidence of supraventricular arrhythmia achieved by selective atrial cooling after aortic valve replacement. Eur J Cardiovasc Thorac Surg. 1987;1:33.

52. Reves JG, Karp RB, Buttner EE, et al. Neuronal and adrenomedullary catecholamine release in response to cardiopulmonary bypass in man. Circulation 1982;66:49.
53. Lamb RK, Prabhakar G, Thorpe JAC, Smith S, Norton R, Dyde JA. The use of atenolol in the prevention of supraventricular arrhythmias following coronary artery surgery. Eur Heart J 1988;9:32.
54. Stephenson LW, MacVaugh H, Tomasello DN, Josephson ME. Propranolol for prevention of post-operative cardiac arrhythmias: a randomized study. Ann Thorac Surg 1980;29:113.
55. Peniston CM, Spence PA, Mihic N, et al. The effects of cardioplegic arrest on right atrial function. Ann Thorac Surg 1986;41:473 .
56. Chen Y-F, Lin Y-T. Comparison of the effectiveness of myocardial preservation in right atrium and left ventricle. Ann Thorac Surg 1985;40:25.
57. Ivey MF, Evey TD, Bailey WW, Williams DB, Hessell EA, Miller DW. Influence of propranolol on supraventricular tachycardia early after coronary artery revascularization. J Thorac Cardiovasc Surg 1983;85:214.
58. Matangi MF, Neutze JM, Graham IC, Hill DG, Kerr AR, Barratt-Boyes BG. Arrhythmia prophylaxis after aorto-coronary bypass. Cardiovasc Surg 1985;89:439.
59. Daudon P, Corcos T, Gardjbakah I, Leuasseur J-P, Cabrol A, Cabrol C. Prevention of atrial fibrillation or flutter by acebutolol after coronary bypass grafting. Am J Cardiol 1986;58:933.
60. Matangi MF, Stickland J, Garbe G, et al. Atenolol for the prevention of arrhythmias following coronary artery bypass grafting. Can J Cardiol 1989; 5:229.
61. Tyras DH, Stother JC, Kaiser GC, Barner HB, Codd JE, Wu L. Supraventricular tachycardia following myocardial revascularization: a randomized trial of prophylactic digitalization. J Thorac Cardiovasc Surg 1979; 77:310.
62. Selzer A, Kelley JJ, Berbode F, Kerth WJ, Osborn JJ, Rapper RW. Case against routine use of digitalis in patients undergoing cardiac surgery. JAMA 1966;195:549.
63. Rose MR, Glassman E, Spencer FC. Arrhythmias following cardiac surgery: relation to serum digoxin levels. Am Heart J 1975;89:288 .
64. Mills SA, Poole GV, Breyer RH. Digoxin and propranolol in the prophylaxis of dysrhythmias after coronary artery bypass grafting. Circulation 1983;68:222.
65. Williams DB, Misbach GA, Kruse AP, Ivey TD. Oral verapamil for prophylaxis of supraventricular tachycardia after myocardial revascularization. J Thorac Cardiovasc Surg 1985;90:592.
66. Davison R, Hertz R, Kaplan K, Parker M, Feiereisel P, Michaelis L. Prophylaxis of supraventricular tachyarrhythmia after cononary bypass surgery with oral verapamil: a randomized, double-blinded trial. Ann Thorac Surg 1985;39:336.
67. Rubin DA, Nieminski KE, Monteferrante JC, Magee T, Reed GE, Herman MV. Ventricular arrhythmias after coronary artery bypass graft surgery: incidence, risk factors and long term prognosis. J Am Coll Cardiol 1985;6:307.
68. Hoie J, Forfang K. Arrhythmias and conduction disturbances following aortic valve implantation. Scand J Thorac Cardiovasc Surg 1979;14:177.

69. Gradman AH, Harbison MA, Berger HJ, et al. Ventricular arrhythmias late after aortic valve replacement and their relation to left ventricular performance. Am J Cardiol 1981;48:824.

70. Abedin Z, Soares J, Phillips DF, Sheldon WC. Ventricular tachycardia following surgery for myocardial revascularization: a follow-up study. Chest 1977;72:426.

71. Foppl M, Hoffman A, Amann FW, et al. Sudden cardiac death after aortic valve surgery: incidence and concomitant factors. Clin Cardiol 1989; 12:2027.

72. Cooperman M, Stinson EB, Griepp RB, Shumway NE. Survival and function after left ventricular aneurysmectomy. J Thorac Cardiovasc Surg 1975;69:321.

73. Topol EJ, Lerman BB, Baughman KL, Platia EV, Griffith LS. De novo refractory ventricular tachyarrhythmias after coronary revascularization. Am J Cardiol 1986;57:57.

74. DeSoyza N, Murphy M, Bissett JK, Kane JJ, Doherty JE. Ventricular arrhythmias in chronic stable angina pectoris with surgical or medical treatment. Ann Intern Med 1978;89:10.

75. Wexelman W, Lichstein E, Cunningham JN, Hollander G, Greengart A, Shani J. Etiology and clinical significance of new fascicular conduction defects following coronary artery bypass surgery. Am Heart J 1986; 111:923.

76. Baerman JM, Kirsh MM, deBuitleir M, et al. Natural history and determinants of conduction defects following coronary artery bypass surgery. Ann Thorac Surg 1987;44:150.

77. Chu A, Califf R, Pryor D, et al. Prognostic effect of bundle branch block related to coronary artery bypass grafting. Am J Cardi ol 1987;59:798.

78. O'Connell JB, Wallis D, Johnson SA, Pifarre R, Gunnar RM. Transient bundle branch block following the use of hypothermic cardioplegia in coronary artery bypass surgery: High incidence without preoperative myocardial infarction. Am Heart J 1982;103:85.

79. Caspi Y, Safadi T, Ammar R, Elamey A, Fishman MH, Merin G. The significance of bundle branch block in the immediate postoperative electrocardiograms of patients undergoing coronary artery bypass. J Thorac Cardiovasc Surg 1987;93:442.

80. Ellis RJ, Mavroudis C, Gardner C, Turley K, Ullyot D, Ebert DA. Relationship between atrioventricular arrhythmias and the concentration of K+ ion in cardioplegia solution. J Thorac Cardiovasc Surg 1980;80:517.

81. Thompson R, Mitchell A, Ahmed M, Towers M, Yacoub M. Conduction defects in aortic valve disease. Am Heart J 1979;98:3.

82. Gannon PG, Sellers RD, Kanjuh VI, Edwards JE, Lillehei CW. Complete heart block following replacement of the aortic valve. Circulation 1966; 33(suppl 4):I152.

83. Krongrad E. Postoperative arrhythmias in patients with congenital heart disease. Chest 1984;85:107.

84. Vetter VL. What every pediatrician needs to know about arrhythmias in children who have had cardiac surgery. Pediatr Ann 1991;20:378.

85. Oh JK, Holmes DR, Hayes DL, Porter CB, Danielson GK. Cardiac arrhythmias in patients with surgical repair of Ebstein's anomaly. J Am Coll Cardiol 1985;6:1351.

86. Yabek SM, Swensson RE, Jarmakani JM. Electrocardiographic recognition of sinus node dysfunction in children and young adults. Circulation 1977;56:235.
87. Gillette PC, Kugler JO, Garson A, Gutgesell HO, Duff DF, McNamara DG. Mechanism of cardiac arrhythmias after the Mustard operation for transposition of great arteries. Am J Cardiol 1980;45:1225.
88. Greenwood RD, Rosenthal A, Sloss LJ, Lacorte M, Nadas A. Sick sinus syndrome after surgery for congenital heart disease. Circulation 1975; 52:208.
89. Goodman M, Roberts N, Izukawa T. Late postoperative conduction disturbances after repair of ventricular septal defect and tetralogy of Fallot. Circulation 1974;49:214.
90. Stanton RE, Lindesmith GG, Meyer BW. Pacemaker therapy in children with complete heart block. Am J Dis Child 1975;129:484.
91. Krongrad E. Prognosis for patients with congenital heart disease and postoperative intraventricular conduction defects. Circulation 1978; 57:856.
92. Steeg CN, Krongrad E, Davachi F, Bowman FO Jr, Gersony WM. Postoperative left anterior hemiblock and right bundle branch block following repair of tetralogy of Fallot: clinical and etiologic considerations. Circulation 1975;51:1026.
93. Quattlebaum TG, Varghese PJ, Neil CA, Donahoo JS. Sudden death among postoperative patients with tetrology of Fallot. A follow-up study of 243 patients for an average of twelve years. Circulation 1976;54:289.
94. Deanfield JE, Ho SY, Anderson RH, McKenna WJ, Allwork SP, Hallidie-Smith KA. Late sudden death after repair of tetralogy of Fallot: a clinico-pathologic study. Circulation 1983;67:626.
95. Deanfield JE, McKenna WJ, Hallidie-Smith KA. Detection of late arrhythmia and conduction disturbance after correction of tetralogy of Fallot. Br Heart J 1980;44:248.
96. Blake R, Chung E, Wesley H, A. H-SK. Conduction defects, ventricular arrhythmias, and late death after surgical closure of ventricular septal defects. Br Heart J 1982;47:305.
97. Harris PD, Malm JR, Bowman FO Jr, Hoffman BF, Kaiser GA, Singer DH. Epicardial pacing to control arrhythmias following cardiac surgery. Circulation 1968;37(suppl 2):178.
98. Waldo AL, Ross SM, Kaiser GA. The epicardial electrogram in the diagnosis of cardiac arrhythmias in the postoperative patient. Geriatrics 1971;26:108.
99. Waldo AL, MacLean WAH. Diagnosis and treatment of cardiac arrhythmias following open heart surgery: emphasis on the use of epicardial wire electrodes. Mt. Kisco, NY: Futura Publishing Company, Inc., 1980.
100. Donovan KD, Power BM, Hockings BE, L K-Y, Barrowcliffe MP, Lovett M. Usefulness of atrial electrograms recorded via central venous catheters in the diagnosis of complex cardiac arrhythmias. Crit Care Med 1993;21:532.
101. Waldo AL, MacLean WAH, Karp RB, Kouchoukos NT, James TN. Continuous rapid atrial pacing to control recurrent or sustained supra ventricular tachycardias following open heart surgery. Circulation 1976; 54:245.

102. Cooper TB, MacLean WAH, Waldo AL. Overdrive pacing for supraventricular tachycardia: a review of theoretical implications and therapeutic techniques. PACE Pacing Clin Electrophysiol 1978;1:196.
103. MacLean WAH, Cooper TB, James TN, Waldo AL. Entrainment and interruption of ventricular tachycardia with rapid ventricular pacing. Circulation 1977;56:III-105.
104. Wells JL Jr, Karp R, Kouchoukos N, MacLean WAH, James TN, Waldo AL. Characterization of atrial fibrillation in man. Studies following open heart surgery. PACE Pacing Clin Electrophysiol 1978;3:426.
105. Fisher JD, Mehra R, Furman S. Termination of ventricular tachycardia with bursts of rapid ventricular pacing. Am J Cardiol 1978;41:94.
106. Waldo AL, Wells JL Jr, Cooper TB, MacLean WAH. Temporary cardiac pacing: applications and techniques in the treatment of cardiac arrhythmias. Prog Cardiovasc Dis 1981;23:451.
107. Kirklin J, Kirklin JW. Management of the cardiovascular subsystem after cardiac surgery: collective review. Ann Thorac Surg 198 3;32:311.
108. Johnson LW, Dickstein RA, Fruehan CT, et al. Prophylactic digitalization for coronary artery bypass surgery. Circulation 1976;5 3:819.
109. Csicsko JF, Schatzlein MH, King RD. Immediate postoperative digitalization in the prophylaxis of supraventricular arrhythmias following coronary artery bypass. Cardiovasc Surg 1981;81:419.
110. Weiner B, Rheinlader HF, Decker EL, Cleveland RJ. Digoxin prophylaxis following coronary artery bypass surgery. Clin Pharm 1986 5:55.

James R. Zaidan

Perioperative Considerations for Rate-Adaptive Implantable Pacemakers

10

Cardiac pacing was, at one time, a therapeutic measure to relieve the symptoms of bradycardia. Patients with sinus bradycardia and sinus arrest as their indications for a permanent pacemaker suffered from a low cardiac output syndrome when the pacemaker stimulated the ventricle. Once the pacemaker controlled the heart rate, the patient could not exercise, because heart rate could not increase the cardiac output to meet the metabolic demand. With these problems in mind, engineers developed pacemakers that afforded patients atrioventricular synchrony. This development helped to eliminate the loss of atrial kick as a clinical problem and simultaneously improved the patients' exercise tolerance.

Atrioventricular synchrony, however, does not account totally for the increased cardiac output associated with exercise. Heart rate remains an important factor. Early pacemakers controlled bradycardia and more recent pacemakers established atrioventricular synchrony. The newest generation of pacemakers adds another level of control: adaptation to metabolic demand and activity of the patient by automatically increasing the pacing rate.

The management of patients with pacemakers in the perioperative setting has become rather sophisticated, and a greater number of complex problems may be associated with rate-adaptive pacing. This chapter serves as an update on the basic principles of pacemakers and

Clinical Cardiac Electrophysiology: Perioperative Considerations
Edited by Carl Lynch III J.B. Lippincott Company, Philadelphia, PA ©1994

a nonengineering view of rate-adaptive pacing. The discussion will start with an explanation of nomenclature then follow to the rationale for rate adaptation. Sensors that are now available for clinical use will receive more detailed attention. It is difficult to determine the anesthetic implications of many of these new pacemakers because they are just now arriving on the clinical scene.

A REVIEW OF IMPLANTABLE PACEMAKER FUNCTION

Pacemaker generators can be implanted in several locations. From one position in the anterior abdominal wall approximately at the belt line, the leads extend from the generator, traverse subcutaneous tissue, and enter the pericardium at the apex of the heart. The electrodes pierce the epicardium, hence their name, finally to lie in the myocardium. When the pacemaker generator lies just below the clavicle, the wire leads are placed through the innominate vein, superior vena cava, right atrium, and via the tricuspid valve into the right ventricle. The electrodes penetrate the right ventricular endocardium to end in the myocardium. The names "endocardial" or occasionally "transvenous" are used to describe this type of electrode.

Nomenclature

It was heretofore unnecessary to learn how pacemakers were named, because all pacemakers were of two varieties. They were either single chamber or dual chamber and all were programmable. With the advent of rate-adaptive pacing, it becomes important to review the code for naming pacemakers. Table 10-1 defines the different letters in the five-letter code.[1]

Some examples of the use of this code follow. VVI pacemakers stimulate the ventricle and sense R waves generated by the ventricle. If the sensing circuit is activated before the programmed pacing interval expires, the pacing circuit is inhibited. AAI pacemakers function like VVI generators, except the electrode is located in the atrium, and are inhibited when P waves are sensed. All pacemakers now implanted are multiprogrammable; therefore, the letter M in the fourth position is usually omitted. The letter R implies multiprogrammability, but it does not describe the type of sensor contained in the pacemaker that controls rate adaptation. VVIR pacemakers include rate

TABLE 10-1. CODES FOR PACEMAKER GENERATORS

First Position: Chamber-Paced	Second Position: Chamber-Sensed	Third Position: Response after Sensing	Fourth Position: Programmability	Fifth Position: Arrhythmia Control
A = atrial	A = atrial	I = inhibited	P = rate and output	P = pacing
V = ventricular	V = ventricular	T = triggered	M = multiprogrammable	S = shock
D = dual (A + V)	D = dual (A + V)	D = dual (I + T)	C = communicating	D = dual (P + S)
	O = none	O = none	R = rate adaptive	O = none
S = single	S = single*		O = none	

*S = single is an industry standard indicating that the generator can be used to pace either the atrium or the ventricle. Once the generator is in place, the letter S must be replaced either by an A or a V. In the programmability column, the letter R always takes precedence over the letters P, M, and C.

adaptation with VVI pacing. VAT pacemakers sense P waves and trigger impulses into the ventricle. More sophisticated pacemakers that track atrial and ventricular rates are the DDDR or fully automatic generators. They pace and sense both the atrium and the ventricle and are also rate adaptive.

Curiously, the atrial output of a DVI pacemaker can inhibit the ventricular output when the R-wave sensitivity of the generator is high (set on a low number) and when the atrial output is high. This inhibition is eliminated by setting the R-wave sensitivity to a higher number and decreasing the atrial output.

Patient Management

The preoperative evaluation of these patients should stress underlying medical diseases. Common findings in the history include coronary and valvular heart disease, hypertension, and diabetes mellitus. The patient probably will be receiving multiple medications that include β-adrenergic and calcium channel-blocking drugs, diuretics, cardiac glycosides, insulin, and possibly anticoagulants. The general rule is to continue these drugs until the time of surgery with special attention paid to the anticoagulants. It might be necessary to stop the anticoagulants and begin a heparin infusion. Insulin administration should be managed as with any other diabetic patient.

While taking the history, determine if the patient experiences a return of prepacemaker symptoms such as dizziness and palpitations while exercising the muscles in the area of the generator. This finding indicates that the patient is experiencing myopotential inhibition of the pacemaker and suggests that shivering or, potentially, succinylcholine-induced muscle fasciculations might inhibit the generator.

The physical examination could reveal chronic congestive heart failure and carotid or abdominal aortic bruits. Note also the location of the pacemaker generator, because this information will reveal the location of the electrodes (epicardial versus transvenous). Observe the chest x-ray to verify the location, type, and continuity of the electrodes. The electrocardiogram will reveal the pacemaker dependency.

Evaluation of the pacemaker's function should assume secondary importance. The general rule is that if the pacemaker is working preoperatively, it should continue to work throughout the procedure. Table 10-2 points out the findings associated with the electrocardiogram when the patient's heart rate is higher, lower, or approximately the same as the pacemaker's rate.

When the patient's rate is faster than the pacemaker's, have the patient perform a Valsalva maneuver to slow the heart rate. Pacemaker spikes and evoked QRS complexes should appear on the electrocardiogram. Instead of the Valsalva maneuver, consider placing the magnet on the generator to convert it to the magnet rate (usually about 90) to determine if capture takes place. An important point is that a pacemaker spike located in the downslope of the R wave, the ST segment, or the T wave indicates an abnormal loss of sensing. Also, a pacemaker spike located in the TP segment or P wave without an associated QRS complex indicates an inability to pace. Both of these findings should be reported to the cardiologist, although in an emergency, it is reasonable to proceed with the surgery.

Monitoring must always include the standard devices plus whatever is necessary to guide the patient safely through surgery. Above all, form the anesthetic management around the patient's underlying disease, not the pacemaker.

Pacemaker Failure

Pacemakers can fail to capture during surgery for several reasons, including: electromagnetic interference, muscle artifacts, and serum electrolytes. Acute changes in the potassium gradient across the cell membrane control myocardial membrane potential and largely determine excitability and the consequent ability of electrical currents to elicit an action potential. Volatile anesthetic agents have no apparent effect on voltage and current threshold when added to a narcotic relaxant anesthetic technique.[2]

Electromagnetic Interference

Electrocautery. Electromagnetic interference in the form of the electrocautery complicates anesthetic management. The VOO generator will not be affected and will continue to emit impulses even while the electrocautery is in use. The VVI nonprogrammable (VVIO) generator probably will be inhibited for one impulse then revert to VOO ac-

TABLE 10-2. ELECTROCARDIOGRAM FINDINGS

Heart Rate	Electrocardiogram Findings
1. Faster than pacemaker	1. No pacemaker impulses observed
2. Slower than pacemaker	2. 1:1 association of pacemaker impulse to peripheral pulse
3. Approximately same as pacemaker	3. a. Sinus beats b. Paced beats c. Fusion beats

tivity or to VVI activity at a different rate. It is reasonable to apply the magnet to the VVIO pacemaker while using the electrocautery to avoid any inhibition.

Programmable generators can reprogram, revert to VOO activity, or temporarily or permanently lose output.[3-6] If they are deliberately reset to VOO activity before the surgical procedure, they can still reprogram. Applying the magnet to a programmable pacemaker will increase the chance of a reprogram; however, the new program will not become manifest until the magnet is removed. Some manufacturers suggest not using the magnet. Always monitor the patient fully when removing the magnet. Avoiding the magnet will minimize the chance of a reprogram. If reprogramming occurs in the absence of a magnet, the new program will be instantly obvious if it involves a change in rate or a decrease in output of the generator below the threshold.

The following guidelines will help to minimize the dangers of using the electrocautery, but they will not eliminate them:

1. If possible, use the bipolar electrocautery.
2. Request that the surgeon use the smallest possible current.
3. Never position the ground plate of the electrocautery so that the generator lies between the ground plate and the surgical site. Try to direct the current away from the generator.
4. Also try to position the ground plate so that a line drawn between the active and ground electrodes of the cautery is perpendicular to a line drawn between the pacemaker's electrode in the ventricle and the generator.
5. Request that the surgeon not activate the electrocautery until he or she is ready to use it. The active electrode of the cautery does not have to touch the patient to reprogram the generator.
6. Do not use the cautery within 5 inches of the pacemaker.

Magnets. Implanted programmable pacemakers can revert to VOO activity, increase the ventricular rate above the high rate limit, or cease activity when they are subjected to strong magnetic fields such as that found in magnetic resonance imaging. The patient with a pacemaker should *never* be placed in an magnetic resonance imaging unit. Normal activity should resume after the magnetic field is eliminated.

Lithotripsy. It is possible for the patient to undergo lithotripsy; however, the circuitry of the pacemaker can be permanently damaged if it is at the focal point of the lithotripsy beam. Some manufacturers suggest programming the pacemaker to a single chamber, non-rate-responsive mode of activity and assuring that the focal point of the beam is more than 2 inches away from the generator. In a large survey by Drach et al.[7] 131 patients with pacemakers underwent 142 lithotripsies. Only 4 patients had problems with the pacemakers, which were easily corrected. Celentano et al.[8] recommend that a cardiologist stand by to treat complications and that emergency pacing equipment be available.

Muscle Artifacts. To determine if inhibition related to skeletal muscle activity occurs, ask the patient if exercising muscles in the area of the generator initiates prepacemaker symptoms.[9,10] One case has been reported in which fasciculations associated with succinylcholine inhibited a pacemaker.[11] Consider using either a defasciculating dose of a neuromuscular blocking drug before administering succinylcholine or a nondepolarizing agent or programming the generator to VOO activity.[12]

Severe shivering may also have several effects depending on the type of pacemaker. VVIO pacemakers could be inhibited. Warming the patient would stop this inhibition. Movement-controlled VVIR and DDDR pacemakers could respond to shivering by increasing the rate. In this case, warming the patient would slow the pacemaker rate. The blood temperature-controlled VVIR or DDDR pacemaker would experience no change in rate during shivering, but would increase its rate as the patient warmed.

Potassium

Acute fluctuations in potassium change the resting membrane potential and can theoretically alter the ability of electrical impulses to stimulate the heart. An acute decrease in serum potassium level will hyperpolarize the myocardial cells away from threshold, but with a counterbalancing decrease in membrane conductance, which will enhance excitability. Acute increases in serum potassium increase the membrane potential toward threshold, but also increase membrane conductance (see Chapter 1). These counterbalancing effects may lead to unpredictable changes in threshold; however, marked changes in pacing threshold are not typically reported. Acute myocardial ischemia has an associated leak of potassium from the cell to the interstitial space, as well as additional cellular changes (calcium overload, decreased potassium conductance), which may also influence pacing threshold and cell excitability.

Other Reports

Nitrous oxide used shortly after placement of a pacemaker and subcutaneous emphysema both have been reported as inhibiting unipolar pacemaker function.[13,14] Apparently, the ground electrode located in the generator's casing did not have good contact with the tissue. In these cases, applying pressure over the pacemaker reestablished pacing.

Andersen et al.[15] have described a method to assure that dorsal column stimulators do not inhibit VVI pacemakers or convert VVI pacemakers to VOO activity. Nerve stimulators have been applied directly over pacemaker generators purposely to inhibit the pacemakers.[16] This inhibition might be possible if the stimulator were used on the arm.

RATIONALE FOR RATE-ADAPTIVE PACING

As people exercise, metabolic demand from muscular activity elevates oxygen consumption. The body responds to meet this demand through several mechanisms. It extracts a greater amount of oxygen from the arterial blood. This change is manifested by a widening of the arterial to venous oxygen difference. Stroke volume increases through enhanced contractility and increased preload. Heart rate increases to meet the demand. Increased stroke volume and heart rate augment the cardiac output. The normal heart maintains a normal atrial to ventricular contraction sequence, or atrioventricular (AV) synchrony, and the PR interval shortens during exercise and sympathetic stimulation.

Unfortunately, in patients with heart disease heart rate cannot always increase to match the metabolic demand. In the presence of conduction disease, they can even lose AV synchrony. They are left with increased preload and oxygen extraction as their only mechanisms of responding to exercise. Adapting the pacing rate to match the metabolic demand and maintaining AV synchrony to augment the preload become important issues in these patients because of their desire to maintain an active lifestyle.

Although rate and synchrony are important, rate seems to be the more important during higher levels of exercise.[17] Comparing VAT, VOO, and rate-adaptive pacing without AV synchrony, Fananapazir et al.[18] found equal improvement in exercise tolerance between VAT or rate-adaptive, asynchronous pacing compared to VOO or asynchronous pacing. This finding suggests that at higher rates AV synchrony could be less important than heart rate in enhancing cardiac output and maintaining maximal exercise capacity. Other investigators found that at rest and at moderate levels of exercise AV synchrony retains its importance in maintaining cardiac output.[19–22]

It seems prudent, therefore, to place either DDD or VAT pacemakers in all patients who require rate augmentation. Since they sense P waves and then trigger an impulse into the ventricle, DDD and VAT pacemakers depend on the patient's intrinsic sinoatrial nodal activity to increase the pacing rate. In patients with disease of the sinoatrial node, the atrial rate cannot increase to the expected level. Chronotropic incompetency, or the failure to achieve the expected atrial rate with exercise, is found in approximately 40% of patients with sick sinus syndrome even when they are in sinus rhythm.[23] Patients with atrial fibrillation, paroxysmal supraventricular tachyarrhythmias, enlarged atria, or persistent ventriculoatrial conduction do not always benefit from P-wave synchronized pacing.[24–26] In these patients, rate-adaptive, VVIR pacemakers play a more important role in maintaining a higher degree of exercise capacity and a better quality of life even without P-wave synchronization.

BASIC DESIGN OF RATE-ADAPTIVE PACEMAKERS

Almost all pacemakers are programmable for rate, output, and various other parameters. Activity-related, rate-adaptive pacing is programmable also for upper and lower rate, activity threshold, rate response, acceleration time, and deceleration time. The upper and lower rates are the maximal and minimal rates that the pacemaker will use when determining the adaptation. Rates are usually programmed between about 60 and 120 to 170 pulses/min. Occasionally, the physician can program a circadian rhythm to allow the pacemaker rate to decrease even more during the sleeping hours.

Activity threshold sets the ease with which the patient can activate rate-adaptive pacing. The pacemaker could require excessive activity or just a little activity to achieve a unit increase in rate. In some models of pacemakers, the number of peaks that pass a threshold are counted, while in other pacemakers, the signals are integrated rather than the peak activity counted.

Rate response is the number of pulses per minute that are in a unit increase in pacing rate. The physician programs into the pacemaker one of 7 to 10 rate-response curves with different slopes (Fig. 10-1). The higher number gives the patient a higher number of pulses per minute per unit increase. The pacemaker can cause a large increase in heart rate (rate-response curve slope = 10) or a small increase in heart rate (rate-response curve slope = 1) for a given level of exercise. Earlier models of activity-related rate adaptation used curvilinear response curves that made it difficult to reach the programmed upper rate. Newer pacemakers have straight line response curves that allow the patient to reach the upper rate limit.

Acceleration time is the final parameter associated with an increase in pacing rate (Fig. 10-2). This setting, by telling the pacemaker how long it should wait until it actually responds to the signal, will allow the patient to perform minimal activity over a very short time span without experiencing an increase in heart rate. Generally, 0.25, 0.50, and 1.0 min are the three choices.

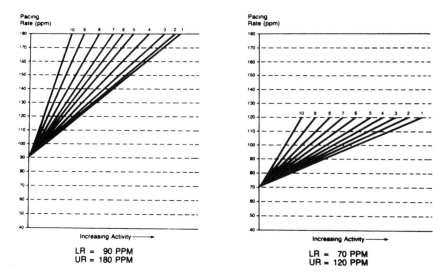

FIGURE 10-1. Rate-response curves tell the pacemaker how much to increase the pacing rate for a given level of exercise. Slope 10 results in a higher increase in pacing rate than does a slope of 1. These two graphs also show a difference in lower (*LR*) and upper rate (*UR*) limits. The rate limits and a choice of a single response curve from the family of curves are programmable. (Reproduced with permission from Medtronic ELITE technical manual. Minneapolis, Medtronic, Inc.)

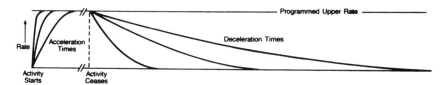

FIGURE 10-2. Programmable acceleration and deceleration times instruct the pacemaker over what period of time it should increase or decrease the pacing rate. The choices generally are 15 sec, 30 sec, or 1 min to reach 90% of the indicated unit change in pacing rate. (Reproduced with permission from Medtronic ELITE technical manual. Minneapolis, Medtronic, Inc.)

Once motion stops, the activity-related sensor will immediately stop sending signals to the control section and pacing rate will decrease to the lower rate. Since the metabolic demand remains high for a few minutes, these pacemakers have another programmable feature called deceleration time (Fig. 10-2). This feature tells the pacemaker how long it should take to lower the automatic rate. It is not related to metabolic demand, but allows the patient some time to recover from the exercise.

The basic design of rate-adaptive pacemakers dictates several components over and above those associated with nonadaptive pacing.

Adaptation requires a sensor that receives the signals from the physical or physiologic change. This signal, which is modified by various mechanisms such as filtering, is then sent through an algorithm to determine if rate modification is necessary. If rate modification is required, then the pacemaker responds by increasing or decreasing its automatic rate. Intermittent increases in the patient's intrinsic heart rate should inhibit rate-adaptive pacing.

The sensed parameter must offer proportional heart rate changes.[27] The sensor must be specific to the change and not respond to other signals. Changes in pacing rate should occur at the beginning of exercise and not have a prolonged lag period. Technically, sensors must be stable and not require excessive programming. They should be an appropriate size and not complicate lead design. Energy required by the sensor should not excessively drain the pacemaker battery.

Closed Loop Systems

When the change in pacing rate reverses the physiologic effect that initiated the change, the pacemaker has a closed loop sensing system (Fig. 10-3). For example, venous oxygen saturation decreases during exercise. An oxygen saturation sensing device located on the pacing lead would function through the algorithm to increase the pacing rate (decrease the automatic rate). Increased heart rate elevates cardiac output. More oxygen is delivered to the tissues and venous saturation increases, which in turn slows the pacing rate. Generally speaking, the physician must program only the lower and upper rates and the rate at which the pacemaker seeks the data.

CLOSED LOOP

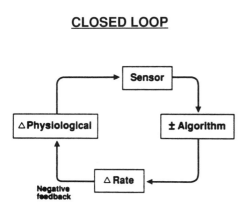

FIGURE 10-3. This diagram represents closed loop sensing. The pacemaker senses an alteration in a parameter such as venous oxygen saturation that changes in response to exercise. Pacing rate increases, the measured parameter returns to its original value, and pacing rate slows to its lower rate limit. (Reproduced with permission from Lau CP, ed. Rate adaptive cardiac pacing: single and dual chamber. Mount Kisco, NY, Futura Publishing Company, Inc., 1993.)

OPEN LOOP

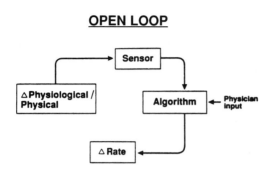

FIGURE 10-4. Shown here is a diagram of an open loop system of sensing in which the change in heart rate does not influence the event which increased the pacing rate. Activity sensing is an example. (Reproduced with permission from Lau CP, ed. Rate adaptive cardiac pacing: single and dual chamber. Mount Kisco, NY, Futura Publishing Company, Inc., 1993.)

Open Loop Systems

An open loop system implies that the change in pacing rate does not reverse the signal that initiated the change (Fig. 10-4). As an example, a sensor that detects motion through vibration or acceleration will induce an increase in pacing rate. The increased heart rate will not negatively feed back to the amount of motion. The open loop system requires more programming because the algorithm will not inherently know the pacing rate required for a given level of exercise. The physician therefore must set not only the lower and upper pacing rates, but also the sensitivity of the system and the rate at which the change takes place. This programming takes place during a series of exercises. The patient relates to the physician subjective feelings of shortness of breath and tiredness, and the physician programs the pacemaker to a higher rate for that level of exercise.

Sensors Used in Pacing

Sensors can be ordered in several different ways.[28] One way is to group them physiologically (Table 10-3). Primary sensors respond to neural activity and endogenous catecholamines and can be considered to be a substitute SA node. These types of experimental sensors are not yet available for clinical use. The secondary group includes sensors that respond to changes brought about by exercise. This group encompasses sensors that respond to changes in oxygen saturation, right ventricular pressure, central venous blood temperature, and electrocardiographic events. Several of these sensors are clinically available. The tertiary group of sensors responds to movement and includes vibration and acceleration of the body or movement of the chest wall during rapid breathing.

TABLE 10-3. TYPES OF SENSORS

Primary	1. Catecholamine Level
	2. Neural activity
Secondary	1. Temperature
	2. Venous oxygen saturation
	3. Evoked QT interval
	4. Ventricular depolarization gradient
	5. Respiratory rate
	6. Minute ventilation
	7. Right ventricular *dp/dt*
	8. Preejection period
Tertiary	1. Body vibration
	2. Body acceleration

The second method of categorizing sensors is to group them according to how they work.[27] Impedance has been used for many years to measure chest wall motion and is now used in pacemakers not only to sense the change in respiratory parameters associated with exercise but also to determine stroke volume and preejection period. Impedance sensing is simple in design, but electrode motion artifacts cause oversensing.

Sensors can measure movement of the body. They use piezoelectricity as their technology in which bending of a crystal is converted into an electric current. Although they function in similar fashion, they are divided according to their ability to sense vibration or acceleration of the casing of the generator.[29] If the piezoelectric crystal is directly attached to the pacemaker casing, then vibrations of the casing associated with body movement create an electrical current which, in turn, changes the pacemaker rate. If the crystal is attached to the circuit board of the pacemaker but not directly to the casing, then a current begins in response to acceleration. Acceleration and vibration sensors are the most widely used. They respond very quickly to movement; however, they tend to slow the pacing rate if the patient is walking slower while ascending an incline.

A third group of sensors are those that detect intracardiac electrical events such as evoked QT interval and the ventricular gradient. The leads that sense electrocardiographic events are the same ones that stimulate the heart; therefore, technology of lead design is available. In this case, when a DDD or VVI generator is due for a change, it can be replaced with a DDDR or a VVIR pacemaker without the need to replace the pacing lead.

The final group encompasses a wide variety of sensors that are located on specialized pacing leads. They include measurement of central body temperature, venous oxygen saturation, and right ventricular *dp/dt* and preejection period. These sensors are more complicated in that they require lead modification; however, once they are in place, they function with minimal input from the physician.

SPECIFIC SENSORS

Body Movement—Vibration

Movement sends vibrations with a frequency below 10 Hz throughout the body.[30–32] External interference usually has frequencies above 10 Hz.[30] All of the vibrations are transmitted to the pacemaker casing. A piezoelectric crystal attached directly to the generator's casing converts these vibrations into an electric current that proceeds through the signal processor to the control section of the pacemaker (Fig. 10-5). By working through the algorithm, the control section determines the final automatic interval and pacing rate of the pacemaker. The mass of the pacemaker casing and to some extent the build and clothing of the patient influence the mechanical vibrations received by the piezoelectric crystal. The sensor should be placed toward the patient's chest wall rather than toward subcutaneous tissue.

Vibration-related sensors have some limitations. Their rate adaptation is not related to metabolic demand. Rate is programmed by the subjective well-being of the patient during the programming phase after implantation. Sometimes the response is not correlated with the activity.[33] Walking an incline results in lower paced rates than walking more quickly on level ground. Isometric exercise might not elicit a rate change. Direct pressure and tapping on the pacemaker casing can result in a higher pacing rate.[34] Some of these sensors use a frequency

FIGURE 10-5. This figure shows the difference between vibration- and acceleration-sensing pacemakers. A vibration-sensing pacemaker has its piezoelectric crystal directly attached to the casing while the accelerometer's crystal is attached to the circuit board but not directly to the casing. (Reproduced with permission from Intermedics technical manual. Angleton, TX, Intermedics, Inc.)

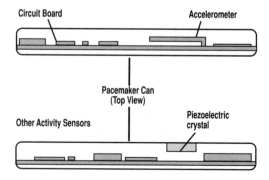

band of up to 70 Hz, which is far outside the range necessary to detect vibrations related to motion. This wide frequency band causes them to detect extraneous artifacts.[35]

One case report exists concerning management of a patient with a vibration-sensing pacemaker scheduled for a cesarean section. The paced heart rate responded appropriately to motion of the patient during movement to the operating room table, intubation, delivery of the fetus, and awakening. Aside from these increases, the paced rate remained at the lower rate limit.[36]

Body Movement—Acceleration

If the piezoelectric crystal is suspended from the circuit board and not directly attached to the pacemaker casing, the sensor is called an accelerometer (Fig. 10-3). This sensor has several advantages over the vibration-sensing device. The accelerometer does not change pacing rate when direct pressure is applied to the casing. It is not affected by orientation, and it senses movement in the 4 Hz range. Acceleration-based pacemakers have a more physiologic response to graded treadmill exercise and descending stairs.

Pacemakers sensing acceleration have the same set of programmable features as the pacemakers using vibration sensors. One difference is the shape of the rate-response curve (Fig. 10-1). It is a triphasic curve that allows a rapid increase in pacing rate early in the exercise period. If exercise level remains constant, pacing rate does not further increase. The last part of the curve permits another increase in heart rate if the patient exercises at a higher level.

Impedance—Respiratory Rate and Minute Ventilation

Respiratory rate and minute volume proportionately increase in response to CO_2 production during exercise and can be used to alter heart rate.[37,38] In the early phase of exercise, tidal volume increases. At high levels of exercise, the respiratory rate begins to increase.[39,40] Pacemakers use both respiratory rate and minute volume to effect changes in pacing rate; however, since it accounts for the early changes in tidal volume, minute ventilation correlates better with heart rate than does respiratory rate.[40]

These sensors work through a tripolar electrode system to detect changes in transthoracic impedance.[41] The pacemaker casing is used as the reference. A current of 1 mA with a pulse duration of 15 µsec is delivered every 50 msec from the distal electrode on the pacing lead. The resulting voltage at the proximal electrode in a bipolar pacing system or a special sensing electrode in a unipolar pacing system is used to calculate the transthoracic impedance. The magnitude of the changes in impedance are related to tidal volume, and the frequency of those changes is related to respiratory rate. Impedance is continuously stored and the measurement from the last minute is compared to the measurement over the last hour. By constantly comparing measurements from minute to hour, the pacemaker will not suddenly increase the pacing rate for short-duration respiratory changes such as a cough or sigh. Just as with activity-related sensors, respiratory-controlled pacemakers are programmable for the ease with which the pacemaker initiates the change and the slope of the rate response.

Although this type of pacemaker has a good correlation between pacing rate and ventilation, impedance sensors measuring respiratory parameters have several limitations. Arm movement can cause an unwanted rate increase.[42,43] However, since tidal volume increases first during exercise, arm motion facilitates an earlier rate response in pacemakers that use respiratory rate in the algorithm. The disadvantage can be used therefore to a somewhat irregular advantage. During exercise, talking will decrease the maximal rate response, because talking initiates an irregular breathing pattern. Cheyne-Stokes breathing has been reported to cause a high pacing rate; however, the pacemaker had a very high rate response slope of 40.[44] Decreasing the rate-response slope to the more commonly used 20 would have decreased the pacing rate.[45] Impedance systems are susceptible to interference caused by electromagnetic interference and myopotentials.[46] Also, electrocardiogram monitors occasionally can count the current impulses used to measure impedance and therefore falsely elevate the heart rate.

There are two case reports relating perioperative experiences with minute ventilation sensing pacemakers. In one case during a cesarean section, the paced rate increased during hyperventilation to preoxygenate the patient.[47] The paced rate remained at the lower rate limit during surgery, but increased again as the patient was awakening. In another case, a patient undergoing a transurethral prostatectomy experienced a heart rate increase during manual ventilation.[48] Decreasing the minute ventilation lowered the paced heart rate.

Electrocardiographic Events—Evoked QT Interval

It is well known that QT interval shortens as heart rate increases. Several formulas have been recommended to standardize the measured QT interval to account for the effect of heart rate. Bazett's formula is

FIGURE 10-6. This figure shows the decrease in evoked QT interval associated with exercise. The time interval is measured from the pacing stimulus to the point of maximal slope on the downslope of the T wave. (Reproduced with permission from Lau CP, ed. Rate adaptive cardiac pacing: single and dual chamber. Mount Kisco, NY, Futura Publishing Company, Inc., 1993.)

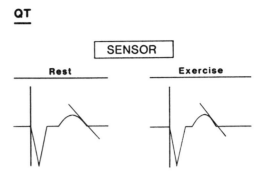

the most commonly used formula.[49] The fact that it does have a relationship to heart rate makes the QT interval measured from the pacing impulse to the point of maximum negative slope of the T wave useful in rate-adaptive pacemakers (Fig. 10-6). This possibility was first suggested by Rickards and Norman,[50] who investigated the QT interval-heart rate relationship in three groups of patients. One group of patients underwent exercise. Patients in the second group were atrially paced without exercise, and the last group were exercising patients who had third degree block and permanent VVI pacemakers. In the first group, the QT interval decreased by 187 msec when the heart rate increased by 100 beats/min. The second group experienced a 66-msec decrease in QT interval with the same increase in heart rate. In the third group the heart rate did not increase, but they still experienced a decrease in stimulus to T-wave interval (evoked QT interval) associated with exercise. Ventricular rate by itself, therefore, is not the only determinant of QT interval. The authors suggested, as later verified by other investigators, that QT interval during exercise is heavily influenced by catecholamines.[51,52]

With QT sensing, the pacemaker blanks for 200 msec after stimulating the ventricle. Then T-wave sensing begins and persists for up to 450 msec. The algorithm sets the timing of the next pacing impulse depending on the duration of the preceding evoked QT interval. If the evoked QT interval is shorter, then the next pacing impulse will be delivered earlier. This pacemaker is actually measuring and then correcting each QT interval to determine if the corrected QT interval correlates with heart rate. If the corrected QT interval is too short for the measured heart rate, then the pacemaker increases the heart rate.

Programmable features of this pacemaker include the rate-response slope at the upper and lower rate limits, upper and lower rates, and the rate of decrease in pacing rate to the lower rate limit. A newer design permits autoprogramming of the rate response slope at the lower rate limit during sleeping hours and at the upper rate limit during exercise.[53]

As expected, this pacemaker has limitations. Some patients are not suitable candidates for evoked QT sensing. Patients receiving class IA or class III antiarrhythmic agents or who have repolarization abnormalities would not benefit, because the T-wave sensing could fall outside of the sensing duration of 450 msec, and the pacemaker would function as a VVI generator. In patients with neuropathies that affect sympathetic stimulation of the heart, QT intervals do not vary, and they also would not benefit. A positive feedback loop is possible. Hedman et al.[54] have reported that a shortened evoked QT interval can increase pacing rate which can, in turn, signal a further increase in pacing rate.

Electrocardiographic Events—Paced Depolarization Integral

The depolarization gradient, or the integral of the intraventricular electrogram, represents the area under the paced R wave (Fig. 10-7). In patients with complete heart block, it decreases with exercise and increases again if the pacing rate is increased.[55] This finding allows the paced depolarization integral (PDI) to be used as a closed loop feedback system to increase the heart rate during exercise.[55] The sensor has the ability to change heart rate within about 10 sec and, like evoked QT interval, increases pacing rate during periods of emotional stress.[56]

By modifying the pacing impulse to minimize electrode polarization, a unipolar electrode system can be used for pacemakers that utilize PDI sensing. Since this parameter must be measured during a paced beat, a patient with a slightly faster intrinsic rate that overrides the pacemaker will not benefit from rate adaptation. To circumvent this

FIGURE 10-7.   The ventricular gradient or paced depolarization integral represents the area under the paced QRS complex. This area decreases with exercise in the patient who cannot intrinsically increase the heart rate. (Reproduced with permission from Lau CP, ed. Rate adaptive cardiac pacing: single and dual chamber. Mount Kisco, NY, Futura Publishing Company, Inc., 1993.)

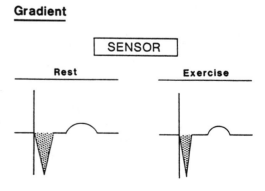

problem, the pacemaker emits one slightly premature impulse every fourth intrinsic beat and measures the PDI from the evoked response.

The PDI is related to ventricular mass and wall thickness.[57] An increase in ventricular mass and a decrease in myocardial wall thickness decrease the PDI. Exercise at a fixed pacing rate will increase end-diastolic volume and decrease wall thickness. The result is a decreased PDI and a faster pacing rate. Lying down from a standing position has the same effect as exercise in the paced patient; therefore, this type of sensor causes paced rate to increase rather then decrease when the patient is supine.[58] Also, administration of β-blocking drugs can increase the paced rate by increasing end-diastolic volume and reducing wall thickness.[59]

Oxygen Saturation

Changes in oxygen saturation can be used as a closed loop sensing system to alter pacing rate during exercise. Exercise results in immediate increases in metabolic demand and extraction of oxygen from the arterial blood. This change is reflected in the venous blood as a decrease in oxygen saturation.[60] In normal subjects heart rates increase to compensate for this increased metabolic demand. In patients with constant rate, implanted pacemakers the heart rate cannot increase.

Two light sources emit signals with wavelengths of 660 and 880 nm. The reflected light is collected by phototransistors, and the ratio of the reflected light is transferred to the timing circuit of the pacemaker. A dual light source has an advantage over light emitted only at 660 nm. Seifert et al.[61] have shown that fibrous tissue layering of a dual light source does not affect the oxygen saturation recording, while a single light source responds with a falsely high saturation. The emission of light requires energy from the battery that could be used for pacing in the non-rate-adaptive mode. To combat this loss of energy the sensors are activated only for approximately 5 msec during diastole. Over a period of 5 years, the sensors, which are active for 100 hr, reduce the pacing life of the generator by 6 months.[62]

Central Venous Temperature

Metabolism associated with exercise generates not only mechanical energy but also heat that increases body temperature. Changes in temperature used to control pacing rate vary from patient to patient.[63] In some patients, temperature decreases at the beginning of exercise as cool blood enters the central circulation, then rises to or above the baseline. Other patients have no initial drop in temperature, but only an increase. Central venous blood temperature can cool either rapidly or slowly. The temperature response remains variable when the patient performs several short periods of exercise. All of these possibilities must be considered when an algorithm is developed .

Temperature sensing requires a thermistor; therefore, a special lead is necessary. Thermistors are very reliable and are able to measure changes of as little as 0.004°C. Temperature is measured in the right ventricle several centimeters proximal to the negative electrode. Since a high impedance is required, breaks in the lead system result in loss of rate-adaptive pacing.[63,64]

Several factors must be considered in the algorithm of temperature-controlled, rate-adaptive pacing. Patients in heart failure experience a decrease in temperature that will not necessarily increase during exercise.[65] Temperature also undergoes a circadian rhythm that the pacemaker's algorithm must consider before initiating a change in pacing rate. Fever can affect the pacing rate. Environment plays an important role in that taking a hot shower or drinking hot liquids could increase the pacing rate. Also, exercising in an environment that facilitates the dissipation of heat such as swimming potentially leads to an attenuated response.

The pacemaker responds by increasing the pacing rate to a predetermined value during the initial decrease in temperature. This initial increase in pacing rate allows the patient to experience a heart rate sufficient to begin exercising without waiting for the blood to warm. The rate will not further increase until the blood temperature has reached a specific, programmed value. At this temperature, the heart rate begins to track the changes in blood temperature according to a programmed response curve. When the patient has finished exercising, the pacing rate decreases as the temperature decreases. Commonly, temperature continues to rise for a few minutes after the exercise is completed. If the patient begins a second exercise period immediately after the first, the blood temperature could decrease again. This time, the pacemaker erroneously could sense the temperature decrease as the end of exercise rather than the beginning and paradoxically decrease the pacing rate just when the patient requires a faster heart rate.[66]

Makes and Models

Over 25 pacemaker manufacturers have developed almost 1000 different models of pacemaker generators. It is obviously impossible to memorize these names. Table 10-4 lists recent rate-adaptive pacemak-

TABLE 10-4. PACEMAKER SENSORS

Model*	Company	Sensor	Toll-Free Phone Number
Activitrax	Medtronic	VIbration	800-328-2518
Legend	Medtronic	Vibration	800-328-2518
ELITE	Medtronic	Vibration	800-328-2518
Synergyst	Medtronic	Vibration	800-328-2518
Sensolog	Siemens	Vibration	800-722-3774
Ergos	Biotronik	Vibration	800-574-0394
Dash	Intermedics	Acceleration	800-231-2330
Relay	Intermedics	Acceleration	800-227-3422
Excel	CPI	Acceleration	800-245-4715
Elvin	Cook	Temperature	800-245-4715
Circadian	Intermedics	Temperature	800-245-4715
	Medtronic	O$_2$ saturation	800-328-2518
Biorate	Biotec	Respiratory rate	
Meta	Telectronics	Minute ventilation	800-525-7042
	Telectronics	PDI	800-525-7042
Rhythmx	Vitatron	Evoked QT	800-328-2518

*Each name in this column is a Trade Mark.

ers, the company that manufacturers them, and the toll-free phone number to access information 24 hr/day.

SUMMARY

Pacemakers are being implanted at the approximate rate of 100,000 per year. In the majority of these patients some type of rate-responsive generator is used. Even though some sensors are just now in clinical trials, within 5 years most of them will be marketed by pacemaker manufacturers. The future will bring not only single sensor devices, but combinations such as evoked QT plus acceleration. With these problems in mind, it behooves anesthesiologists to learn the technology of these sensing devices so that we will understand the responses that occur in the perioperative period.

References

1. Bernstein AD, Camm AJ, Fletcher RD, et al. The NASPE/BPEG generic pacemaker code for antibradyarrhythmia and adaptive-rate pacing and antitachyarrhythmia devices. PACE Pacing Clin Electrophysiol 1987;10:794.

2. Zaidan JR, Curling PE, Craver JM. Effect of enflurane, isoflurane, and halothane on pacing stimulation thresholds in man. PACE Pacing Clin Electrophysiol 1985;8:32.
3. Dresner DL, Lebowitz PW. Atrioventricular sequential pacemaker inhibition by transurethral electrosurgery. Anesthesiology 1988;68:599.
4. Domino KB, Smith TC. Electrocautery-induced reprogramming of a pacemaker using a precordial magnet. Anesth Analg 1983;62:609.
5. Mangar D, Atlas GM, Kane PB. Electrocautery-induced pacemaker malfunction during surgery. Can J Anaesth 1991 38:616.
6. Levine PA, Balady GJ, Lazar HL, Belott PH, Roberts AJ. Electrocautery and pacemakers: management of the paced patient subject to electrocautery. Ann Thorac Surg 1986;41:313.
7. Drach GW, Weber C, Donovan JM. Treatment of pacemaker patients with extracorporeal shock wave lithotripsy: experience from two continents. J Urol 1990;143:895.
8. Celentano WJ, Jahr JS, Nossaman BD. Extracorporeal shock wave lithotripsy in a patient with a pacemaker. Anesth Analg 1992;74:770.
9. Echeverria HJ, Luceri RM, Thurer RJ, Castellanos A. Myopotential inhibition of unipolar AV sequential (DVI) pacemaker. PACE Pacing Clin Electrophysiol 1982;5:20.
10. Iesaka Y, Pinakatt T, Gosselin AJ, Lister JW. Bradycardia dependent ventricular tachycardia facilitated by myopotential inhibition of a VVI pacemaker. PACE Pacing Clin Electrophysiol 1982;5:23-29.
11. Finfer SR. Pacemaker failure on induction of anaesthesia. Br Anaesth 1991;66:509.
12. Sidhu VS. Interference with pacemaker function [letter]. Br J Anaesth 1991;67:664.
13. Lamas GA, Rebecca GS, Braunwald NS, Antman EM. Pacemaker malfunction after nitrous oxide anesthesia. Am J Cardiol 1985;56:995.
14. Beaudoin M. An unusual case of failure to pace [letter]. Anaesth Intens Care 1989;17:235.
15. Andersen C, Oxhøl H, Arnsbo P. Management of spinal cord stimulators in patients with cardiac pacemakers. PACE Pacing Clin Electrophysiol 1990;13:574.
16. Ducey JP, Fincher CW, Baysinger CL. Therapeutic suppression of a permanent ventricular pacemaker using a peripheral nerve stimulator. Anesthesiology 1991;75:533.
17. Jutzy RV, Isaeff DM, Bansal RC, et al. Comparison of VVIR, DDD and DDDR pacing. J Electrophysiol 1989;3:194.
18. Fananapazir L, Bennett DH, Monks P. Atrial synchronized ventricular pacing: contribution of the chronotropic response to improved exercise performance. PACE Pacing Clin Electrophysiol 1983;6:601.
19. Markewitz A, Hemmer W. What's the price to be paid for rate response: AV sequential versus ventricular pacing? PACE Pacing Clin Electrophysiol 1991;14:1782.
20. Sulke N, Chambers J, Dritsas A, et al. A randomized double-blind crossover comparison of four rate-responsive pacing modes. J Am Coll Cardiol 1991;17:696-706.
21. Lau CP, Wong CK, Leung WH, et al. Superior cardiac hemodynamics of atrioventricular synchrony over rate responsive pacing at submaximal exercise: observations in activity-sensing DDDR pacemakers. PACE Pacing Clin Electrophysiol 1990;13:1832.

22. McMeekin JD, Lautner D, Hanson S, et al. Importance of heart rate response during exercise in patients using atrioventricular synchronous and ventricular pacemakers. PACE Pacing Clin Electrophysiol 1990; 13:59.
23. Rosenqvist M. Atrial pacing for sick sinus syndrome. Clin Cardiol 1990;13:43.
24. Greenspon AJ, Greenberg RM, Franke WS. Tracking of atrial flutter during DDD pacing: another form of pacemaker-mediated tachycardia. PACE Pacing Clin Electrophysiol 1984;7:955.
25. Derr C, Mason MA. Amplitude of atrial electrical activity during sinus rhythm and during atrial flutter-fibrillation. PACE Pacing Clin Electrophysiol 1985;8:348.
26. Gross J, Moser S, Benedek M, et al. Clinical predictors and natural history of atrial fibrillation in patients with DDD pacemakers. PACE Pacing Clin Electrophysiol 1990;13:1828.
27. Lau CP. Design of a rate adaptive pacing system: an overview of sensors and algorithms. In: Lau CP, ed. Rate adaptive cardiac pacing: single and dual chamber. Mount Kisco, NY, Futura Publishing Company, Inc., 1993.
28. Gillette P. Critical analysis of sensors for physiological responsive pacing. PACE Pacing Clin Electrophysiol 1984;7:1263.
29. Alt E, Matula M. Comparison of two activity-controlled rate-adaptive pacing principles: acceleration versus vibration. Cardiol Clin 1992; 10:635.
30. Alt E, Matula M, Theres H, et al. The basis for activity controlled rate variable cardiac pacemakers: an analysis of mechanical forces on the human body induced by exercise and environment. PACE Pacing Clin Electrophysiol 1989;12:1667.
31. Lau CP, Stott JR, Toff WD, et al. Selective vibration sensing: a new concept for activity-sensing rate-responsive pacing. PACE Pacing Clin Electrophysiol 1988;11:1299.
32. Lau CP, Stott RJ, Toff WD, et al. Selective vibration sensing: a new concept for activity-sensing rate-responsive pacing. PACE Pacing Clin Electrophysiol 1988;11:1299.
33. Lau CP, Mehta D, Toff WD, et al. Limitations of rate response of an activity-sensing rate-responsive pacemaker to different forms of activity. PACE Pacing Clin Electrophysiol 1988;11:141.
34. Lau CP, Tai YT, Fong PC, et al. Clinical experience with an activity sensing DDDR pacemaker using an accelerometer sensor. PACE Pacing Clin Electrophysiol 1992;15:334.
35. Snoeck J, Berkhof M, Claeys M, et al. External vibration interference of activity based rate responsive pacemakers. PACE Pacing Clin Electrophysiol 1992;15(part II):1841.
36. Andersen C, Oxhoj H, Arnsbo P, Lybecker H. Pregnancy and cesarean section in a patient with a rate-responsive pacemaker. PACE Pacing Clin Electrophysiol 1989;12:386.
37. Rossi P, Plicche G, Canducci G, et al. Respiratory rate as a determinant of optimal pacing rate. PACE Pacing Clin Electrophysiol 1983;6:502.
38. Rossi P, Plicche G, Canducci G, et al. Respiration as a reliable physiological sensor for controlling cardiac pacing rate. Br Heart J 1984;51:7.
39. Cummin ARC, Iyawa VI, Mehta N, et al. Ventilation and cardiac output during the onset of exercise, and during voluntary hyperventilation in humans. J Physiol (Lond) 1986;370:567.

40. Vai F, Bonnet JL, Ritter PH, ert al. Relationship between heart rate and minute ventilation, tidal volume and respiratory rate during brief and low level exercise. PACE Pacing Clin Electrophysiol 1988;11:1860.
41. Nappholtz T, Valenta H, Maloney J, et al. Electrode configurations for a respiratory impedance measurement suitable for rate responsive pacing. PACE Pacing Clin Electrophysiol 1986;9:960.
42. Webb SC, Lewis LM, Morris-Thurgood JA, et al. Respiratory-dependent pacing: a dual response from a single sensor. PACE Pacing Clin Electrophysiol 1988;11:730.
43. Lau CP, Ritchie D, Butrous GS, et al. Rate modulation by arm movements of the respiratory dependent rate responsive pacemaker. PACE Pacing Clin Electrophysiol 1988;11:744.
44. Scanu P, Guilleman D, Grollier G, et al. Inappropriate rate response of the minute ventilation rate responsive pacemaker in a patient with Cheyne-Stokes dyspnea (letter). PACE Pacing Clin Electrophysiol 1989;12:1963.
45. Kertes PJ, Mond HG. (letter). PACE Pacing Clin Electrophysiol 1990; 13:948.
46. Lau CP, Linker NJ, Butrous GS, et al. Myopotential interference in unipolar rate responsive pacemakers. PACE Pacing Clin Electrophysiol 1989; 12:1324.
47. Lau CP, Lee CP, Wong CK, et al. Rate responsive pacing with a minute ventilation sensing pacemaker during pregnancy and delivery. PACE Pacing Clin Electrophysiol 1990;13;158.
48. Madsen GM, Andersen C. Pacemaker-induced tachycardia during general anaesthesia: a case report. Br J Anaesth 1989;63:360.
49. Bazett HC. An analysis of the time-relations of electrocardiograms. Heart 1920;7:353.
50. Rickards AF, Norman J. Relation between QT interval and heart rate: new design of physically adaptive cardiac pacemaker. Br Heart J 1981;45:56.
51. Fananapazir L, Bennett DH, Faragher EB. Contribution of heart rate to QT interval shortening during exercise. Eur Heart J 1983;4:265.
52. Jordaens L, Backers J, Moerman E, et al. Catecholamine levels and pacing behavior of QT-driven pacemakers during exercise. PACE Pacing Clin Electrophysiol 1990;13:603.
53. Boute W, Gebhardt U, Begemann MJS. Introduction of an automatic QT interval drive rate responsive pacemaker. PACE Pacing Clin Electrophysiol 1988;11:1804.
54. Hedman A, Hjemdahl P, Nordlander R, et al. Effects of mental and physical stress on central hemodynamics and cardiac sympathetic nerve activity during QT interval-sensing rate-responsive and fixed rate ventricular inhibited pacing. Eur Heart J 1990;11:903.
55. Callaghan F, Vollmann W, Livingston A, et al. The ventricular depolarization gradient: effects of exercise, pacing rate, epinephrine, and intrinsic heart rate control on the right ventricular evoked response. PACE Pacing Clin Electrophysiol 1990;12:1115.
56. Singer I, Brennan AF, Steinhaus B, et al. Effects of stress and beta 1 blockade on the ventricular depolarization gradient of the rate modulating pacemaker. PACE Pacing Clin Electrophysiol 1991;14:460.
57. Lau CP. The sensing of ventricular depolarization gradient and output pulse parameters. In: Lau CP, ed. Rate adaptive cardiac pacing: single and dual chamber. Mount Kisco, NY, Futura Publishing Company, Inc., 1993, p. 142.

58. Paul V, Garrett C, Ward DE, et al. Closed loop control of rate adaptive pacing: clinical assessment of a system analyzing the ventricular depolarization gradient. PACE Pacing Clin Electrophysiol 1989;12:1896.
59. Lasaridis K, Paul VE, Katritsis D, et al. Influence of propranolol on the ventricular depolarization gradient. PACE Pacing Clin Electrophysiol 1991;14:787.
60. Wirtzfeld AL, Goedel-Mienen L, Bock T, et al. Central venous oxygen saturation for the control of automatic rate responsive pacing. Circulation 1981;64(suppl. IV):299.
61. Seifert GP, Moore AA, Graves KL, it al. In vivo and in vitro studies of a chronic oxygen saturation sensor. PACE Pacing Clin Electrophysiol 1991; 14:1514.
62. Lau CP. Central venous oxygen saturation and other sensors. In: Lau CP, ed. Rate adaptive cardiac pacing: single and dual chamber. Mount Kisco, NY, Futura Publishing Company, Inc., 1993, p. 184.
63. Fearnot NE, Smith HJ, Sellers D, et al. Evaluation of the temperature response to exercise testing in patients with single chamber, rate adaptive pacemakers: a multicenter study. PACE Pacing Clin Electrophysiol 1989; 12:1806.
64. Arakawa M, Kambara K, Hiroyasu I, et al. Intermittent oversensing due to internal insulation damage of temperature sensing rate responsive pacemaker lead in subclavian venipuncture method. PACE Pacing Clin Electrophysiol 1989;12:1312.
65. Shellock FG, Rubin SA, Ellrodt AG, et al. Unusual core temperature decrease in exercising heart-failure patients. J Appl Physiol 1983;52:544.
66. Lau CP. Central venous temperature. In: Lau CP, ed. Rate adaptive cardiac pacing: single and dual chamber. Mount Kisco, NY, Futura Publishing Company, Inc., 1993, p. 158.

Robert Gow
Frederick A. Burrows

Cardiomyopathy Associated Arrhythmias: A Review of the Problem in Children

11

The term "arrhythmia" is not limited to irregularities of the heart beat but is applied also to disturbances of rate and of conduction. Arrhythmias are being identified as having an important relationship to the prognosis of children with heart disease of many etiologies. It is well recognized that patients who have undergone previous repair of structural heart disease have a predisposition to certain arrhythmias. In particular, previous Mustard repair for transposition of the great arteries is associated with sinus node dysfunction, atrial arrhythmias, and sudden death. Similarly, ventricular arrhythmias have been noted in other patients, e.g., after repair of tetralogy of Fallot. Heart block is also a known complication of the natural history of congenitally corrected transposition of the great arteries.

It is the purpose of this chapter to examine the role of rhythm abnormalities in pediatric patients with nonstructural heart muscle disease. An arrhythmia may develop in a patient with a well-defined disorder such as idiopathic dilated cardiomyopathy, hypertrophic cardiomyopathy, or arrhythmogenic right ventricular dysplasia. Less frequently, an incessant tachycardia may cause a secondary cardiomyopathy, which is completely reversible once the arrhythmia is controlled. Uncommon disorders (connective tissue disease or neuromuscular disease) may affect the heart and have a component of either myocardial or conduction tissue dysfunction as a late consequence.

Clinical Cardiac Electrophysiology: Perioperative Considerations
Edited by Carl Lynch III. J.B. Lippincott Company, Philadelphia, PA ©1994

UNDIAGNOSED HEART MUSCLE DISEASE AND SUDDEN CARDIAC DEATH

It is important to appreciate the fact that unrecognized heart disease may be responsible for sudden death in previously healthy children. Although there have been a number of studies that reviewed the autopsy findings in young people who died suddenly, it is not always possible to be certain whether cardiac disease was suspected prior to the event.[1-6]

Driscoll[6] reviewed all the deaths in individuals 1 to 22 years of age in a small community and concluded that 4 of the 12 sudden unexpected deaths were of cardiac origin. The analysis included a further review of 13 studies that had examined sudden and unexpected deaths in children and adolescents. Hypertrophic cardiomyopathy accounted for more than half the diagnoses and death occurred during exercise in 60% of these. In the study of Corrado and colleagues[4] 6 of 17 apparent arrhythmic deaths were caused by arrhythmogenic right ventricular dysplasia, a disorder characterized by partial replacement of the right ventricular myocardium with fibrous and fatty tissue.[7] All of these individuals died during physical exertion.

Clinically unrecognized myocarditis is the other diagnosis that is disproportionately represented in some series of unexpected sudden death in young people.[1,2,5] Neuspeil and Kuller[2] identified cardiac lesions in 51 (25%) of 207 children and adolescents who died suddenly and unexpectedly. Myocarditis was present in 14 (27%) of the 51, and although 11 had prodromal symptoms only 1 had prior physician contact. Myocarditis was found in 17% of children with sudden unexpected death by Noren and colleagues.[8] In their study, 4% of children who experienced sudden, violent death also had histologic evidence of myocarditis. What is not clear from these studies is whether or not the deaths were caused by arrhythmias.

These studies highlight the association between sudden death and heart muscle disease. However, current medical technologies make it possible for individuals to survive an episode of out-of-hospital cardiac arrest, allowing a diagnosis to be established. Silka and colleagues[9] reviewed the finding in 15 such patients and identified dilated cardiomyopathy in 3 and hypertrophic cardiomyopathy in 1. The majority of children had undergone previous surgery for congenital heart abnormalities or had "electrical" problems such as Wolff-Parkinson-White syndrome or long QT syndrome. Similarly, myocarditis, dilated cardiomyopathy, or restrictive cardiomyopathy was diagnosed in 5 of 11 patients in another study of young patients with ventricular tachycardia or fibrillation.[10]

HYPERTROPHIC CARDIOMYOPATHY

Hypertrophic cardiomyopathy (HCM) has been recognized as an important clinical entity in only the last 20 years. The most feared complication is sudden death, which is said to occur in 2 to 4% of adult patients[11-13] and in 6 to 8% of children[14-17] with this disease per annum. Tachyarrhythmias (both ventricular and supraventricular) have been implicated as some of the most likely causes of sudden death in these individuals; however, other mechanisms, such as conduction abnormalities or hemodynamic collapse, may be important in other cases.

Ventricular tachycardia degenerating to fibrillation has only rarely been observed at the time of sudden death in ambulatory patients with HCM[18]; however, the prevalence of ventricular arrhythmias on 24-hr ambulatory electrocardiographic monitoring of adults with HCM has been shown. Savage and colleagues[19] described the results of monitoring in 100 patients and showed that 65% had either single polymorphic ventricular extrasystoles, couplets, or ventricular tachycardia. The prognostic role of ventricular arrhythmias was first shown by Maron and colleagues.[20] In their study, the occurrence of a sudden cardiac catastrophe was significantly more common in patients with asymptomatic ventricular tachycardia (24%) than those without (3%). The finding of ventricular tachycardia on 24-hr monitoring identified a high-risk group with a yearly mortality of 8.6%. It is important to emphasize that a causal relationship cannot be established from these data.

There are some similarities between pediatric and adult patients in regard to the presence of ventricular arrhythmias, but also some important differences. Unfortunately, the amount of available literature is quite small, and many of the studies that describe the adult population include some pediatric patients. McKenna and colleagues[16] reported an analysis of arrhythmias in 53 children aged from 6 months to 21 years with HCM. Twenty-five patients had occasional ventricular extrasystoles, but only 1 had more than 20 in 24 hr. Ventricular tachycardia was found in 4 adolescents at initial evaluation, and in 1 other during follow-up. These last 5 patients were older, had more symptoms, and had echocardiographic evidence of greater degrees of right and left ventricular hypertrophy. Five patients died suddenly and another 2 had documented out-of-hospital ventricular fibrillation; yet none of these 7 individuals had been shown to have frequent ventricular extrasystoles or ventricular tachycardia on routine monitoring. It was concluded that the absence of ventricular arrhythmia during ambulatory electrocardiographic monitoring does not indicate a low risk of sudden death.

Spontaneous day-to-day variability of ventricular arrhythmias is recognized in HCM. Mulrow et al.[21] performed 48 to 168 hr of ambulatory electrocardiographic monitoring in 16 patients with documented ventricular tachycardia. The probability of failing to detect ventricular tachycardia in this group was 48% for 24 hr of monitoring, 23% for 48 hr, and 11% for 72 hr. It was concluded that 48 to 72 hr of monitoring was a reasonable compromise for follow-up; however, 5 to 6 days of monitoring may be required at the initial evaluation to exclude the presence of significant arrhythmias.

The invasive electrophysiologic study is an important part of the investigation of selected patents with ventricular arrhythmias, syncope, and out-of-hospital cardiac arrest. There is currently a great interest in whether these studies can identify individuals at increased risk of sudden death, and whether they can predict a successful therapeutic outcome in HCM.[22] Although it is speculated that in children the precipitating event for sudden death is most often hemodynamic,[23] arrhythmia cannot be excluded. An effective diagnostic test in the pediatric age group appears to be the signal-averaged electrocardiogram. An abnormal signal-averaged electrocardiogram was found to be associated with clinical nonsustained ventricular tachycardia in adults with HCM by Cripps and colleagues,[24] but not in young patients (<25 years of age). However, only 2 of the 20 young patients had documented nonsustained ventricular tachycardia on ambulatory electrocardiography, with neither showing delayed potentials on their signal-averaged electrocardiograms. Of note is the fact that 2 of the 3 young patients with out-of-hospital cardiac arrest and ventricular fibrillation had abnormal signal-averaged electrocardiograms. This test may be a significant additional method for noninvasive assessment of young patients with HCM who may be at increased risk for sudden death. Systematic application of this technique needs to be applied in this population.

A wide range of supraventricular arrhythmias have been documented in patients with HCM.[25] McKenna et al.[16] found an 8% incidence of nonsustained atrial tachycardias in 53 children with HCM. One other child had preexcitation and recurrent atrioventricular reentry tachycardia, and 1 developed atrial fibrillation during follow-up. Frequent supraventricular extrasystoles were not mentioned. There has been a widely held belief that atrial fibrillation heralds a poor prognosis[26] and that symptomatic deterioration occurs with the onset of rapid atrial rates. Indeed, Stafford and colleagues[27] reported a youth who developed ventricular fibrillation as a consequence of rapid ventricular conduction during atrial fibrillation, while Wigle et al.[26] have shown marked elevation of left atrial pressure and pulmonary edema during atrial fibrillation. Others have shown that this arrhythmia does not represent an ominous development in the natural history of patients with HCM.[28]

Complete heart block, although very uncommon, has been associated with sudden death in the patient with HCM.[29] Postmortem studies have described the anatomic substrate as fibrosis in the sinus node, atrioventricular node, and His bundle.[30] However, much of the literature that describes ambulatory electrocardiography and exercise testing in this population does not make any specific reference to the sinus node, atrioventricular node, or distal conduction system disease in either the adult or child. Conduction abnormalities after surgical resection in the left ventricular outflow tract are well recognized with left bundle branch block being the most common.[31] Changes to the surgical approach, in particular a more anterior resection, have decreased the incidence of this complication[32] although left anterior hemiblock may still occur.

DILATED CARDIOMYOPATHY

Dilated cardiomyopathy (DCM) is an uncommon cause of cardiac disease in the pediatric age group, representing only 1 to 2.5% of hospital admissions.[33] It is well recognized, however, that the mortality in this patient population can be substantial,[34–37] with the reported 5-year survival ranging from 20 to 80%. Death in DCM may be from progressive heart failure, embolism, or sudden death (presumed arrhythmic).

Ventricular arrhythmias may occur in as many as 49% of adults with DCM.[38] The situation is unclear in the pediatric patient with DCM, and there is no overall consensus of the prognosis for a child in whom the condition has been recently diagnosed. Griffin and associates[34] reported their experience with 32 children and concluded that if a diagnosis of DCM is made in a child older than 2 years of age, the prognosis is poor. The development of significant arrhythmias was a risk factor for poor outcome. In particular, all children who died had complex ventricular ectopy with 2 having nonsustained ventricular tachycardia. Treatment of the arrhythmia was not associated with an improved outcome. Such arrhythmias were found to be very frequent in infants and children with DCM by Lewis and Chabot.[37] In fact, rhythm disturbances were found in only 16% of survivors compared with 53% of those who died. Ventricular arrhythmias included frequent extrasystoles, nonsustained ventricular tachycardia, and ventricular fibrillation. Sudden death occurred in 11 individuals, with 73% of these resulting from an arrhythmia. However, two other recent studies were not able to identify arrhythmias as independent predictors of poor outcome.[35,36] Complex ventricular arrhythmias, including ventricular tachycardia, were evenly distributed between survivors and nonsurvivors in the report of Chen and colleagues.[35] Similarly, Friedman et al.[36] specifically examined the significance of arrhythmias in relation to prognosis . Ventricular arrhythmias were present in 34% of those who had arrhythmias found on 24-hr ambulatory electrocardiograms, with 10% having ventricular tachycardia. The presence of ventricular arrhythmias did not predict a worse outcome, and treatment of the arrhythmia did not influence the mortality.

Although all types of supraventricular arrhythmias have been described in DCM, atrial fibrillation is particularly frequent and occurs in 10 to 20% of patients. Hofmann and colleagues[39] showed that atrial fibrillation increased the risk of sudden death and death from heart failure in adult patients. Atrial arrhythmias were diagnosed in 22% of the pediatric patients described by Friedman and associates,[36] and were more common than ventricular arrhythmias (22% versus 34%).

ARRHYTHMOGENIC RIGHT VENTRICULAR DYSPLASIA

Arrhythmogenic right ventricular dysplasia (ARVD) is a condition of unknown etiology that predominantly affects the right ventricle and, in its typical form, is characterized by fibro-fatty replacement of the myocardium.[7] An evolving concept suggests that ARVD represents a part of the spectrum of a generalized cardiomyopathic process.[40] Left ventricular abnormalities are being recognized more frequently and may be evidenced by latent dysfunction[40] or profound histologic involvement.[41] The clinical spectrum of ARVD is also variable. Patients may have minimal symptoms and occasional ventricular arrhythmias[42] or sudden death.[4] One report revealed that ARVD was the most common cardiac cause of sudden arrhythmic death in young athletes[4]; however, it was not found in any of 15 pediatric survivors of out-of-hospital cardiac arrest in the study of Silka and colleagues.[9]

The familial occurrence of ARVD is also well recognized,[43,44] and the genetic pattern is consistent with an autosomal dominant condition with variable penetrance and expression. Sporadic forms occur and appear to be clinically indistinguishable from the familial form. Buja and colleagues[44] found clinical or subclinical ARVD in 35% of family members after a diagnosis was made in an index case. The polymorphism of the clinical picture has been stressed,[42,45] and the severity of the clinical arrhythmias may not be related to the extent of the right ventricular abnormality found on investigation.[45]

The prevalence of ARVD in pediatric patients who are seen with ventricular arrhythmias is unclear. Arrhythmogenic right ventricular dysplasia was found in 35% of 24 children without overt heart disease who had clinical ventricular tachycardia at presentation.[46] Ventricular tachycardia on exercise testing has been suggested as an important indicator of the presence of ARVD in children with no apparent heart disease and has been diagnosed in up to 75% of these patients.[46–48] Similarly, Fauchier et al.[49] diagnosed ARVD in 14 of 20 young patients who had a range of clinical ventricular arrhythmias. Only 3 of these individuals were younger than 22 years old, and ARVD was found in 1. What is important from this study is the analysis of the clinical arrhythmias, and their response to exercise testing. Seven patients had clinical ventricular tachycardia on ambulatory electrocardiography, with 6 of these having exercise-induced ventricular tachycardia. Another 7 patients had only frequent isolated ventricular extrasystoles on monitoring, and none of these had exercise-induced ventricular tachycardia. In fact, the extrasystoles disappeared during exercise in all these patients—a so-called "normal" response. The echocardiogram was abnormal in only 2 of this group. From these data it appears that in patients with frequent isolated ventricular extrasystoles that disappear with exercise and a normal echocardiogram the prognosis may not have be benign.

Both the standard 12-lead electrocardiogram and the signal-averaged electrocardiogram may be abnormal in patients with ARVD. The typical electrocardiographic abnormality in ARVD is T-wave negativity in the precordial leads; this may be found in up to 85% of cases.[50] In mild forms of the disease the T-wave abnormalities may be restricted to the right chest leads, and a relationship has been demonstrated between right ventricular volume and the number of precordial leads with negative T waves.[51] For the pediatric patient the sensitivity of this finding is problematic because of the influence of age-related changes in the electrocardiogram. Most children have milder (or earlier) forms of ARVD, and positive T waves in all the precordial leads may not be normally be seen until adolescence. The signal-averaged electrocardiogram may be positive for late potentials in 75 to 85% of patients with ARVD and ventricular tachycardia.[52–54] Haissaguerre and colleagues[52] demonstrated that late potentials occur more frequently in ARVD patients with clinical sustained ventricular tachycardia than in those with nonsustained ventricular tachycardia, frequent extrasystoles, or a family history of sudden death. Similarly, late potentials were found in nearly all patients in whom programmed stimulation induced sustained ventricular tachycardia compared with 38% of those in whom programmed stimulation induced either nonsustained ventricular tachycardia or no arrhythmia.

The 24-hr ambulatory electrocardiogram almost invariably identifies ventricular arrhythmias in these patients. As has already been shown, the severity of the arrhythmia is quite variable. Children with ventricular tachycardia demonstrated on ambulatory monitoring are more likely to have exercise-induced ventricular tachycardia,[46] and further investigation frequently reveals ARVD. However, because of the lack of sensitivity of the echocardiogram, children with mild (or early) disease who have isolated ventricular extrasystoles and a "normal" exercise test are likely to be followed sporadically, if at all. Therefore, it will be some time before it becomes clear in which children arrhythmias that have benign features will progress to clinically diagnosable ARVD in their 20s or 30s. It is to be hoped that further information which has better predictive capabilities will become available.

An interesting approach to the noninvasive diagnosis of arrhythmias in ARVD was taken by Haissaguerre et al.[52] They have proposed that there is a distinctive response to an infusion of isoproterenol in patients with ARVD. It was found that 8 to 30 µg/m in induced ventricular tachycardia in 88% of their ARVD population, compared to 2% in control patients. The pattern was polymorphic in the majority of patients, and this was considered to be the distinctive response. There was a good correlation with findings on a signal-averaged electrocardiogram and programmed stimulation, and the tests were considered complementary.

Although the overwhelming interest in ARVD has been its association with ventricular arrhythmias, Tonet and associates[55] have described supraventricular arrhythmias in 24% of 72 consecutive patients with ARVD. The arrhythmias were atrial tachycardia, a trial flutter, and atrial fibrillation. There did not appear to be any relationship to underlying ventricular involvement or atrial dimension on the echocardiogram. Clinical sick sinus syndrome has also been reported in these patients, and biopsy material showed fatty replacement of the myocardium in both the atrium and ventricle.[56] No comparable information is available for children.

MYOCARDITIS

Myocarditis is an inflammatory process involving the heart and may be caused by an infectious or noninfectious process. The infections that cause myocarditis are usually viral, but any organism may be responsible. Noninfectious myocarditis can be caused by physical agents, toxins, or generalized inflammatory processes.

The arrhythmic consequences of myocarditis may be acute, chronic, or subclinical and can be out of proportion to the other clinical manifestations. Sudden unexpected death occurs infrequently in the young population, and autopsy examination may be required to reveal evidence of myocarditis. Topaz and Edwards[5] found histologic evidence of myocarditis in 24% of postmortem examinations performed on young people from a nonselected population who died suddenly. The mechanisms of sudden death are probably arrhythmic if no or minimal symptoms were present prior to the event and would be due to ventricular arrhythmias or complete heart block. Histologic evidence of myocarditis is found in 4 to 10% of routine autopsies.[57] Other series have confirmed the increased incidence of myocarditis in these individuals.[8,58] In competitive athletes, postmortem evidence of myocarditis is less frequent than evidence of HCM[3] or ARVD.[4]

Ventricular arrhythmias occurring during clinical episodes of myocarditis are well described. Dec and colleagues[59] examined the relation between histologic and clinical features in 27 myocarditis patients who exhibited heart failure. Most patients had 24-hr ambulatory electrocardiograms during the acute illness, and ventricular tachycardia was found in 11. Sudden death occurred in 4 individuals, 2 of whom had documented ventricular tachycardia. In this entire group, 4 patients were younger than 25 years of age, and ventricular tachycardia was present in 1. Although the overall prevalence of ventricular tachycardia was high in this study, the mean left ventricular ejection fraction at presentation was only 22%, indicating extensive disease. Ventricular tachycardia has recently been described in Lyme carditis.[60] This is an important consideration in endemic areas because the cardiac involvement can progress if it is not treated, and antibiotics may be an effective treatment.

A population-based study of young military recruits with electrocardiographic changes during acute infection and the diagnosis of mild myocarditis revealed ventricular extrasystoles in only 13%.[61] In the majority of these patients the extrasystoles either persisted or increased with exercise. These authors also performed electrocardiograms on 68 consecutive individuals with infectious symptoms. In this group, findings consistent with myocarditis were noted in 6%, and ventricular extrasystoles in 1%. Similar population studies in children with viral illnesses are needed.

The importance of myocarditis as a cause of severe arrhythmias in young patients has been shown. Myocarditis was diagnosed in 3 of 38 children who had ventricular tachycardia.[62] One died acutely and the other 2 recovered. Interestingly, the 2 who survived had slow ventricular tachycardia with rates of 130 to 150/min while the other child had severe cardiovascular collapse and tachycardia rates up to 300/min. Myocarditis may also be a cause of neonatal ventricular arrhythmias[63] and needs to be distinguished from accelerated ventricular rhythm, which frequently has a good prognosis.[64] In the pediatric survivor of cardiac arrest, myocarditis is an infrequent diagnosis and was seen in only one patient in the report of Benson and colleagues[10] and in none of the patients investigated by Silka et al.[9] Isolated case reports of children with severe ventricular arrhythmias and autopsy-proven myocarditis have been published.[65]

Lymphocytic myocarditis has been described as a low-grade, chronic inflammation. Vignola and colleagues[66] reported cardiac biopsy results from 17 patients who demonstrated no evidence of heart disease but who had severe ventricular arrhythmias; lymphocytic myocarditis was diagnosed in 6. After therapy with immunosuppressive agents, the inflammation resolved or improved in all, and ventricular arrhythmias were noninducible in 5. These results, although from an uncontrolled study, suggest that a chronic inflammatory process may be responsible for severe arrhythmias in patients previously thought to have 'primary electrical disease." The resolution of the arrhythmias with immunosuppressive therapy is encouraging; however, a larger controlled study is needed to confirm these results. Recently, similar findings in children have indicated that subclinical myocarditis is present in some children with complex ventricular arrhythmias.

In general, the prognosis for patients with acute myocarditis and ventricular arrhythmias is determined by the extent of the myocardial involvement. In most patients whose myocardial function returns to normal arrhythmias have resolved, and if death occurred during the acute phase, the myocardial damage was probably extensive. Arrhythmias may persist if a chronic cardiomyopathic state persists .

The other important arrhythmia associated with acute myocarditis is sudden complete heart block. Predominantly single case reports[67] were published until Lim et al.[68] reported on 10 children and young adults with Stokes-Adams attacks caused by acute myocarditis. Initial electrocardiography revealed complete heart block with ventricular rates from 28 to 48/min. Broad complex escape rhythms were noted in 80%. Ventricular tachycardia was also present in 2 patients. Initial management included temporary transvenous pacing in 9. Normal conduction returned in 1 to 12 hr, although relapses occurred in 3 individuals. By hospital discharge, 8 patients showed normal atrioventricular conduction. Complete heart block has been described in children with viral infections such as mumps[69] and in Mycoplasma *pneumoniae* myocarditis.[70] The patient with mumps myocarditis had permanent complete heart block. Lyme carditis may also cause heart block. In a prospective study of 61 patients with Lyme disease, one developed acute heart block which resolved within 2 weeks.[71] An invasive electrophysiologic study revealed that the block was at the level of the atrioventricular node, but the distal conduction tissues were involved as well. The overall prognosis for patients with heart block caused by myocarditis appears to be favorable; however, recovery is not universal. Dysrhythmias and conduction abnormalities associated with these various myopathic states are summarized in Table 11-1.

TACHYCARDIA-INDUCED CARDIOMYOPATHY

In addition to their occurrence as a consequence of cardiomyopathies, it is now accepted that chronic elevation of the heart rate can cause deterioration of normal ventricular function. The etiologic role of atrial arrhythmias in the production of a cardiomyopathy in children was suggested by Keane and associates.[72] In their report, heart size normalized in four of five children with atrial ectopic tachycardia following digoxin therapy. Since then the relationship between chronic tachycardia and a secondary cardiomyopathy has been established in both children and adults and in experimental models designed to elucidate the pathophysiology of this process.

Supraventricular arrhythmias that have been implicated in tachycardia-induced cardiomyopathy are atrial fibrillation, atrial flutter, atrial ectopic tachycardia, and the permanent form of junctional reciprocating tachycardia, which may utilize a decremental accessory pathway or be "intranodal."[73–78] Pharmacologic management of the arrhythmias has not been uniformly successful. Amiodarone and propranolol were found to be equally effective pharmacologic treatments for children demonstrating atrial ectopic tachycardia with reduced ventricular function.[78] If medical therapy fails, a surgical cure is frequently possible[74–79] with resolution of a dilated cardiomyopathy after surgical ablation of the ventricular tachycardia.[80] Similar results have now been described following radiofrequency ablation of atrial ectopic tachycardia.

Although recovery from the cardiomyopathy can be quite rapid, most studies have shown that at least 1 week is required before any substantial improvement.[74,81] Although continued improvement over many months is often seen, not all patients will show complete normalization of left ventricular function.[74]

In summary, chronic supraventricular or ventricular tachycardia can produce a cardiomyopathy that is frequently reversible. The diagnosis needs to be considered in all children who are seen with newly diagnosed ventricular dysfunction before a diagnosis of idiopathic DCM is made. In many cases differentiation from sinus tachycardia is difficult,[82] and invasive electrophysiologic studies with mapping may be required. Consideration should be given to radiofrequency ablation of the focus if it can be adequately localized. Successful pharmacologic therapy may include propafenone, sotalol, amiodarone, or flecainide, while surgical ablation remains an option if other techniques are ineffective.

TABLE 11-1. DYSRHYTHMIAS ASSOCIATED WITH CARDIOMYOPATHIES IN THE PEDIATRIC POPULATION

| | Complication | | | | |
Disease Process	Supraventricular Tachycardia	Ventricular Arrhythmias	Conduction Abnormalities	Late Potentials on Signal-Averaged ECG	Sudden Cardiac Death in Children and Adolescents
Hypertrophic cardiomyopathy	8%	47–65% (?nonpredictive)	Rare	Yes	~50% (6–8% of patients per year) Probable cause
Dilated cardiomyopathy	22% (10–20% AF)	34% (10% VT) (?nonpredictive)			
Arrhythmogenic right ventricular dysplasia	24%	75% with exercise; 85% with isoproterenol		75–85%	Probable cause (~20%)
Myocarditis		13–50% (increases with severity)	CHB occasional		4–24%

ECG, electrocardiogram; AF, atrial fibrillation; VT, ventricular tachycardia; CHB, complete heart block.

SYSTEMIC DISORDERS WITH CARDIAC MANIFESTATIONS

There are a variety of disorders in which cardiac involvement is an important manifestation of disease, but the clinical expression is dominated by other features. Examples are neuromuscular, connective tissue, and hematologic disease. This section will focus on the rhythm abnormalities that can be found in some of these conditions.

Neuromuscular Disorders

Duchenne's muscular dystrophy is an hereditary (X-linked) myopathy in which dystrophin, an intrinsic muscle membrane protein, is absent. The typical electrocardiogram of patients with Duchenne's muscular dystrophy shows tall R waves in the right chest and deep, narrow Q waves in the left chest.[83] Short PR intervals have been found and could represent accelerated atrioventricular node conduction or atriofascicular bypass tracts although classical Wolff-Parkinson-White syndrome is rarely seen. There have been a number of studies that document the presence of rhythm and conduction abnormalities in Duchenne's muscular dystrophy.

Perloff[84] examined 20 patients aged 4 to 24 years with standard and 24-hr electrocardiography. Abnormal interatrial conduction was shown in 20% of individuals, one of whom developed atrial flutter. A short PR interval was found in 35%, right ventricular conduction delay was found in 35%, and left anterior hemiblock was suspected in 20%. It was not clear whether the majority of these changes were present in the same individuals. Inappropriate sinus tachycardia was present in 65% of patients. Frequent atrial and ventricular extrasystoles were common, and ventricular tachycardia occurred in one patient. The cause of the sinus tachycardia is unknown; however, speculation includes abnormalities of autonomic function, sinoatrial node fibrosis, or as a compensatory change for decreased myocardial function. Histologic evidence has been provided that shows abnormalities of the specialized conducting tissues. A morphologic examination of hearts from three patients who manifested conduction abnormalities on an electrocardiogram revealed degenerative changes in the sinoatrial node, atrioventricular node, bundle of His, and bundle branches.[85] While dystrophin is normally present in cardiac Purkinje fibers, the absence of this critical protein in conduction tissues of patients with Duchenne's muscular dystrophy has been recently documented.[86]

Frequent arrhythmias were also noted by D'Orsogna and colleagues.[87] In particular, sinus tachycardia was present in all 18 of their patients and frequent and complex ventricular ectopy were both noted in 17% and were related to age. Interestingly, no patients had conduction abnormalities. Signal-averaged electrocardiograms are abnormal in up to 50% of patients with advanced Duchenne's muscular dystrophy,[88] and an association has been found between the frequency of ventricular extrasystoles and the presence of late potentials. It is likely that late potentials are a manifestation of diffuse cardiac involvement, and it remains to be seen if any correlation with sudden death is found.

Late onset, slowly progressive (Becker) dystrophy has an associated cardiomyopathy in up to 80% of individuals. The cardiac features are of a dilated cardiomyopathy with frequent involvement of the atrioventricular node and distal conduction system. Some forms of the disease may manifest primarily as atrial arrhythmias and heart block.[89]

The cardiac involvement in patients with *fascioscapulohumeral muscular dystrophy* was initially characterized by permanent atrial paralysis. Stevenson and colleagues[90] investigated 30 patients with electrocardiographic monitoring, and a third had invasive electrophysiology studies. Sinus rhythm was predominant, and atrial paralysis was not observed. Minor electrocardiographic changes were noted in 24%; however, significant infranodal conduction delay was not observed. Atrial flutter and fibrillation were easily induced and sinus node dysfunction was noted in 10%. It is probable that the cases of atrial paralysis were observed in a phenotypically similar disorder, *Emery-Dreifuss dystrophy.* The cardiac involvement in Emery-Dreifuss dystrophy is now characterized by atrial paralysis, atrial arrhythmias, and infranodal conduction defects.[91]

The cardiac manifestations of *myotonic muscular dystrophy* are mostly related to the conduction system and can cause Stokes-Adams attacks from sudden heart block.[92] Atrial and ventricular arrhythmias are frequently present. Invasive electrophysiologic studies reveal distal conduction disease in 80% of patients.[93] Histologic findings are consistent with the clinical features, and atrophy of the atrioventricular node and bundle branches is common.[94] In contrast, the heart does not appear to be involved in *myotonia congenita*.

Progressive distal conduction block is an inherent feature of the *Kearns-Sayre syndrome.*. The syndrome is possibly caused by mitochondrial deoxyribonucleic acid deletion.[95] Clinically, the electrocardiographic appearances progress from left anterior hemiblock and right bundle branch block to complete heart block. A short PR interval is frequently present and, from electrophysiology studies, has been shown to be caused by enhanced atrioventricular node conduction.[96] Because of the risk of sudden death, elective cardiac pacing should be considered in these patients.[97]

Inflammatory and Connective Tissue Disorders

Autopsy reports from individuals with systemic lupus erythematosus have shown histologic evidence of myocarditis in 25 to 50%.[98] The incidence has decreased since the universal use of steroids. Clinically, myocarditis may manifest itself as conduction abnormalities or arrhythmias; atrial fibrillation and frequent ventricular extrasystoles may occur. Heart block may cause Stokes-Adams attacks.[99] *Systemic sclerosis* has long been known as a cause of conduction and rhythm disorders. Electrocardiographic abnormalities were reported in 75% of patients with visceral involvement compared with 25% of those without.[100] The standard electrocardiogram shows evidence of distal conduction disease in 5 to 10% of patients, and the extent of the conduction disorder seems to correlate with the degree of myocardial involvement. The 24-hr ambulatory electrocardiogram identifies more abnormalities than the standard electrocardiogram. Premature atrial extrasystoles and supraventricular tachycardia occur in up to 60% of patients, while complex ventricular ectopy or ventricular tachycardia are found in 30%.[101] These latter arrhythmias are strong independent risk factors. Forty to 60% of all deaths in systemic sclerosis are sudden and presumably caused by ventricular tachycardia or fibrillation.[102] Atrioventricular block occurs in other connective tissue diseases. *Rheumatoid arthritis* may be complicated by atrial conduction abnormalities and heart block.[103] Electrocardiographic abnormalities were present in 52% of patients with *polymyositis* and included evidence of atrioventricular block and atrial and ventricular arrhythmias.[104]

ANESTHETIC IMPLICATIONS AND CONCLUSIONS

In this chapter we have focused on the arrhythmic manifestations of cardiomyopathies present in the pediatric population. The proper preoperative preparation and anesthetic management of patients with known arrhythmogenic cardiac disease requires close cooperation between the surgeon, cardiologist, and anesthesiologist. There is no place for last minute cardiac or anesthesia consultations in these patients, and this requires that surgeons be aware of the need for preadmission cardiac and possibly anesthetic consultations to enable the specific concerns for the particular patient to be properly addressed. Preoperative echocardiograms, Holter monitoring, and exercise data can be gathered as needed to assess myocardial function and predisposition to arrhythmia development. This enables planning for any special precautions. Assessment of myocardial function is as important as assessment of arrhythmogenic potential as they both may affect the anesthesiologist's decision as to which anesthetic technique to use. Unless patients are undergoing arrhythmia procedures, it is suggested that antiarrhythmic medications be continued before surgery and recommended immediately after surgery. This may require the conversion of some oral medications to an intravenous administration until normal oral intake is reestablished.

In studies of adult patients with underlying cardiac disease, frequent premature ventricular contractions and premature atrial contractions as well as rhythms other than normal sinus or atrial fibrillation have been implicated as risk factors,[105,106] although they are not as important in asymptomatic persons without any underlying heart disease.[107] However, there are few data regarding such preoperative arrhythmias as a cardiac risk factor in pediatric patients. What is apparent from the literature is that there is still an enormous uncertainty surrounding many of the reported observations. Unlike the adult studies,[107] the presence of arrhythmias in otherwise clinically normal pediatric patients may indicate a previously undiagnosed underlying cardiomyopathy with reduced myocardial reserve, and such patients are a priority because of the potential for exaggerated responses to myocardial depressant and arrhythmogenic anesthetic agents. Currently the preoperative preparation is complicated by the controversies surrounding appropriate drug management of these conditions. Innovations such as programmed stimulation of the ventricle and signal-averaged electrocardiograms will assist in the identification, assessment, and preoperative preparation of these patients.

References

1. Molander N. Sudden natural death in later childhood and adolescence. Arch Dis Child 1982;57:572.
2. Neuspeil DR, Kuller LH. Sudden and unexpected natural death in competitive athletes. JAMA 1985;254:1321.
3. Maron BJ, Epstein SE, Roberts WE. Causes of sudden death in competitive athletes. J Am Coll Cardiol 1986;7:204.
4. Corrado D, Thiene G, Nava A, Rossi L, Pennelli N. Sudden death in young competitive athletes: clinicopathologic correlations in 22 cases. Am J Med 1990;89:588.

5. Topaz O, Edwards JE. Pathologic features of sudden death in children, adolescents, and young adults. Chest 1985;87:476.
6. Driscoll DJ. Sudden unexpected deaths in children and adolescents. J Am Coll Cardiol 1985;5(suppl B):118.
7. Fontaine G, Guiraudon G, Frank R, Cabrol C, Grosgogeat Y. Arrhythmogenic right ventricular dysplasia: a previously unrecognized syndrome. Circulation 1979;52(suppl II):65.
8. Noren GR, Staley NA, Bandt CM, Kaplan EL. Occurrence of myocarditis in sudden death in children. J Forensic Sci 1977;22:188.
9. Silka MJ, Kron J, Walance CG, Cutler JE, McAnulty JH. Assessment and follow-up of pediatric survivors of sudden cardiac death. Circulation 1990;82:341.
10. Benson DW, Benditt DG, Anderson RW, et al. Cardiac arrest in young, ostensibly health patients: clinical, hemodynamic, and electrophysiologic findings. Am J Cardiol 1983;52:65.
11. Shah PM, Adelman AG, Wigle ED, et al. The natural (and unnatural) history of hypertrophic obstructive cardiomyopathy. Circ Res1 973;34/35 (suppl II):179.
12. Swan DA, Bell B, Oakley CM, Goodwin JF. Analysis of symptomatic course and prognosis and treatment of hypertrophic obstructive cardiomyopathy. Br Heart J 1971;33:671.
13. Adelman AG, Wigle ED, Ranganathan N, et al. The clinical course in muscular subaortic stenosis: a retrospective and prospective study of 60 hemodynamically proved cases. Ann Intern Med 1972;77:515.
14. Maron BJ, Henry WL, Clark CE, Redwood DR, Roberts WC, Epstein SE. Assymetric septal hypertrophy in childhood. Circulation 1976;5 3:9.
15. Maron BJ, Roberts WC, Epstein SE. Sudden death in hypertrophic cardiomyopathy: a profile of 78 patients. Circulation 1982;67:1388.
16. McKenna WJ, Franklin RCG, Nihoyannopoulos P, et al. Arrhythmia and prognosis in infants, children and adolescents with hypertrophic cardiomyopathy. J Am Coll Cardiol 1988;11:147.
17. McKenna WJ, Deanfield JE. Hypertrophic cardiomyopathy: an important cause of sudden death. Arch Dis Child 1984;59:971.
18. Nicod P, Polikar R, Peterson KL. Hypertrophic cardiomyopathy and sudden death. N Engl J Med 1988;318:1255.
19. Savage DD, Seides SF, Maron BJ, Myers DJ, Epstein SE. Prevalence of arrhythmias during 24-hour electrocardiographic monitoring and exercise testing in patients with obstructive and nonobstructive hypertrophic cardiomyopathy. Circulation 1979;59:866.
20. Maron BJ, Savage DD, Wolfson JK, Epstein SE. Prognostic significance of 24 hour ambulatory electrocardiographic monitoring in patients with hypertrophic cardiomyopathy: a prospective study. Am J Cardiol 1981; 48:252.
21. Mulrow JP, Healy MJ, McKenna WJ. Variability of ventricular arrhythmias in hypertrophic cardiomyopathy and implications for treatment. Am J Cardiol 1986;58:615.
22. Fananapazir L, Tracey CM, Leon MB, et al. Electrophysiologic abnormalities in patients with hypertrophic cardiomyopathy: a consecutive analysis in 155 patients. Circulation 1989;80:1259.
23. McKenna WJ, Camm AJ. Sudden death in hypertrophic cardiomyopathy: assessment of patients at high risk. Circulation 1989;80:1489.

24. Cripps TR, Counihan PJ, Frenneaux MP, Ward DE, Camm AJ, McKenna WJ. Signal-averaged electrocardiography in hypertrophic cardiomyopathy. J Am Coll Cardiol 1990;15:956.

25. Krikler DM, Davies MJ, Rowland E, Goodwin JF, Evans RC, Shaw DB. Sudden death in hypertrophic cardiomyopathy: associated accessory atrioventricular pathways. Br Heart J 1980;43:245.

26. Wigle ED, Sasson Z, Henderson MA, et al. Hypertrophic cardiomyopathy: the importance of the site and the extent of hypertrophy: a review. Prog Cardiovasc Dis 1985;28:1.

27. Stafford WJ, Trohman RG, Bilsker M, Zaman L, Castellanos A, Myerburg RJ. Cardiac arrest in an adolescent with atrial fibrillation and hypertrophic cardiomyopathy. J Am Coll Cardiol 1983;15:70.

28. Robinson K, Frenneaux MP, Stockins B, Karatasakis G, Poloniecki JD, McKenna WJ. Atrial fibrillation in hypertrophic cardiomyopathy: a longitudinal study. J Am Coll Cardiol 1990;15:1279.

29. Tajik AJ, Giuliani ER, Frye RL. Muscular subaortic stenosis associated with complete heart block. Am J Cardiol 1973;31:101.

30. James TN, Marshall TK. De subitaneis mortibus. XII. Asymmetric hypertrophy of the heart. Circulation 1975;51:1149.

31. Maron BJ, Epstein SE, Morrow AG. Symptomatic status and prognosis of patients after operation for hypertrophic obstructive cardiomyopathy: efficacy of ventricular septal myotomy and myectomy. Eur Heart J 1983; 4(suppl F):175.

32. Williams WG, Mehta H, Wigle ED, et al. Results of operation for obstructive idiopathic subaortic stenosis. Proc Can Cardiovasc Soc Clin Invest Med 1982;5(suppl 1):30.

33. Colan SD, Spevak PJ, Parness IA, Nadas AS. Cardiomyopathies. In: Fyler DC, ed. Nadas' pediatric cardiology. Philadelphia: Hanley & Belfus, 1992:329.

34. Griffin ML, Hernandez A, Martin TC, et al. Dilated cardiomyopathy in infants and children. J Am Coll Cardiol 1988;11:139.

35. Chen S-C, Nouri S, Balfour I, Jureidini S, Appleton RS. Clinical profile of congestive cardiomyopathy in children. J Am Coll Cardiol 1990;15:189.

36. Friedman RA, Moak JP, Garson A Jr. Clinical course of idiopathic dilated cardiomyopathy in children. J Am Coll Cardiol 1991;18:152.

37. Lewis AB, Chabot M. Outcome of infants and children with dilated cardiomyopathy. Am J Cardiol 1991;68:365.

38. Meinertz T, Hoffmann T, Kasper W, et al. Significance of ventricular arrhythmias in idiopathic dilated cardiomyopathy. Am J Cardiol 1984; 53:902.

39. Hofmann T, Meinertz T, Kasper W, et al. Mode of death in idiopathic dilated cardiomyopathy: a multivariate analysis of prognostic determinants. Am Heart J 1988;116:1455.

40. Manyari DE, Klein GJ, Gulamhusein S, et al. Arrhythmogenic right ventricular dysplasia: a generalized cardiomyopathy? Circulation 1983; 68:251.

41. Shoji T, Kaneko M, Onodera K, et a. Arrhythmogenic right ventricular dysplasia with massive involvement of the left ventricle. Can J Cardiol 1991;7:303.

42. Rizzon P. Polymorphism of the clinical picture. Eur Heart J 1989;10(suppl D):10.

43. Nava A, Thiene G, Canciani B, et al. Familial occurrence of right ventricular dysplasia: a study involving nine families. J Am Coll Cardiol 1988;12:1222.

44. Buja GF, Nava A, Martini B, Canciani B, Thiene G. Right ventricular dysplasia: a familial cardiomyopathy? Eur Heart J 1989;10(suppl D):13.

45. Nava A, Scognamiglio R, Thiene G, et al. A polymorphic form of familial arrhythmogenic right ventricular dysplasia. Am J Cardiol 1987;59:1405.

46. Deal B, Miller S, Scagliotti D, et al. Ventricular tachycardia in the young without overt heart disease. Circulation 1985;72(suppl III):340A.

47. Palileo E, Ashley W, Swiryn S, et al. Exercise provocable right ventricular outflow tract tachycardia. Am Heart J 1982;104:185.

48. Soloman S, Osdol V, Massumi A, Warda M, Hall RJ. Exercise induced ventricular tachycardia and arrhythmogenic right ventricular dysplasia: electrophysiologic and therapeutic considerations. Texas Heart Inst J 1983;10:351.

49. Fauchier JP, Desveaux B, Cosnay P, Raynaud P, Philippe L, Itti R. Troubles du rhthme ventriculaire complexes du sejet jeune apparemment sain. Arch Mal Coeur 1985;9:1333.

50. Marcus FK, Fontaine MD, Fuiraudon G, et al. Right ventricular dysplasia: a report of 24 adult cases. Circulation 1982;65:384.

51. Nava A, Canciani B, Buja G, et al. Electrovectorcardiographic study of negative T waves on precordial leads in arrhythmogenic right ventricular dysplasia: relationship with right ventricular volumes. J Electrocardiol 1988;21:239.

52. Haissaguerre M, Le Metayer P, D'Ivernois C, Barat JL, Montserrat P, Warin J. Distinctive response of arrhythmogenic right ventricular disease to high dose isoproterenol. PACE Pacing Clin Electrophysiol 1990; 13:2119.

53. Lemery R, Brugada P, Janssen J, Cherieux E, Dugernier T, Wellens HJJ. Nonischemic sustained ventricular tachycardia: clinical outcome in 12 patients with arrhythmogenic right ventricular dysplasia. J Am Coll Cardiol 1989;1989:96.

54. Blomstrom-Lundqvist C, Olsson S, Edvardsson N. Follow-up by repeated signal-averaged surface QRS in patients with the syndrome of arrhythmogenic right ventricular dysplasia. Eur Heart J 1989;10(suppl D):54.

55. Tonet JL, Castro-Miranda R, Iwa T, Poulain F, Frank R, Fontaine GH. Frequency of supraventricular tachyarrhythmias in arrhythmogenic right ventricular dysplasia. Am J Cardiol 1991;67:1153.

56. Nogami A, Adachi S, Nitta J, et al. Arrhythmogenic right ventricular dysplasia with sick sinus syndrome and atrioventricular conduction disturbance. Jpn Heart J 1990;31:417.

57. Stevens PJ, Underwood Ground KE. Occurrence and significance of myocarditis in trauma. Aerospace Med 1970;41:776.

58. Wentworth P, Jentz LA, Croal AE. Analysis of sudden unexpected death in southern Ontario, with emphasis on myocarditis. Can Med Assoc J 1979;120:676.

59. Dec GW Jr, Palacios IF, Fallon JT, et al. Active myocarditis in the spectrum of acute dilated cardiomyopathies: clinical features, histologic correlates, and clinical outcome. N Engl J Med 1985;121:1558.

60. Vlay SC, Dervan JP, Elias J, Dattwyler R. Ventricular tachycardia associated with Lyme carditis. Am Heart J 1991;121:1558.

61. Heikkila J, Karjalainen J. Evaluation of mild acute infectious myocarditis. Br Heart J 1982;47:381.
62. Rocchini AP, Chun PO, Dick M. Ventricular tachycardia in children. Am J Cardiol 1981;47:1091.
63. Stevens DC, Schreiner RL, Hurwitz RA, Gresham EL. Fetal and neonatal ventricular arrhythmia. Pediatrics 1977;63:771.
64. Van Hare GF, Stranger P. Ventricular tachycardia and accelerated ventricular rhythm presenting in the first month of life. Am J Cardiol 1991;67:42.
65. Woolf PK, Chung T-S, Stewart J, Lialios M, Davidian M, Gewitz MH. Life-threatening dysrhythmias in varicella myocarditis. Clin Pediatr 1987;26:480.
66. Vignola PA, Aonuma K, Swayze PS, et al. Lymphocytic myocarditis presenting as unexplained ventricular arrhythmias: diagnosis with endomyocardial biopsy and response to immunosuppression. J Am Coll Cardiol 1984;4:812.
67. Johnson JL, Phang Lee L. Complete atrioventricular heart block secondary to acute myocarditis requiring intracardiac pacing. J Pediatr 1971; 78:312.
68. Lim C-H, Toh CCS, Chia B-N, Low L-P. Stokes-Adams attacks due to acute nonspecific myocarditis. Am Heart J 1975;90:172.
69. Arita M, Ueno Y, Masuyama Y. Complete heart block in mumps myocarditis. Br Heart J 1981;46:342.
70. Agarwala B, Ruschhaupt DG. Complete heart block from mycoplasma pneumoniae infection. Pediatr Cardiol 1991;12:233.
71. Rubin DA, Sobera C, Nikitin P, McAllister A, Wormser GP, Nadelman RB. Prospective evaluation of heart block complicating early Lyme disease. PACE Pacing Clin Electrophysiol 1992;15:252.
72. Keane JF, Plauth WH, Nadas AS. Chronic ectopic tachycardia of infancy and childhood. Am Heart J 1972;84:748.
73. Phillips E, Levine SA. Auricular fibrillation without other evidence of heart disease: a cause of reversible heart failure. Am J Med 1949;7:479.
74. Packer DL, Bardy GH, Worley SJ, et al. Tachycardia-induced cardiomyopathy: a reversible form of left ventricular dysfunction. Am J Cardiol 1986;57:563.
75. McLaren CJ, Gersh BJ, Sugrue DD, Hammill SC, Seward JB, Holmes DR Jr. Tachycardia induced myocardial dysfunction: a reversible phenomenon? Br Heart J 1985;53:323.
76. Gillette PC, Smith RT, Garson A Jr, et al. Chronic supraventricular tachycardia: a curable cause of congestive cardiomyopathy. J AMA 1985; 253:391.
77. Koike K, Hesslein PS, Finlay CD, Williams WG, Izukawa T, Freedom RM. Atrial automatic tachycardia in children. Am J Cardiol 1988 ;61:1127.
78. Mehta AV, Sanchez GR, Sacks EJ, Casta A, Dunn JM, Donner RM. Ectopic automatic atrial tachycardia in children. J Am Coll Cardiol 1988;11:379.
79. Giorgi LV, Hartzler GO, Hamaker WK. Incessant focal atrial tachycardia: a surgically remedial cause of cardiomyopathy. J Thorac Cardiovasc Surg 1984;87:466.
80. Fyfe DA, Gillette PC, Crawford FA Jr, Kline CH. Resolution of dilated cardiomyopathy after surgical ablation of ventricular tachycardia in a child. J Am Coll Cardiol 1987;9:231.

81. Rabbani LE, Wang PJ, Couper GL, Friedman PL. Time course of improvement in ventricular function after ablation of incessant automatic atrial tachycardia. Am Heart J 1991;121:816.
82. Gelb BD, Garson A Jr. Noninvasive discrimination of right atrial ectopic tachycardia from sinus tachycardia in "dilated cardiomyopathy." Am Heart J 1990;120:886.
83. Perloff JF, DeLeon AC Jr, O'Doherty D. The cardiomyopathy of progressive muscular dystrophy. Circulation 1966;33:625.
84. Perloff JK. Cardiac rhythm and conduction in Duchenne's muscular dystrophy: a prospective study of 20 patients. J Am Coll Cardio l 1984;3:1263.
85. Sanyal SH, Johnson WW. Cardiac conduction abnormalities in children with Duchenne's progressive muscular dystrophy: electrocardiographic features and morphologic correlates. Circulation 1982;66:853.
86. Bies RD, Friedman D, Roberts R, Perryman MB, Caskey CT. Expression and localization of dystrophin in human cardiac Purkinje fibers. Circulation 1992;86:157.
87. D'Orsogna L, O'Shea JP, Miller G. Cardiomyopathy of Duchenne muscular dystrophy. Pediatr Cardiol 1988;9:205.
88. Yotsukura M, Ishizuka T, Shimada T. Late potentials in progressive muscular dystrophy of the Duchenne type. Am Heart J 1991;121: 1137.
89. Perloff JK. Neurological disorders and heart disease. In: Braunwald E, ed. Heart disease: a textbook of cardiovascular medicine. Philadelphia: WB Saunders, 1988:1782.
90. Stevenson WG, Perloff JK, Weiss JN, Anderson TL. Fascioscapular muscular dystrophy: evidence for selective, genetic electrophysiologic involvement. J Am Coll Cardiol 1990;15:292.
91. Waters DD, Nutter DO, Hopkins LC, Dorney D. Cardiac features of an unusual X-linked humero-peroneal neuromuscular disease. N Engl J Med 1975;293:1017.
92. Clements Jr SD, Colmers RA, Hurst JW. Myotonia dystrophia: ventricular arrhythmias, intraventricular conduction abnormalities, atrioventricular block and Stokes-Adams attacks successfully treated with permanent transvenous pacemaker. Am J Cardiol 1976;37:933 .
93. Perloff JK, Stevenson WG, Roberts NK, Cabeen W, Weiss J. Cardiac involvement in myotonic dystrophy (Steinert's disease): a prospective study of 25 patients. Am J Cardiol 1984;54:1074.
94. Nguyen JJ, Wolfe JT III, Holmes DR Jr, Edwards WD. Pathology of the cardiac conduction system in myotonic dystrophy: a study of 12 cases. J Am Coll Cardiol 1988;3:662.
95. Moraes CT, DiMauro S, Zeviani M, et al. Mitochondrial DNA deletions in progressive external ophthalmoplegia and Kearns-Sayre syndrome. N Engl J Med 1989;320:1293.
96. Roberts NK, Perloff JK, Kark P. Cardiac conduction in Kearns-Sayre syndrome. Am J Cardiol 1979;44:1396.
97. Schwartzkopff B, Frenzel H, Breithardt G, et al. Ultrastructural findings in endomyocardial biopsy of patients with Kearns-Sayre syndrome. J Am Coll Cardiol 1988;12:1522.
98. Ansari A, Larson PH, Bates HD. Cardiovascular manifestations of systemic lupus erythematosus: current perspectives. Prog Cardiovasc Dis 1985;27:421.

99. Moffat GR Jr. Complete atrial ventricular dissociation with Stokes-Adams attacks due to disseminated lupus erythematosus, report of a case. Ann Intern Med 1965;63:508.
100. Escudero J, McDevitt E. The electrocardiogram in scleroderma: analysis of 60 cases and review of the literature. Am Heart J 195 8;56:846.
101. Roberts NK, Cabeen WR, Moss J, et al. The prevalence of conduction defects and cardiac arrhythmias in progressive systemic sclerosis. Ann Intern Med 1981;94:38.
102. Kostis JB, Seibold JR, Turkevich D, et al. Prognostic importance of cardiac arrhythmias in systemic sclerosis. Am J Med 1988;84:1007.
103. Rokas S, Mavrikakis M, Iliopoulou A, Moulopoulos S. Electrophysiologic studies of the heart in patients with rheumatoid arthritis. Int J Cardiol 1990;26:75.
104. Gottdeiner JS, Sherber HS, Hawley RJ, Engel WK. Cardiac manifestations of polymyositis. Am J Cardiol 1978;41:1141.
105. The Multicenter Postinfarction Research Group: Risk stratification and survival after myocardial infarction. N Engl J Med 1983;309:331.
106. Califf RM, McKinnis RA, Burks J, et al. Prognostic implications of ventricular arrhythmias during 24-hour ambulatory monitoring in patients undergoing catheterization for coronary artery disease. Am J Cardiol 1982;50:23.
107. Kennedy HL, Whitlock JA, Sprague MK, et al. Long-term follow-up of asymptomatic healthy subjects with frequent and complex ventricular ectopy. N Engl J Med 1985;312:193.

Terry W. Latson

Principles and Applications of Heart Rate Variability Analysis

12

The term "heart rate variability" (HRV) refers to small oscillations in heart rate associated with the activity of homeostatic autonomic reflexes.[1–11] The cyclical change in heart rate associated with respiration (i.e., the respiratory sinus arrhythmia) is one component of HRV known to most physicians. Cyclical changes in heart rate at frequencies lower than the respiratory frequency comprise other distinct components of HRV (Fig. 12-1). Analysis of these cyclical changes in heart rate has been shown to provide information regarding the integrity and "balance" of sympathetic and parasympathetic reflexes. The current resurgence of interest in HRV is based upon the evidence that analysis of HRV may: 1) help elucidate the role of alterations in autonomic reflex activity in multiple pathophysiologic conditions; 2) provide a noninvasive diagnostic tool for detecting alterations in autonomic reflex activity in specific patient populations. This chapter will provide a brief overview of HRV studies, with specific emphasis on measurement techniques, clinical applications, and current limitations of this technology.

Clinical Cardiac Electrophysiology: Perioperative Considerations
Edited by Carl Lynch III. J.B. Lippincott Company, Philadelphia, PA ©1994

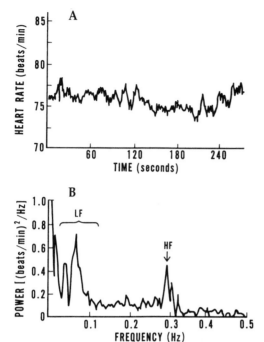

FIGURE 12-1. A, beat-to-beat heart rate signal of a patient presenting for elective cardiac surgery. B, power spectrum (i.e., plot of power spectral density as a function of frequency) of the heart rate signal in A. The peak denoted HF represents the dominant high-frequency HRV component associated with respiration (i.e., the respiratory sinus arrhythmia). The peaks denoted by LF are caused by cyclical changes in heart rate at frequencies lower than the respiratory frequency and represent low-frequency HRV components. (Reproduced with permission from Komatsu T, Kimura T, Sanchafa V, et al. Effects of fentanyl-diazepam-pancuronium anesthesia on heart rate variability: a spectral analysis. J Cardiothorac Vasc Anesth 1992;6:444.)

BRIEF HISTORICAL BACKGROUND

Cyclical variations in heart rate synchronized with respiration were noted in ancient times. These variations were formally described by Ludwig in 1846[12] using the newly invented kymograph. A description of corresponding respiratory variations in arterial blood pressure was documented by Hales as early as 1733.[13] Additional oscillations in blood pressure occurring at frequencies of 6 to 9 cycles/min (0.1 to 0.15 Hz), which correspond to heart rate variations at this same frequency, were subsequently described by Mayer[14] in 1876 and are still referred to as "Mayer waves" in the current medical literature. With respect to clinical applications of HRV, attenuation of the respiratory sinus arrhythmia with advanced age was taught to medical students as early as 1871.[15] Probably the most notable early clinical investigation of HRV was that by Hon and Lee in 1965[16] describing the use of fetal HRV as a qualitative measure of fetal distress. In the same year, McCrady et al.[17] suggested that changes in the amplitude of the respiratory sinus arrhythmia might provide an index of "anesthetic depth."

The application of power spectral analysis to the study of HRV was pioneered in the 1970s by Hyndman,[18] Sayers,[19] and Kitney and Rompelman.[20] Power spectral analysis of HRV provides quantitative measurements of the amplitude and frequency of heart rate oscillations. This early work described three peaks in the HRV power spectrum, typically located at 0.04 Hz, 0.10 to 0.12 Hz, and the respiratory frequency (0.3 Hz). Studies by Hyndman et al.[18] and Kitney and Rompelman[20] suggested that the low-frequency peak was related to rhythmic variations in peripheral vasomotor tone associated with thermal regulation and that the mid-frequency peak was related to homeostatic baroreceptor reflex control. In 1981, Akselrod et al.[4] presented evidence that activities of the sympathetic, parasympathetic, and renin-angiotensin systems made frequency-specific contributions to the HRV power spectrum. Parasympathetic blockade was shown to essentially abolish the peak at the respiratory frequency, as well as diminish the amplitude of the lower frequency peaks. Combined sympathetic and parasympathetic blockade dramatically attenuated heart rate variations at all frequencies. Numerous other investigations since 1985 have further established the effects of autonomic blockade, experimental interventions, and various disease states on heart rate variability in both animals and humans.

ETIOLOGY

Research is still ongoing to define both the etiology of HRV and the proper interpretation of changes in HRV. In general terms, HRV results from the interplay of multiple homeostatic reflexes involved with respiration, blood pressure, venous return, regional vasomotor tone, thermal regulation, cognitive function, and emotions. Essential components of this interplay include peripheral inhibitory reflex mechanisms (with negative feedback characteristics), peripheral excitatory reflex mechanisms (with positive feedback characteristics), and central neural integration.[1] Both endogenous perturbations (e.g., respiration and autoregulatory adjustments in the vasomotor tone of different tissue beds) and exogenous perturbations (e.g., changes in posture and mental stress) provoke these homeostatic reflex mechanisms, which then leads to cyclical fluctuations characteristic of feedback control systems. Some insights into the reflex mechanisms involved with HRV have been provided by pharmacologic blocking studies, surgical and/or anesthetic "denervation" studies, and mathematical models of physiologic control systems.

Both pharmacologic blocking studies and surgical denervation studies have established that high-frequency HRV (i.e., heart rate oscillations associated with respiration) is mediated almost exclusively by parasympathetic reflexes.[1,4,5,21-25] These high-frequency variations are almost completely abolished by the administration of atropine. Although rhythmic discharges in phase with respiration have been described in central sympathetic outflows,[26] sympathetic reflexes are thought to have minimal contribution to the genesis of high-frequency HRV. This probably reflects the multistep stimulus-response time of sympathetic reflexes at the sinus node, which are mediated by the receptor-G protein-adenylyl cyclase-cAMP-dependent protein kinase cascade (see Chapter 1). In contrast, the more rapid parasympathetic muscarinic actions are mediated by the direct receptor-G protein-potassium channel interaction.[4,27] However, a central inhibitory influence by sympathetic reflexes on central parasympathetic reflex centers cannot be ruled out.[1,28] The primary etiology of high-frequency HRV is believed to involve reflex responses to physiologic changes associated with respiration. Postulated stimuli causing this cyclical increase in heart rate with inspiration include changes in venous return, pulmonary venous and/or left atrial stretch, arterial pressure, and lung distention.[25,29,30] Some evidence suggests that "central" stimuli originating from the medullary respiratory centers may also be involved.[30,31]

The source of low-frequency HRV (i.e., HR oscillations at frequencies below 0.15 Hz) is less well understood than the etiology of high-frequency HRV. Pharmacologic blocking studies and surgical denervation studies have suggested that low-frequency HRV is jointly mediated by the sympathetic and parasympathetic systems,[1,4,5,8,32] with possible additional effects by the renin-angiotensin system.[4,5] Postulated stimuli for low-frequency HRV include: cyclical changes in regional blood flows, thermoregulatory responses, baroreceptor feedback mechanisms, and the "buffering" activity of neurohumoral reflex systems. Whether low-frequency HRV should be subdivided into two or more frequency bands is debated. Although controversial, some evidence suggests that HRV below 0.03 to 0.04 Hz is mediated by thermoregulatory reflexes and/or slower acting neurohumoral reflexes (e.g., the renin-angiotensin system).[1,3,4,18,19,34-36] Many current studies exclude these lower frequency oscillations due to difficulties involved with measuring and interpreting HRV below 0.03 Hz.

HRV in the frequency range between 0.05 and 0.15 Hz is usually attributed to baroreceptor reflex activity in conjunction with autoregulatory adjustments in vasomotor tone.[1,3,4,32,37,38] Although these oscillations typically have a "center frequency" of about 0.1 Hz in healthy young subjects, recent studies suggest that this center frequency is decreased in patients with autonomic dysfunction (Fig. 12-2).[38,39] Studies using epidural blockade suggest that vasomotor reflex-induced oscillations in blood pressure may be the primary stimulus for these heart rate oscillations.[32] Other evidence suggests that there may be a "central oscillator" involved with these low-frequency cyclical fluctuations in both heart rate and blood pressure.[40] Another hypothesis, supported

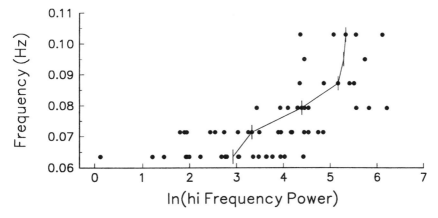

FIGURE 12-2. Central frequency of the low-frequency HRV peak (i.e., HRV power between 0.05 and 0.125 Hz) versus high-frequency HRV power. Vertical lines represent the median value of high-frequency HRV power at each central frequency. The observed decrease in central frequency with decreasing HRV power (indicative of parasympathetic reflex dysfunction) is consistent with a recently proposed model of baroreceptor function.[38] Data obtained from 60 elective surgical patients ranging in age from 20 to 80 years.

by models of cardiovascular homeostasis, is that these oscillations result from the input of "broad-band noise" (i.e., multiple inputs at varying frequencies) into a highly integrated reflex control system. Such a control system exhibits limited oscillatory activity in response to a wide range of endogenous and exogenous perturbations.[38]

MODELS OF CARDIOVASCULAR HOMEOSTASIS

Models of cardiovascular homeostasis usually involve either electrical or mathematical simulations of defined input-output relationships for specific cardiovascular reflexes. Model inputs have included physiologic changes associated with respiration, venous return, autoregulatory adjustments in regional vasomotor tone, and beat-to-beat changes in blood pressure. Model outputs typically include changes in systemic vascular resistance and beat-to-beat changes in heart rate.[38,41–44] Essential elements in these modeled input-output relationships include detection of the input stimulus (e.g., by baroreceptors or stretch receptors), propagation of this stimulus detection to the central nervous system, and subsequent alteration of steady-state and/or phasic influences by the central nervous system on effector organs (e.g., systemic vascular resistance or heart rate). Since clinical application of HRV is frequently focused on changes in the "balance" of sympathetic and parasympathetic reflexes, model identification of the separate influences of these two reflex systems and how they interact is an essential component of more recent models.[38] More advanced models also include parameters to account for different response times of the sympathetic and parasympathetic reflex systems, as well as the different "gains" (both linear and nonlinear) of various feedback reflex systems. An extensive review of the multiple physiologic models previously proposed is beyond the scope of this chapter. For the interested reader, the thesis by Ben ten Voorde is highly recommended.[38] This reference provides a detailed review of prior studies and also describes the development of a highly integrated model, which appears capable of describing many important aspects of HRV.

MEASUREMENT TECHNIQUES

There is no question that additional information is needed regarding how various reflex mechanisms interact for the genesis of HRV under both normal and pathophysiologic conditions. Although this lack of complete information regarding mechanisms is a limitation, it should not preclude appropriate observational studies. An appropriate analogy is that we have limited understanding of the genesis of electroencephalogram (EEG) wave forms, yet EEG analysis has had immense clinical application. This analogy between observational studies with the EEG and HRV can be extended to "justify" the multiple measurement techniques used to study both phenomena. The multiplicity of techniques developed (and still being developed) reflects a lack of consensus involving: 1) the best method for mathematically describing a given signal; and 2) what informational aspects of that signal are most important with regard to the observation under study.

Three major methodologies are currently used for analysis of HRV. These methodologies differ with respect to the mathematical "domain" used to describe the measured variations in beat-to-beat heart rate. These three different domains are: 1) time; 2) frequency; and 3) chaos.

Time-Domain Techniques

Time-domain techniques involve measurements of statistical variability in R to R intervals between normal QRS complexes. Two of the most frequently used measurements are: 1) standard deviation of all RR intervals over a specified time period (e.g., 24 hr); and 2) the proportion of adjacent RR intervals which differ by more than 50 msec.[45-51] The advantage of this methodology is that it is the most "repeatable" among different investigators because the measurement algorithms are simple and well-defined. These methods have been used in large clinical studies establishing correlations between reductions in HRV and mortality after myocardial infarction.[45,49,51,52] The disadvantage of time-domain techniques is that information regarding the characteristics of cyclical fluctuations in heart rate (e.g., repetition frequency and amplitude) is not easily appreciated in time-domain measurements. Some generalizations concerning the relative influence of specific frequency components of HRV on different time-domain measurements can be derived.[48,51] However, these time-domain techniques are generally less well-suited than frequency-domain techniques for analyzing specific alterations in both sympathetic and parasympathetic reflex activity.

Frequency-Domain Techniques

Frequency-domain techniques measure both the frequency and amplitude of oscillations in heart rate. These are the same techniques commonly used for power spectral analysis of the EEG. However, the frequencies of interests are much lower in HRV analysis compared to EEG analysis (i.e., up to 30 Hz for beta waves in the EEG versus up to 0.5 Hz for high-frequency HRV measurements). The advantage of these frequency-specific HRV measurements is that they are believed to provide additional information about sympathetic versus parasympathetic reflex activity (see below). A disadvantage of these techniques is that there are multiple algorithms by which these measurements can be made, and there is no consensus regarding which algorithms are best (i.e., there are no "standards" for either analyzing or reporting power spectral measurements of HRV). Significant differences between the methods of different investigators include: 1) whether the "wave form" analyzed is based on RR intervals (units of milliseconds) or heart rate (units of beats per minute); 2) the length of the data segment analyzed (which can influence the ability to detect lower frequency oscillations); 3) whether the data analyzed are "preconditioned" prior to spectral analysis (e.g., removal of linear trends that can introduce low-frequency artifacts); 4) whether the spectral analysis is performed using fast Fourier transform (FFT) or autoregressive algorithms, described below; 5) the frequency ranges over which spectral power is calculated (see below, Frequency Range of HRV Measurements); and 6) whether the power spectral measurements are reported in normalized units or absolute units (see below, Normalized Versus Absolute HRV Power Measurements).

The two major techniques used for spectral analysis of HRV are FFT algorithms and autoregressive algorithms. Because of their computational efficiency and popular application in other medical and engineering fields, the FFT algorithms are the most commonly used at present. Disadvantages of FFT algorithms include the need to "window" the data (i.e., mathematically taper the ends of the data segment to avoid "leakage") and greater sensitivity to nonperiodic "noise" in the analyzed data.[1,53] Newer spectral analysis techniques using autoregressive (AR) algorithms involve characterizing the observed data using mathematical models of oscillatory phenomena.[1,8,54] After selection of an appropriate model (usually an "AR" model) and the model "order" (dependent upon the number of oscillatory components in the model), one of several algorithms is applied to determine the model parameters that most closely "fit" the measured data using various statistical criteria. The differences between the resultant model predictions and the observed data (termed the "residuals") can also be analyzed in order to assess the presence of any nonrandom signal components not accounted for in the model predictions. The frequency-domain characteristics of the measured data (i.e., the number, amplitude, and frequency of oscillatory components) can then be derived directly from the determined model parameters. Potential advantages of autoregressive algorithms include greater "noise immunity" to nonperiodic signal components, the ability to define spectral components using shorter data segments, and avoidance of the need for data windowing.[1] Potential disadvantages include the different approaches taken by various investigators to determine the best-fit model parameters and the model order (which can significantly affect the model-derived power spectrum).

Although not yet commonly used for HRV analysis, the technique of "complex demodulation" is another signal analysis technique that can be used to measure time-dependent changes in the amplitude of signals at known frequencies. This technique was recently used by Hayano et al.[55] to examine changes in HRV during postural tilt. The main advantage of this technique is that it allows more rapid assessment of changes in HRV. Hayano et al. report a time of <15 sec for measuring the changes in HRV amplitude of both low-frequency and high-frequency HRV components using this approach.

Chaos- or Entropy-Domain Techniques

Chaos-domain techniques are used to detect subtle differences in the "regularity" of data. Using the example of Pincus et al.,[56] note the similarities and differences in the following data sets:

Data Set 1: 90, 70, 90, 70, 90, 70, 90, 70, 90, 70, 90, 70, 90, 70, ...
Data Set 2: 90, 70, 70, 90, 90, 90, 70, 70, 90, 90, 70, 90, 70, 70, ...

Classical statistical measures, such as mean and variance, will not distinguish between these two series (e.g., the mean of both series is 80; the standard deviation of both series is 10). However, the two series are distinctly different with respect to the "regularity" of the data, and thus the two series can be distinguished using "regularity statistics" such as entropy and chaos. Compared to frequency-domain techniques, chaos-domain techniques provide a measure of the "broadbandedness" or "whiteness" of the data in the frequency domain.[56] Several HRV studies have now been conducted using a recently developed entropy statistic called approximate entropy (ApEn).[56-60] These studies have demonstrated that loss of irregularity or complexity (i.e., decreases in ApEn) is associated with compromised physiologic status.

INTERPRETATION OF FREQUENCY-DOMAIN MEASUREMENTS

Current research is dominated by studies using frequency-domain measurements of HRV as a "noninvasive probe" of sympathetic and parasympathetic reflex activity. The multiplicity of approaches used by various investigators for both deriving and reporting frequency-domain measurements has significantly complicated interpretation of the many studies in this area. The differences between commonly reported frequency-domain measurements, and the current "conventional" interpretation of these measurements will now be briefly discussed.

Frequency Ranges of HRV Measurements

HRV parameters are typically reported as HRV spectral power contained within specific frequency ranges (similar to EEG spectral power measurements in specific frequency ranges such as α, β, and θ powers). These frequency ranges are based on the locations of characteristic peaks in the HRV power spectrum associated with the activity of different reflexes. Although early HRV studies suggested the presence of three distinct peaks at approximately 0.04 Hz, 0.10 Hz, and the respiratory frequency, latter studies have suggested that not all peaks are present in all patients or even in the same patient at different times.

Almost all investigators now report a "low-frequency" and a "high-frequency" HRV power measurement, with some investigators also defining "mid-frequency" and/or "very low-frequency" measurements. High-frequency HRV power predominantly measures the variability in heart rate induced by respiration (i.e., the respiratory sinus arrhythmia), and thus the defined frequency band for this parameter usually encompasses the frequency range corresponding to the frequency of normal respiration. Typical frequency ranges are between 0.10 to 0.15 Hz for a lower cutoff frequency and between 0.30 to 0.50 Hz for an upper cutoff frequency. The upper cutoff frequency for low-frequency HRV power is usually equal to the lower cutoff frequency for high-frequency HRV power (i.e., between 0.10 and 0.15 Hz). The lower cutoff frequency for low-frequency HRV power has varied between 0 and 0.05 Hz, with a typical value of 0.03 Hz. The term "mid-frequency" HRV has sometimes been used to describe this low-frequency HRV (e.g., 0.05 to 0.15 Hz). Alternatively, this term may be used to describe a more narrow frequency band usually beginning at 0.06 to 0.07 Hz and extending up to 0.10 to 0.15 Hz.

Rather than using predefined frequency bands for calculating HRV parameters, some investigators have used specific criteria for selecting variable frequency bands. One technique involves simultaneously analyzing another physiologic variable and then defining an HRV frequency band to correspond with the dominant frequency of this other variable (using either peak-detection or cross-correlation techniques). This approach has been used by Estafanous et al.[61] and others with measurements of respiratory movements for defining high-frequency HRV parameters. Recent studies by Muzi et al.[3,62] have extended this technique using measurements of arterial pressure variability to help define mid-frequency HRV parameters. When autoregressive spectral analysis techniques are used, the derived AR model parameters can also be used to define the number, central frequency, and power of dominant oscillatory components. Some investigators have advocated using these model-derived parameters (rather than using predefined frequency bands) for calculating the power of low- and high-frequency HRV components.[1,6]

Normalized Versus Absolute HRV Power Measurements

In reading the HRV literature, it is also important to distinguish whether HRV power measurements are being reported in "absolute" or "normalized" power measurements. Measurements of absolute HRV power are usually obtained by integration of the area under the power spectral density curve between the predefined frequency limits of the specific HRV parameter (e.g., 0.15 to 0.40 Hz as a typical range for high-frequency HRV). These absolute power measurements represent the variance (i.e., standard deviation squared) of the measured signal about its mean value. Absolute power measurements are important for quantitating the integrity of autonomic reflexes, but as discussed later, they may not always be an accurate index of absolute changes in sympathetic or parasympathetic "tone" when the defined study condition dramatically affects total HRV power.

HRV parameters are now often also reported in terms of normalized power measurements. These measurements represent HRV power in the predefined frequency range expressed as a proportion of total HRV power (e.g., area under the power spectral density curve for a specific frequency range divided by the area under the curve for all frequencies studied). Normalized measurements thus quantitate the fraction of total HRV power contained within defined frequency ranges, irrespective of variations in total HRV power. Normalized measurements of HRV are commonly used in clinical studies to adjust for variations in total HRV power among different individuals and/or between different study conditions. As discussed later, normalized power measurements are particularly useful when examining changes in sympathetic-parasympathetic "balance" by comparing the components of HRV at low versus high frequencies.

Interpretation of High-Frequency HRV Measurements

Since both pharmacologic blocking studies and surgical denervation studies have demonstrated that high-frequency HRV is mediated almost exclusively by parasympathetic reflexes, measurements of high-frequency HRV power are conventionally interpreted in clinical studies as an index of parasympathetic activity or tone.[1,3,7] Additional evidence from both experimental intervention studies and observations in disease conditions supports this interpretation. Infusions of Neo-Synephrine and assumption of the recumbent position, both of which are associated with an increase in vagal activity, cause a corresponding increase in high-frequency HRV.[6,63] Conversely, both head-up tilt and the infusion of vasodilators cause a decrease in high-frequency HRV.[6] High-frequency HRV is also reduced in pathophysiologic conditions (e.g., congestive heart failure,[64] hypertension,[65] and myocardial infarction[66]), which have been shown to be associated with decreased parasympathetic reflex activity using other autonomic testing techniques. When changes in high-frequency HRV measurements are interpreted, it is important to remember that significant alterations in ventilatory pattern can also affect this measurement independent of changes in vagal tone.[8,67]

Interpretation of Low-Frequency HRV Measurements

Despite the evidence from pharamacologic blocking studies that multiple reflex system may be involved in the genesis of low-frequency HRV, measurements of low-frequency HRV are conventionally interpreted in clinical studies as an index of sympathetic activity or tone.[1,7] Many studies have confirmed that low-frequency HRV is increased by experimental interventions that increase sympathetic activity (e.g., head-up tilt, infusion of vasodilators, moderate exercise, and mental stress).[6,22,68,69] Low-frequency HRV is also increased in pathophysiologic conditions characterized by increased sympathetic tone (e.g., congestive heart failure[64] and recent myocardial infarction[66]) and decreased in conditions characterized by decreased sympathetic tone (e.g., quadriplegia).[70] When experimental and clinical HRV studies are reviewed, it is important to recognize that: 1) dogs usually exhibit minimal low-frequency sympathetic HRV under normal resting conditions;[1] and 2) the relative proportion of low-frequency HRV in "normal" humans can be significantly influenced by age.[3]

Do HRV Measurements Reflect Changes in Sympathetic Tone, Parasympathetic Tone, or Sympathetic-Parasympathetic Balance?

Many of the original studies of HRV used only absolute power measurements of HRV.[4,5,23] These early studies suggested that absolute power measurements of HRV correlated well with changes in sympathetic and parasympathetic tone during defined experimental interventions. However, more recent studies have suggested that interpretation of these changes in absolute HRV power measurements may not be as straightforward as initially presumed. In particular, this interpretation may not be valid when the defined intervention is accompanied by significant alterations in total HRV power.[1] For example, under some circumstances of sympathetic activation (e.g., exercise or mental stress), absolute measurements of low-frequency "sympathetic" HRV are reduced. This "paradoxical" decrease in low-frequency HRV power is thought to result from the decrease in total HRV power that accompanies these interventions. However, even under conditions where total HRV power is altered, changes in the frequency distribution of HRV power (e.g., low-frequency HRV expressed in normalized units or ratios of low-frequency HRV to high-frequency HRV) appear to remain a qualitative measure of changes in sympathetic-parasympathetic balance.[1] Thus, during moderate exercise and mental stress, although absolute measurements of low-frequency HRV may be reduced, normalized measurements of low-frequency sympathetic HRV are found to be increased (in accordance with the expected shift of sympathetic-parasympathetic balance toward sympathetic dominance).[68,71,72]

More recent HRV studies thus suggest that examination of normalized power measurements is important when defined study conditions are accompanied by significant changes in total HRV power. Changes in these normalized measurements are believed to reflect changes in sympathetic-parasympathetic balance, rather than absolute measurements of sympathetic or parasympathetic tone. Other relevant examples where changes in HRV expressed in normalized units (rather than absolute units) appear to be more consistent with known autonomic reflex alterations include: 1) the effect of chronic β-blockade therapy, where absolute measurements of low-frequency sympathetic HRV may increase, but normalized measurements are decreased[6]; and 2) the effect of sufentanil induction, where absolute measurements of high-frequency "parasympathetic" HRV are decreased, but normalized measurements are increased.[73]

CORRELATION OF HRV MEASUREMENTS WITH OTHER TESTS OF AUTONOMIC FUNCTION

Muscle Sympathetic Nerve Activity

Measures of sympathetic activity in nerves to skeletal muscle have also been used to examine alterations in sympathetic reflex activity. Saul et al. compared power spectral measures of HRV with measurements of muscle sympathetic activity in the peroneal nerve of human subjects.[73a] Autonomic activity was manipulated by step-wise infusions of nitroprusside and Neo-Synephrine. No spectral measurement of HRV correlated significantly with muscle sympathetic activity during either control or Neo-Synephrine infusion. However, a significant correlation was found between normalized measurements of low-frequency HRV power and muscle sympathetic nerve activity during nitroprusside infusion. The authors explained their findings by a model of heart rate control in which low-frequency heart rate fluctuations are due to changing levels of both sympathetic and parasympathetic inputs to the sinoatrial node. Thus, during periods of sympathetic activation and parasympathetic withdrawal (e.g., nitroprusside infusion), correlation between low-frequency sympathetic HRV and muscle sympathetic activity exists. However, during periods of parasympathetic activation (e.g., Neo-Synephrine infusion), the parasympathetic influence on low-frequency HRV is not reflected in measurements of muscle sympathetic nerve activity, and hence no correlation was detected.

Baroreflex Function

Under normal resting conditions, baroreflex changes in heart rate are believed to be primarily mediated by vagal reflex mechanisms. Measures of baroreflex sensitivity have thus been correlated with parasympathetic reflex integrity and function. Bigger et al.[47] examined the correlations between baroreflex sensitivity (assessed with phenylephrine injection) and three measures of parasympathetic HRV (high-frequency HRV power, the percent of successive normal RR intervals greater than 50 msec, and the root-mean-square successive difference of normal RR intervals). Weak correlations (r values of 0.57 to 0.63) were found between baroreflex sensitivity and the three measures of HRV; the highest correlation was with high-frequency HRV power ($r = 0.63$, $p < 0.01$).

Correlations between baroreflex sensitivity (heart rate responses to phenylephrine and nitroglycerin injections) and HRV were also investigated by Mancia et al.[74] in 82 ambulatory hypertensive subjects. Both HRV and arterial blood pressure variability were examined. Baroreflex sensitivity showed a weak positive correlation with measures of HRV ($r = 0.32$ to 0.47, $p < 0.05$) and a weak negative correlation with arterial blood pressure variability ($r = -0.28$ to -0.50, $p < 0.05$). These authors concluded that: "because of the low correlation between baroreflex sensitivity and blood pressure and heart rate variabilities, other factors (probably central in nature) are important in determining the size of these variations."

Other investigators have combined measurements of HRV and arterial pressure variability in an attempt to measure baroreflex sensitivity without the need for exogenous blood pressure manipulations (e.g., phenylephrine injection). These studies are based on the concept that heart rate variability represents baroreceptor-mediated changes in heart rate, which occur in response to spontaneous variations in blood pressure. Therefore, by quantitating the relationship between HRV and arterial pressure variability, a measure of baroreflex sensitivity (i.e., change in heart rate per incremental change in blood pressure) can be derived. The necessary quantitative relationship between HRV and arterial pressure variability can be derived using the technique of "cross-spectral analysis." This analysis provides measures of amplitude relationships (i.e., gain), temporal relationships (i.e., phase), and "linearity" relationships (i.e., coherence—a measure analogous to the "r value" in linear regression) between these two variability signals. In an initial study by Robbe et al.,[36] very high correlations ($r = 0.94$) were found between baroreflex sensitivity measured using phenylephrine injections as compared to measurements using cross-spectral analysis of heart rate and blood pressure variabilities in healthy subjects at rest and during mental activity. Subsequent studies by Pagani et al.[68] in hypertensive subjects also revealed highly significant but weaker correlations (r = 0.6 to 0.61) between these two methods of assessing baroreflex sensitivity.

Other Standardized Tests of Autonomic Reflex Integrity

Prior studies in diabetics have described qualitative correlations between measures of HRV and the presence of autonomic neuropathy by other tests (e.g., changes in heart rate with a Valsalva maneuver and forced breathing).[75,76] More quantitative relationships between HRV measurements and these other tests, in both diabetics and nondiabetics, have recently been described.[77,77a,77b] Latson et al. correlated measures of low, mid-, and high-frequency HRV (obtained during supine rest) with the Valsalva Index, Strain Index, Valsalva Ratio, change in heart rate with forced breathing, and change in heart rate with standing. HRV parameters were expressed in terms of both HRV power (units of [beats per minute]^2) and HRV standard deviation (units of beats per minute), since the units of the latter measurements are more analogous to the units of the other autonomic tests studied. Highly significant and moderately strong correlations (r values between 0.5 and 0.8) were found between all HRV parameters and other tests except for the change in heart rate with standing. For all Valsalva parameters, the strongest correlations were observed with high-frequency HRV expressed in units of standard deviation (r values of 0.76 to 0.81, $p < 0.001$).

CLINICAL APPLICATIONS

Cardiovascular Disease and Parasympathetic Reflex Dysfunction

Multiple studies beginning as early as 1971[78] have documented impaired parasympathetic control of heart rate in patients with a variety of cardiac diseases. Impaired parasympathetic responsiveness has been documented by decreased heart rate responses to atropine, decreased baroreceptor modulation of heart rate, and attenuated cardiac slowing during facial immersion in cold water.[78–83] The etiology for this decreased parasympathetic influence on heart rate is not well-defined. Several animal studies using direct vagal stimulation have suggested that the decreased parasympathetic/baroreceptor responsiveness in cardiovascular disease is not due to alterations in the sinus node or cardiac muscarinic receptor responsiveness (in fact, muscarinic responsiveness may even be increased).[82,84,85] Proposed mechanisms include alterations in myocardial mechanoreceptors, myocardial chemoreceptors, structural changes in high (arterial)- and low (cardiopulmonary)-pressure baroreceptors, altered interactions between high- and low-pressure baroreceptors, sympathetically mediated inhibition of parasympathetic reflexes, and effects of various neurohumoral alterations (e.g., elevated levels of atrial natriuretic peptide, norepinephrine, β-endorphins, and angiotensin II).[28,51,85–91] Studies in heart transplant patients suggest that this impaired vagal responsiveness is fairly rapidly reversible following restoration of normal cardiac function (as measured by baroreflex responsiveness and amplitude of the respiratory sinus arrhythmia of the native atria).[87,88]

Coronary Artery Disease and Sudden Death

A possible link between impaired parasympathetic reflexes and mortality following myocardial infarction was first reported by Wolf et al. in 1978.[92] These authors reported that reduced HRV (i.e., reduced amplitude of the respiratory sinus arrhythmia) measured in a 60-sec electrocardiographic recording in patients with acute myocardial infarction was an indicator of long-term mortality. A probable causal relationship between parasympathetic dysfunction and post-myocardial infarction (MI) mortality has been explained by the influence of autonomic activity on the vulnerability of the heart to malignant ventricular arrhythmias.[93–100] Multiple animal studies have demonstrated that decreased vagal activity and/or increased sympathetic activity decrease the threshold for ventricular fibrillation and thus predispose the heart to malignant arrhythmias during myocardial ischemia. Conversely, a protective effect of vagal stimulation has also been demonstrated.[94–97,100]

Subsequent studies by several investigators between 1984 and 1987 suggested that analysis of HRV might be a useful tool for identifying persons at increased risk following myocardial infarction and/or at high risk of sudden cardiac death.[48,49,101] The hypothesis behind these studies was that analysis of HRV would allow noninvasive detection of the abnormalities in autonomic reflex function that predispose to malignant ventricular arrhythmias. Myers et al. examined both time-domain and frequency-domain measurements of HRV in a group of patients at high risk for sudden cardiac death.[48] Decreases in parasympathetic HRV, measured by either technique, were the most sensitive for distinguishing patients at increased risk.

A subsequent study by Kleiger et al.[49] in over 800 patients who survived acute myocardial infarction demonstrated that decreased

HRV (measured as the standard deviation of all normal RR intervals in a predischarge 24-hr Holter recording) was a very significant, independent predictor of mortality. Patients with a standard deviation of RR intervals less than 50 msec had a mortality of 34% during the 31-month follow-up period, compared to a mortality of 9% in patients with an RR interval standard deviation over 100 msec (Fig. 12-3). Reduced HRV remained an independent predictor of mortality even after multivariate analysis of multiple other clinical and Holter-derived variables (reduced ejection fraction, New York Heart Association Functional Class, occurrence of rales and/or postinfarction angina, and frequency/type of ventricular ectopy). Decreased HRV provided stratification of patient risk even within defined subsets of high-risk patients (e.g., within the subgroup of patients with an ejection fraction of less than 30%, low HRV was associated with a more than 2-fold increase in risk). A follow-up study documented that reduced HRV also remained an independent predictor of mortality after controlling for exercise testing results.[45]

The potential role of increased sympathetic reflex activity in the autonomic imbalance following myocardial infarction was investigated by Lombardi, Pagani, Malliani, and co-workers using spectral analysis of HRV.[66] In patients studied 2 weeks after acute myocardial infarction, low-frequency sympathetic HRV was significantly greater (and high-frequency parasympathetic HRV was significantly less) than in age-matched control subjects. In post-MI patients, the increase in low-frequency sympathetic HRV that normally accompanies head-up tilt was significantly attenuated, which also suggests enhanced

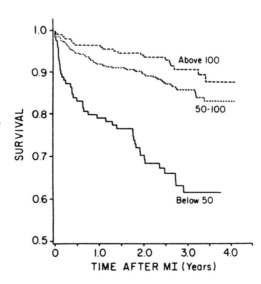

FIGURE 12-3. Cumulative survival over total the follow-up period as a function of heart rate variability (standard deviation of all normal RR intervals in an predischarge 24-hr Holter recording). MI, myocardial infarction. (Reproduced with permission from Kleiger RE, Miller JP, Bigger JT, Moss AJ. Decreased heart rate variability and its association with increased mortality after acute myocardial infarction. Am J Cardiol 1987;59:256.)

sympathetic activity in the post-MI setting. These abnormalities in HRV (both at rest and with tilt) progressively resolved over a 1-year period. A subsequent study demonstrated that post-MI patients also did not show the normal increase in low-frequency HRV with mental stress.[69]

These abnormalities in HRV following myocardial infarction are consistent with other studies assessing autonomic abnormalities in post-MI patients via analysis of changes in baroreflex sensitivity.[91,98,102] These studies have demonstrated marked impairment of baroreflex sensitivity in the early stages of myocardial infarction and variable recovery of baroreflex sensitivity between 10 days and 3 months after myocardial infarction. Similar to patients with greater depression in HRV, those patients demonstrating greater depression of baroreflex sensitivity are at increased risk for subsequent death.

Bigger et al.[47] subsequently investigated whether measures of HRV and baroreflex sensitivity provided redundant information in the post-MI setting (i.e., since both tests measure parasympathetic reflex integrity, and since both tests correlate with post-MI mortality, would the results from one test accurately predict the results of the other?). Although baroreflex sensitivity was significantly correlated with measures of HRV, the strength of these correlations was limited (r values of 0.57 to 0.63). Both baroreflex sensitivity and HRV were reduced more in patients with inferior myocardial infarction compared to those with anterior infarction. These authors suggest that measures of HRV primarily represent a measure of tonic vagal activity, whereas measures of baroreflex sensitivity represent a marker of phasic responsiveness to vagal reflexes and that the two measures are not redundant. The relative value of these different measures of parasympathetic reflex activity in predicting sudden cardiac death continues to be evaluated.[101] Other recent studies investigating the prognostic value of HRV measurements include those by Cripps et al.[103] and Odemuyiwa et al.[52]

Additional studies in patients with coronary artery disease without recent myocardial infarction have examined possible relationships between measures of HRV, angiographic severity of coronary artery disease, and late mortality. In a study by Hayano et al.,[104] a heart rate-adjusted measurement of high-frequency HRV was significantly associated with advancing angiographic severity (decreasing HRV for increasing angiographic severity). In contrast, Rich et al.[105] did not find any significant association between HRV and angiographic variables. However, these latter authors did demonstrate increased mortality in patients with coronary artery disease and decreased HRV, even in the absence of recent myocardial infarction.

Latson et al.[106] investigated whether these HRV abnormalities observed in patients with coronary artery disease were correlated with altered neurohumoral responses to orthostatic stress (e.g., attenuated or absent increases in norepinephrine with head-up tilt). Such altered neurohumoral responses have been demonstrated in several patient groups (e.g., those with congestive heart failure or hypertension), but were never directly related to alterations in HRV. These investigators reported a significant association between reduced power spectral measures of HRV and altered neurohumoral responses to head-up tilt in a group of patients undergoing elective coronary revascularization. Both reduced HRV and altered norepinephrine responses were also associated with an increased incidence of hypotension following anesthesia induction.[107,108]

Several investigators have also examined changes in HRV preceding cardiac events. These studies have shown an increase in low-frequency HRV (and/or the ratio between low-frequency and high-frequency HRV) prior to both asymptomatic nocturnal ischemia[109] and spontaneous episodes of ventricular tachycardia.[110] Studies by Rimoldi et al.[22] have also documented a similar increase in low-frequency HRV during experimental myocardial ischemia in dogs.

Analysis of heart rate variability has also been used to investigate whether circadian variations in autonomic activity correlate with the well-known circadian variations in the incidence of myocardial ischemia and infarction (i.e., the increased incidence of ischemia, infarction, and sudden cardiac death during the morning hours).[111] Furlan et al.[7] examined the circadian variation in both heart rate and arterial pressure variabilities. Low-frequency sympathetic variability in both heart rate and arterial pressure decreased during the night and then subsequently increased upon waking up in the morning (even while subjects were still lying in bed). These changes were accompanied by reciprocal changes in high-frequency "vagal" HRV. These authors concluded that "the ominously increased rate of cardiovascular events in the morning hours may reflect the sudden rise of sympathetic activity and the reduction of vagal tone."

Congestive Heart Failure

Patients with congestive heart failure also exhibit significant alterations in HRV.[64,88,112-118] Frequency-domain measurements are decreased over all frequencies, and the residual variability is shifted toward a relative increase in low-frequency sympathetic components. This pattern of change in frequency-domain measurements is consistent with preservation and/or augmentation of sympathetic control of heart rate, along with significantly diminished vagal modulation of heart rate (Fig. 12-4).[64,116,118] In particular, patients with congestive heart failure have a significantly higher percentage of HRV power in the very low-frequency range (0.01 to 0.04 Hz).[64]

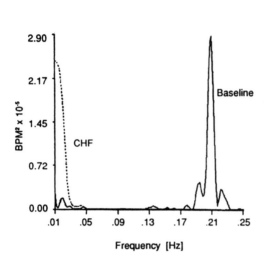

FIGURE 12-4. Power spectrum of HRV in at dog at baseline and after induction of congestive heart failure (CHF) by rapid ventricular pacing. The baseline spectrum (*solid line*) demonstrates a prominent high-frequency peak consistent with the predominance of parasympathetic tone normally present in dogs. With congestive heart failure (*dotted line*) this high-frequency peak is markedly diminished, along with dramatic augmentation of low-frequency HRV. BPM^2, = (beats/min)^2. (Reproduced with permission from Binkley PF, Nunziata E, Haas GJ, Nelson SD, Cody RJ. Parasympathetic withdrawal is an integral component of autonomic imbalance in congestive heart failure: demonstration in human subjects and verification in a paced canine model of ventricular failure. J Am Coll Cardiol 1991;18:464.)

As a prognostic test, Saul et al.[64] were unable to demonstrate any significant relation between spectral analysis of HRV and survival in patients with severe congestive heart failure (median survival time of 5 months). However, these authors did demonstrate a limited correlation (r value of 0.5) between HRV measurements and elevated plasma norepinephrine levels in these patients. HRV measurements may have some prognostic value regarding the development of congestive heart failure in patients treated with doxorubicin, where reductions in high-frequency HRV (i.e., respiratory sinus arrhythmia) were found to be predictive for the development of doxorubicin-induced cardiomyopathy.[117]

Hypertension

Alterations in HRV in hypertensive patients have been less well-studied than in patients with myocardial infarction and congestive heart failure. Although one study demonstrated no significant difference in HRV between normotensive patients and patients with mild essential hypertension,[119] another study demonstrated relative enhancement of sympathetic HRV in patients with moderate essential hypertension.[65] In this latter study, the normal increase in low-frequency HRV with passive tilt was also attenuated in hypertensive patients. These findings are thus qualitatively similar to the changes observed with recent myocardial infarction and congestive heart failure and suggest a shift in sympathetic-parasympathetic balance toward more sympathetic dominance in patients with central hypertension. In another study by Pagani et al.,[68] the "gain" of the relationship between heart rate and arterial pressure variabilities was found to be reduced in hypertensive patients, consistent with other findings of decreased baroreflex sensitivity in this patient group. The normal pattern of circadian changes in HRV may also be altered in hypertensive subjects.[1]

Therapeutic Interventions for Cardiovascular Disease

Analysis of HRV has also been used to investigate the effects of various therapeutic interventions on the autonomic imbalance associated with cardiovascular disease. Since physical training may enhance vagal activity and baroreflex sensitivity, the effects of training and post-MI physical rehabilitation on HRV are under investigation. Physical training in patients without myocardial infarction has been shown to increase both HRV and the gain between heart rate and arterial pressure variabilities.[67,120,121] However, studies in patients after myocardial infarction have been less convincing that physical training is able to significantly enhance vagal HRV in this group of patients in whom reduced vagal HRV has been shown to be a mortality risk factor.[122]

In healthy subjects, chronic β-blocker administration has been shown to increase total HRV and to shift the frequency distribution of HRV toward greater parasympathetic high-frequency HRV.[6,123] This shift in autonomic balance toward increased parasympathetic influence may explain the improved mortality observed with β-blocker administration following myocardial infarction. Potentially detrimental increases in low-frequency HRV have also been demonstrated after propranolol withdrawal,[124] consistent with the known increased incidence of cardiac events following acute β-blocker withdrawal. In contrast to the effects of β-blockers, vagal measures of HRV either decrease[125] or do not change[123] with administration of calcium channel blockers, consistent with the lack of beneficial effect of calcium channel blockers on post-MI mortality.

Another recent study has documented increases in vagal measures of HRV with transdermal scopolamine and suggested a possible therapeutic role for this drug in the post-MI setting.[126] This increase in vagal HRV by a "vagolytic" agent most likely represents a central "vagotonic" effect previously described with lower doses of belladonna alkaloids. Prior studies examining the effects of graded doses of atropine have demonstrated that low doses of atropine can also cause an increase in HRV (in contrast to the depressant effects of higher doses of atropine).[3,25]

Recent investigations by Muzi et al.[127] examined the effects of clonidine on HRV. As a result of presumed central sympatholytic effects, as well as potential vagotonic effects, this drug may also have significant therapeutic potential for attenuating the autonomic imbalance associated with several forms of cardiovascular disease. Clonidine was found to reduce normalized measurements of low-frequency sympathetic HRV, to increase high-frequency parasympathetic HRV, and to significantly decrease the ratio of low-frequency to high-frequency HRV.

HRV analysis has also been used to investigate the autonomic reflex effects of several drugs used in the treatment of heart failure. Similar to the autonomic imbalance following myocardial infarction, heart failure is accompanied by increased sympathetic activity and decreased parasympathetic activity. Using HRV analysis, a beneficial increase in parasympathetic activity relative to sympathetic activity has been observed with administration of angiotensin-converting enzyme inhibitors,[128] digoxin,[129] and flosequinan.[130]

Cardiac Transplantation

Since the transplanted heart is completely denervated, it is not surprising that heart transplant patients show very minimal HRV. Existing studies present conflicting results regarding the presence of any discrete frequency components in the residual HRV of transplant patients.[131–135] Studies by several investigators suggest that heart rate variations at the respiratory frequency can still be detected using precise analytical techniques.[132–134] On the basis of these studies, investigators have suggested that changes in myocardial wall stretch induced by ventilation may have some influence on intrinsic heart rate independent of autonomic reflex effects. Another study by Fallen et al.[135] did not demonstrate any discrete HRV peaks in eight of nine posttransplant patients, but did document both distinct low- and high-frequency HRV peaks in one patient 33 months after transplantation. Furthermore, administration of atropine abolished the high-frequency vagal peak, suggesting that some transplanted hearts may undergo functional reinnervation (consistent with another recent study demonstrating cardiac release of norepinephrine in transplant patients).[136]

Analysis of heart rate variability has also been suggested as a noninvasive method for detection of myocardial rejection. However, whereas Sands et al.[131] found that patients who develop allograft rejection demonstrated significantly more variability across all frequency bands, Zbilut et al.[133] reported that rejection was associated with a decrease in high-frequency HRV. Additional studies are needed to interpret these varied results.

Diabetes

Analysis of HRV may provide a sensitive, noninvasive measurement for the early detection and subsequent progression of autonomic dysfunction in diabetics.[75,76,134,137,138] Diabetics demonstrate alterations in both low-frequency sympathetic HRV as well as high-frequency vagal HRV. Alterations in HRV have been shown to correlate with measurements of autonomic dysfunction by other autonomic tests, with the presence of concomitant peripheral neuropathy, and with the severity of disease.

Advanced Age

Advanced age is known to impair both sympathetic and parasympathetic reflex responsiveness. Corresponding changes in HRV with advancing age include decreased low-frequency HRV, decreased high-frequency HRV, and attenuation of the expected shift toward greater low-frequency HRV with orthostatic stress.[8,139–142] Studies by Shannon et al.[140] and Korkushko et al.[142] document that high-frequency parasympathetic HRV declines at a faster rate with advancing age than low-frequency sympathetic HRV, although this has not been a uniform finding.[8]

Mental Stress

Analysis of HRV in normal subjects during mental stress results in: 1) a reduction in total HRV power; and 2) a shift in sympathetic-parasympathetic balance toward greater sympathetic dominance and parasympathetic withdrawal.[11,20,37,69,71] This withdrawal of parasympathetic influence is consistent with the decrease in baroreflex sensitivity during mental stress demonstrated by other investigators.[37] While multiple studies have suggested that analysis of HRV may be useful as a physiologic indicator of mental load, such measurements must be interpreted with caution. Although measurements during sedentary activities may be useful, it must be remembered that concomitant alterations in the pattern of respiration and/or physical activity may also significantly influence HRV measurements.[11,20,69,71]

Neonates and Infants

Analysis of fetal HRV was the first major clinical application of the study of HRV and remains the most widely used clinical application today. Reductions in fetal HRV, usually assessed from visual inspection of strip chart recordings of beat-to-beat heart rate, may reflect fetal distress indicative of the need for rapid delivery.[143] This reduction in fetal HRV is believed to be caused by depression of central nervous system reflexes due to hypoxia or other fetal insults. Several types of drugs administered during labor and delivery may also affect fetal HRV, including central depressants (e.g., narcotics, tranquilizers, and anesthetics), and parasympatholytics (e.g., atropine and scopolamine).[143]

Additional studies in infants have examined changes in HRV associated with sudden infant death syndrome,[11,144] cardiac arrest following cardiac surgery,[145] and neonatal "well-being."[56] In eight infants who subsequently died from sudden infant death syndrome, spectral analysis of HRV revealed increases in low-frequency HRV power.[144] In contrast, patients who sustained cardiac arrest after cardiac surgery demonstrated a relative decrease in low-frequency HRV (i.e., a reduction in the ratio of low-frequency to high-frequency HRV power) prior to the arrest.[145] Another study using approximate entropy measurements of HRV (ApEn) demonstrated reduced ApEn in a group of nine neonates in the intensive care unit and suggested that ApEn might be a useful marker of general well-being in this setting.[56]

Neurologic Dysfunction

Studies in patients with neurologic dysfunction have shown that central integration of cardiorespiratory reflexes are required for the genesis of HRV. Lowenshoh et al.[146] demonstrated that changes in HRV parallel changes in cognitive function (i.e., level of consciousness and orientation) in patients with episodic increases of intracranial pressure. These observations suggest that cerebral function and perhaps consciousness per se may play an important role in HRV. In a study of six quadriplegic men (traumatic spinal cord transections at cervical levels 6 or 7), Inoue et al.[70] documented disappearance of the low-frequency component of HRV with preservation of high-frequency HRV. These authors suggested that disappearance of the low-frequency HRV component in quadriplegic patients was caused by interruption of the spinal pathways linking supraspinal cardiovascular centers with peripheral sympathetic outflow.

APPLICATIONS IN ANESTHESIOLOGY

Effects of General Anesthetics

Early studies examining the effects of anesthesia on HRV documented dose-dependent reductions in the amplitude of the respiratory sinus arrhythmia (i.e., high-frequency HRV) with inhalational anesthetics (halothane,[17] isoflurane,[147,148] and enflurane[149]). These investigators suggested that this depression of high-frequency HRV might serve as an indicator of anesthetic depth and postoperative recovery. More recent studies have examined the effects of inhalational and intravenous anesthetics on both high- and low-frequency components of HRV.

With the possible exceptions of pure etomidate[33] and N_2O,[150] all general anesthetics studied have been shown to cause reductions in HRV. Potential etiologies for this reduction in HRV include nonspecific effects of general anesthetics on cerebral function (i.e., unconsciousness) as well as specific alterations in autonomic reflex function. Based on HRV studies in nonanesthetized subjects, potential effects of anesthetics on autonomic reflex function that might influence HRV include: 1) alterations in sympathetic and parasympathetic tone; 2) changes in baroreflex sensitivity; 3) depressant effects on reflex pathways at both peripheral and ganglionic sites; and 4) altered integration of reflex activity at central sites. To the extent that reductions in HRV may be caused by nonspecific effects of general anesthesia on cerebral function (i.e., unconsciousness), all anesthetics might be expected to have similar effects on HRV. However, to the extent that reductions in HRV are caused by agent-specific effects on autonomic reflex function, different anesthetics would be expected to have different effects on HRV.

Preliminary studies by Latson et al.[73] suggest that anesthetic effects on both cerebral and autonomic reflex function may be involved with anesthetic depression of HRV. These investigations demonstrated that an anesthetic technique known to cause significant depression of autonomic reflex activity (thiopental-N_2O) caused a significantly greater reduction in HRV than an anesthetic technique known to have minimal effects on autonomic reflex activity (etomidate-N_2O) (Fig. 12-5). Despite reductions in total HRV by all techniques studied, changes in the normalized measurements of HRV (which adjusts for the reduction in total HRV) appeared to correlate with known changes in sympathetic-parasympathetic balance following anesthesia induction with: 1) thiopental-N_2O (reduced high-frequency parasympathetic HRV); 2) sufentanil (increased parasympathetic HRV); and 3) etomidate-N_2O (no change in sympathetic-parasympathetic balance).

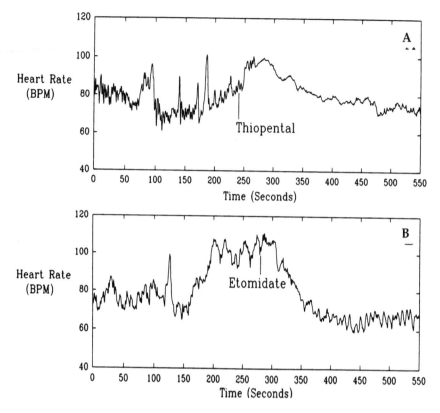

FIGURE 12-5. Beat-to-beat heart rate signals from a patient induced with thiopental (A) and a patient induced with etomidate (B). Note the reduction in amplitude of heart rate oscillations after drug administration in both patients. The greater reduction in amplitude after thiopental is consistent with the known depressant effects of this drug on autonomic reflexes. (Reproduced with permission from Latson TW, Mirhej A, McCarroll SM, et al. Effects of three anesthetic techniques on heart rate variability. J Clin Anesth 1992;4:265.)

Intravenous Anesthetics

Studies with intravenous anesthetic agents have examined the effects of potent narcotics,[61,73,151,152] thiopental,[33,73,153] propofol,[33,154,155] and etomidate.[33,73] All studies with potent narcotics have shown reductions in HRV and a shift in parasympathetic-sympathetic balance toward greater parasympathetic influence.[61,73,151] Loss of consciousness during sufentanil infusion is accompanied by a dramatic decrease in HRV (Fig. 12-6),[73] and this decrease in HRV is correlated with concomitant decreases in blood pressure.[152] All studies with thiopental have shown large reductions in HRV with a shift toward lesser parasympathetic influence.[32,73,153] Studies with propofol have also shown large reductions

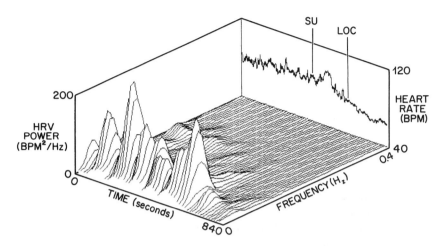

FIGURE 12-6.   Beat-to-beat heart rate signal and trended HRV power spectra in a patient anesthetized with a sufentanil infusion. Power spectra are spaced 15 sec apart. Each spectrum was derived from a 64-sec data epoch of the heart rate signal (32 sec on each side of the plotted spectrum). SU, start of sufentanil infusion; LOC, loss of consciousness. HRV power at all frequencies decreases dramatically following loss of consciousness. BPM, beats per minute. (Reproduced with permission from Latson TW, Mirhej A, McCarroll SM, et al. Effects of three anesthetic techniques on heart rate variability. J Clin Anesth 1992;4:265.)

in HRV with either no change in sympathetic-parasympathetic "balance"[155] or a shift toward lesser parasympathetic influence.[33] In contrast to the findings by Latson et al.[73] using etomidate-N_2O, Scheffer[33] found that pure etomidate (a single dose of 2.5 mg/kg) had no significant effect on HRV. This latter study would suggest that anesthetic-induced loss of consciousness does not necessarily cause reductions in HRV. Alternatively, etomidate may directly or indirectly cause transient cortical stimulation, which offsets any possible depressant effects of loss of consciousness on HRV. Subsequent studies by Latson (unpublished data) have shown that while initial unconsciousness with a pure etomidate induction is not always accompanied by reduced HRV (some patients even show an initial increase), HRV is subsequently reduced by a second dose of etomidate and/or a maintained infusion. Whether this represents a time-dependent or dose-dependent effect will require further study.

Inhalational Anesthetics

Studies with inhalational anesthetics also suggest that different inhalational techniques have different effects on HRV. Recent studies by Ebert and Muzi[150] compared the dose-dependent effects of isoflurane, halothane, and enflurane on HRV. These investigators reported a graded fall in HRV power from 0.5 to 1 minimal alveolar concentration anesthesia, which was more pronounced with enflurane than with isoflurane or halothane. They also demonstrated an increase in HRV power when N_2O was added to 0.5 minimal alveolar concentration anesthesia and postulated that this may represent cortical stimulation by this agent. Different inhalational techniques also appear to have different effects on the frequency distribution of residual HRV. Ishikawa et al.[156] reported a shift in the frequency distribution of residual HRV toward greater parasympathetic high-frequency HRV with sevoflurane-N_2O,[156] and a similar shift was reported by both Galletly et al.[154] and Latson and O'Flaherty[155] with isoflurane-N_2O (Fig. 12-7). In contrast, Kato et al.[157] found no significant change in the frequency distribution of residual HRV under pure isoflurane anesthesia except when EEG bursting activity was accompanied by transient heart rate accelerations (which acted to increase low-frequency HRV).

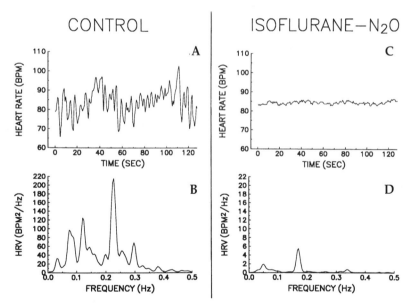

FIGURE 12-7. Beat-to-beat heart signals and derived HRV power spectra from a patient anesthetized with isoflurane and N_2O. A and B, control measurements. C and D, postinduction measurements. Note the change in scale between B and D. Isoflurane-N_2O caused a marked reduction in HRV at all frequencies and a shift in the frequency distribution of the residual HRV toward relatively greater high-frequency HRV power. (Reproduced with permission from Latson TW, O'Flaherty D. Effects of surgical stimulation on autonomic reflex function: assessment by changes in heart rate variability. Br J Anaesth 1993;70:301.)

Factors That May Influence HRV During General Anesthesia

As more studies investigating the effects of anesthetics on HRV appear, it will become important to recognize factors that may contribute to varied results. One important factor is the use of ancillary anesthetic agents, both as part of the anesthetic technique per se (e.g., N_2O, potent narcotics, or muscle relaxants) and as premedicants (e.g., benzodiazepines, scopolamine, or clonidine). Recent studies demonstrating synergistic interactions between intravenous agents would suggest that even relatively small doses of ancillary anesthetics might have significant influence on drug potency[158] and hemodynamics[159] (the latter probably reflecting an effect on autonomic reflex function). These drug interactions would also be expected to influence HRV. In fact, alterations in HRV may provide a valuable method for evaluating these interactions.

Another factor that may influence HRV measured during general anesthesia is alterations in ventilatory patterns. Ventilatory pattern can affect both high- and low-frequency HRV in awake subjects even when the respiratory frequency is well above 0.1 Hz.[8,67] If the respiratory frequency is lowered under anesthesia (e.g., 6 to 8 breaths/min), potential effects of ventilation on low-frequency HRV can be even more pronounced. Lastly, the potential effects of any recent or concurrent noxious stimulation (e.g., intubation or surgery) must be considered (see below).

Depth of Anesthesia

Results of recent studies would suggest that use of HRV for monitoring anesthetic depth may not be as straightforward as initially hoped.[17,147–149] As detailed above, different anesthetic techniques appear to have different effects on HRV, such that drug-specific criteria would be required when HRV measurements are used for monitoring anesthetic depth. In addition, the effects of surgical stimulation on HRV may vary with anesthetic technique.[155] In healthy subjects undergoing laparoscopic surgery with a propofol-fentanyl anesthesia technique, Latson et al.[155] found that surgical stimulation significantly increased total HRV power and increased the percent of total power in the low-frequency range (suggesting a shift in autonomic balance toward sympathetic dominance). In contrast, surgical stimulation had no significant effect on HRV (measured frequency range 0.03 to 0.50 Hz) in a similar group of patients anesthetized with an isoflurane-N_2O-fentanyl technique (Fig. 12-8). From clinical observations, as well as from what little information

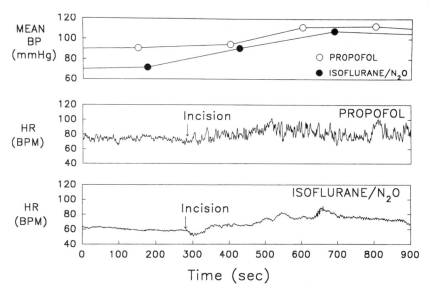

FIGURE 12-8. Change in mean blood pressure and beat-to-beat heart rate with surgical incision in a patient anesthetized with propofol and a patient anesthetized with isoflurane-N₂O. Despite similar increases in mean blood pressure, note the much greater augmentation in the amplitude of heart rate oscillations after incision in the patient anesthetized with propofol. Also note the initial transient heart rate deceleration after incision in the patient anesthetized with isoflurane. This initial heart rate deceleration (which was common in patients anesthetized with isoflurane-N₂O but absent in patients anesthetized with propofol) is consistent with changes in the HRV power spectrum of these patients, demonstrating a shift in sympathetic-parasympathetic balance toward greater parasympathetic dominance. (Reproduced with permission from Latson TW, O'Flaherty D. Effects of surgical stimulation on autonomic reflex function: assessment by changes in heart rate variability. Br J Anaesth 1993;70:301.)

is available concerning the effects of noxious stimulation on autonomic reflex function,[155,160,161] it seems probable that the effects of surgical stimulation on HRV will vary significantly with anesthetic technique, anesthetic dose, stimulus intensity, and stimulus location.

Regional Anesthesia

Scheffer[32] examined changes in HRV caused by controlled lumbar epidural anesthesia to a sensory level of T5/T6. He reported that epidural anesthesia to this level caused a loss of low-frequency HRV with preservation of high-frequency HRV. Scheffer postulated that the observed loss of low-frequency HRV was due to loss of sympathetic modulation of vasomotor tone (and hence loss of blood pressure fluctuations that may act as the stimulus for low-frequency HRV) rather than loss of sympathetic innervation to the heart (mean heart rate did not change). Pruett et al.[162] have shown generalized decreases in HRV accompanying high levels of spinal anesthesia. Clinical application of this finding (e.g., as a continuous monitor of the level of spinal blockade) awaits further studies.

Preoperative Assessment

Preoperative differences in autonomic function among surgical patients may play a significant role in the hemodynamic effects[61,107,108,163–166] of anesthetics. This was first systematically investigated by Burgos et al. in 1989.[163] These investigators demonstrated an increased incidence of hypotension in diabetics with autonomic neuropathy. More recent investigations using spectral analysis of HRV have extended these findings to autonomic dysfunction in nondiabetic patients undergoing both cardiac[61,108] and noncardiac[164,165] surgery. In patients scheduled for elective coronary bypass surgery, Latson et al.[108] demonstrated an increased incidence of hypotension after sufentanil-midazolam induction in patients who demonstrated decreased mid-frequency HRV and/or a lack of the normal increase in low-frequency HRV with head-up tilt. A subsequent study by Estafanous et al.[61] in cardiac surgery patients demonstrated that preoperative differences in autonomic function may influence the effects of sufentanil induction on mean heart rate.[61] Patients who developed a greater than 20% decrease in mean heart rate with sufentanil induction had decreased high-frequency HRV prior to induction and an increased ratio of low-frequency to high-frequency HRV. Additional studies by Latson et al. in noncardiac surgical patients have documented an increased incidence of hypotension in patients with decreased HRV following both thiopental[77a,164] and propofol[165] induction techniques (Table 12-1). These latter studies also suggest that some degree of autonomic reflex dysfunction (sufficient to influence the hemodynamic response to induction) is not uncommon in elective surgery patients over the age of 40, and that the occurrence of hypotension may be better predicted by measures of autonomic function (e.g., HRV measurements, Valsalva maneuver) than a preoperative history of specific medical conditions (e.g., hypertension, diabetes, coronary artery disease, advanced age).

Donchin et al.[167] examined alterations in high-frequency HRV in patients scheduled for elective neurosurgical procedures. They demonstrated that preoperative vagal dysfunction was associated with poor surgical outcome. In contrast, age, heart rate, Glasgow Coma Scale scores, and tumor location, size, and malignancy were not related to outcome in these patients. Additional studies are needed to

TABLE 12-1. INCIDENCE OF HYPOTENSION IN 70 CONSECUTIVELY CONSENTING DAY SURGERY PATIENTS INDUCED WITH A PROPOFOL-NARCOTIC TECHNIQUE

HRV Measurement	SBP <85 mm Hg (%)	SBP <75 mm Hg (%)
HRVmid <2.3 BPM² ($n = 26$)	50	27
HRVmid >2.3 BPM² ($n = 44$)	7	2
p by Fisher exact test	<0.001	<0.003
HRVhi <1.1 BPM² ($n = 26$)	46	31
HRVhi >1.1 BPM² ($n = 44$)	9	0
p by Fisher exact test	<0.001	<0.001

Patients (age range 21 to 73 years) with decreased HRV had a significantly increased incidence of hypotension. SBP, systolic blood pressure; HRVmid, mid-frequency HRV power; HRVhi, high-frequency HRV power; BPM, beats per minute. (Adapted from Latson TW, Liu J, Kellin K. Increased incidence of hypotension after propofol narcotic induction in day surgery patients with autonomic reflex dysfunction (abstract). Anesth and Analg 1992;76:S210.)

determine whether preoperative evaluation of the autonomic nervous system may be predictive of other perioperative complications.[166,168]

Postoperative Assessment

Several authors have suggested that measurements of HRV may provide useful criteria for assessing recovery from inhalational or intravenous anesthesia. In patients anesthetized with isoflurane-N_2O, Donchin et al.[147] reported return of the amplitude of the respiratory sinus arrhythmia to approximately 90% of control within 30 min after arrival in the recovery room. Studies with enflurane[149] and propofol[169] suggest that recovery of HRV may be more prolonged with these agents. Additional studies correlating HRV measurements with other measures of anesthesia recovery (e.g., sedation and motor skills) are needed.

Other studies have assessed changes in heart rate variability following major surgery. Komatsu et al.[170] documented prolonged reduction in high-frequency HRV following esophagectomy and postulated that this reduction may have been due to vagal injury. These same investigators also documented reductions in all frequency components of HRV following cardiac surgery.[171] Although some recovery of HRV was evident in these patients during the 28-day follow-up period, HRV remained reduced even at the end of this period. The authors postulated that myocardial ischemia, use of myocardial preservation solutions, and use of iced slush may have contributed to the observed autonomic dysfunction.

Fleisher et al.[57] examined changes in the ApEn of HRV in 23 high-risk patients undergoing elective noncardiac surgery (Fig. 12-9). Nine

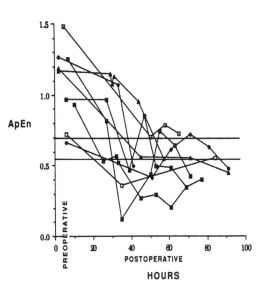

FIGURE 12-9. Hourly approximate entropy (ApEn) values during the preoperative and postoperative periods for patients who demonstrated postoperative ventricular dysfunction. Each point represents the average value during a 1-hour period, with each line representing one patient. (Reproduced with permission from Fleisher LA, Pincus SM, Rosenbaum H. Approximate entropy of heart rate as a correlate of postoperative ventricular dysfunction. Anesthesiology 1993;78:683.)

of these patients demonstrated postoperative ventricular dysfunction, whereas 14 had an uncomplicated course. Decreased postoperative ApEn had a sensitivity of 88% and a specificity of 71% as a potential predictor of postoperative ventricular dysfunction. These authors concluded that chaos-domain measurements of HRV may provide valuable prognostic information in the postoperative period and that additional studies are warranted.

Conclusions

Although significant insight into the etiology of HRV has been gained in recent years, many questions remain. In terms of physiologic control systems, HRV represents one physiologic "output" resulting from the interaction of multiple homeostatic reflex systems. Additional research is needed to define all of the relevant "inputs" to these reflex systems, as well as to determine how these multiple systems interact under both normal and pathophysiologic conditions. Because of the large number of reflex systems that can influence HRV, the sensitivity and specificity of HRV measurements with relation to a specific phenomenon may be limited (e.g., aging, recent myocardial infarction, and ventricular dysfunction all appear to have qualitatively similar effects on HRV). Additional limitations include the lack of "standards" by which HRV is either measured or reported. Because of these limitations, the measurement of HRV should not be considered established technology. Despite these limitations, mounting evidence from HRV studies and other studies suggests that alterations in autonomic reflex function may play an important role in the pathophysiology and clinical outcome of multiple medical conditions. HRV analysis has clearly been a valuable research tool to examine pharmacologic and pathophysiologic alterations in the autonomic nervous system. Since measurements of HRV offer considerable potential as a noninvasive monitor of reflex alterations, this developing technology may some day also make important contributions to patient care.

References

1. Malliani A, Pagani M, Lombardi F, Cerutti S. Cardiovascular neural regulation explored in the frequency domain. Circulation 1991;84:482.
2. Appel ML, Berger RD, Saul JP, Smith JM, Cohen RJ. Beat to beat variability in cardiovascular variables: noise or music? J Am Coll Cardiol 1989; 14:1139.
3. Muzi M, Ebert TJ. Quantification of heart rate variability with power spectral analysis. Curr Opin Anaesthesiol 1993;6:3.
4. Akselrod S, Gordon D, Ubel FA, Shannon DC, Barger AC, Cohen RJ. Power spectrum analysis of heart rate fluctuation: a quantitative probe of beat-to beat cardiovascular control. Science 1981;213:220.
5. Akselrod S, Gordon D, Madwed JB, Snidman NC, Shannon DC, Cohen RJ. Hemodynamic regulation investigation by spectral analysis. Am J Physiol 1985;249:H867.
6. Pagani M, Lombard F, Guzzetti S, et al. Power spectral analysis of heart rate and arterial pressure variabilities as a marker of sympathovagal interaction in man and conscious dog. Circ Res 1986;59:178.
7. Furlan R, Guzzetti S. Crivellaro W, et al. Continuous 24-hour assessment of the neural regulation of systemic arterial pressure and RR variabilities in ambulant subjects. Circulation 1990;81:537.
8. Pagani M, Lombardi F, Guzzetti S, et al. Power spectral analysis of heart rate and arterial pressure variabilities as a marker of sympathovagal interaction in man and conscious dog. Circ Res 1986;59:178.
9. Latson TW. Heart rate variability and anesthesiology: reasons for cautious optimism. J Cardiothorac Vasc Anesth 1992;6:647.
10. O'Flaherty D. Heart rate variability and anaesthesia. Eur J Anesthesiol 1994;IV (in press).
11. Kitney RI, Rompelman O. The beat-by-beat investigation of cardiovascular function. Oxford, Clarendon Press, 1987.
12. Ludwig C. Arch Anat Physiol Wissenschofl Med 1847;242.
13. Hales S. Statistical essays. Containing haemastaticks. London: Manby and Woodward, 1773; vol 2.
14. Mayer S. Studien zur physiologie des herzens und der blutgefasse: 5. Abhandlung: Uber spontane blutdruckschwankungen. Sber Akad Wiss Wien 1876;74:281.
15. Hering E. Sitzungsber. Akad Wiss Wien Math Naturwiss Kl Abt 1871; 64:333.

16. Hon EH, Lee ST. Electronic evaluation of the fetal heart rate patterns preceding fetal death, further observations. Am J Obstet Gynecol 1965;87:814.
17. McCrady JD, Vallbona C, Hoffe HE. The effect of preanesthetic and anesthetic agents on the respiration-heart rate response of dogs. Am J Vet Res 1965;26:710.
18. Hyndman BW, Kitney RI, Sayers BM. Spontaneous rhythms in physiological control systems. Nature 1971;233:339.
19. Sayers B. Analysis of heart rate variability. Ergonomics 1973;16:17.
20. Kitney RI, Rompelman O. The study of heart rate variability. Oxford, Clarendon Press, 1980.
21. Randall DC, Brown DR, Raisch RM, Yinglin JD, Randall WC. SA nodal parasympathectomy delineates autonomic control of heart rate power spectrum. Am J Physiol 1991;260:H985.
22. Rimoldi O, Pierini S, Ferrari A, Cerutti S, Pagani M, Malliani A. Analysis of short-terms oscillations of R-R and arterial pressure in conscious dogs. Am J Physiol 1990;258:H967.
23. Pomeranz B, McCaulay RJB, Caudill MA, et al. Assessment of autonomic function in humans by heart rate spectral analysis. Am J Physiol 1985; 248:H151.
24. Hayano J, Sakakibara Y, Yamada A, et al. Accuracy of assessment of cardiac vagal tone by heart rate variability in normal subjects. Am J Cardiol 1991;67:199.
25. Porges SW. Respiratory sinus arrhythmia: physiological basis, quantitative methods and clinical implications. New York, Plenum Press, 1987:101.
26. Lombardi F, Montano N, Finocchiaro ML, et al. Spectral analysis of sympathetic discharge in decerebrate cats. J Auton Nerv Syst 1990;30 (suppl):S97.
27. Berger RD, Saul JP, Cohen RJ. Transfer function analysis of autonomic regulation. I. Canine atrial rate response. Am J Physiol 1989;25:H142.
28. Schwartz PJ, Pagani M, Lombardi F, Malliani A, Brown AM. A cardiocardiac sympathovagal reflex in the cat. Circ Res 1973;32:215.
29. Melcher A. Respiratory sinus arrhythmia in man. A study in heart rate regulation mechanisms. Acta Physiol Scand 1976;435(suppl):3.
30. Smith JJ, Kampine JP. Circulatory physiology—the essentials. 3rd ed. Baltimore, Williams & Wilkins, 1990.
31. Joel N, Samuelhoff M. The activity of the medullary centres in diffusion respiration. J Physiol (Lond) 1956;133:360.
32. Scheffer GJ. Neuro-cardiovascular control during anesthesia. Doctoral thesis, University of Amsterdam, Amsterdam, The Netherlands, 1990.
33. Scheffer GJ, ten Voorde BJ, Karemaker JM, Ross HH, DeLange JJ. Effects of thiopeutone, etomidate and propofol on beat-to-beat cardiovascular signals in man. Anaesthesia 1993;48:849.
34. Kitney RI, Rompelman O. Thermal entrainment patterns in heart rate variability. J Physiol (Lond) 1977;270:41.
35. Kobayashi M, Musha T. 1/f fluctuation of heartbeat period. IEEE Trans Biomed Eng 1982;29:456.
36. Lindqvist A, Parviainen P, Jalonen J, Tuominen J, Valimaki I, Lattinen LA. Clinical testing of thermally stimulated cardiovascular oscillations in man. Cardiovasc Res 1991;25:666.

37. Robbe HWJ, Mulder LJM, Ruddel H, Langewitz WA, Beldman JBP, Mulder G. Assessment of baroreceptor reflex sensitivity by means of spectral analysis. Hypertension 1987;10:538.
38. Jan ten Voorde B. Modeling the baroreflex—a system analysis approach. Academic thesis, University of Amsterdam, Department of Medical Physics, Amsterdam, The Netherlands, 1992.
39. Van Den Akker TJ, Koeleman ASM, Hogenhuis LAH, Rompelman O. Heart rate variability and blood pressure oscillations in diabetics with autonomic neuropathy. Automedica 1983;4:201.
40. Preiss G, Polosa C. Patterns of sympathetic neuron activity associated with Mayer waves. Am J Physiol 1974;226:724.
41. Baselli G, Cerutti S, Civardi S, Malliani A, Pagani M. Cardiovascular variability signals: towards the identification of a closed-loop model of the neural control mechanism. Trans Biomed Eng 1988;35:1033.
42. deBoer RW, Karemaker JM, Strackee J. Hemodynamic fluctuations and baroreflex sensitivity in humans: a beat-to-beat model. Heart Circ Physiol 1987;22:680.
43. Saul JP, Berger RD, Albrecht P, Stein SP, Chen MH, Cohen RJ. Transfer function analysis of the circulation: unique insights into cardiovascular regulation. Am J Physiol 1991;261:H153.
44. Faes Th JC, Lanting P, TenVoorde BJ, Rompelman O. The origin of respiratory sinus arrhythmia: towards a closed-loop identification of autonomic regulation in the cardiovascular system using respiratory induced fluctuations in heart rate and arterial blood pressure. Automedica 1990; 13:33.
45. Kleiger RE, Miller JP, Krone RJ, Bigger JT. The independence of cycle length variability and exercise testing on predicting mortality of patients surviving acute myocardial infarction. Am J Cardiol 1990;65:408.
46. Bigger JT, Kleiger RE, Fleiss JL, Rolnitzky LM, Steinman RC, Miller J. Components of heart rate variability measured during healing of acute myocardial infarction. Am J Cardiol 1988;61:208.
47. Bigger JT, La Rovere MT, Steinman RC, et al. Comparison of baroreflex sensitivity and heart period variability after myocardial infarction. J Am Coll Cardiol 1989;14:1511.
48. Myers GA, Martin GJ, Magid NM, et al. Power spectral analysis of heart rate variability in sudden cardiac death: comparison to other methods. Trans Biomed Eng 1986;BME-33:1149.
49. Kleiger RE, Miller JP, Bigger JT, Moss AJ. Decreased heart rate variability and its association with increased mortality after acute myocardial infarction. Am J Cardiol 1987;59:256.
50. Hull SS, Evans AR, Vanoli E, et al. Heart rate variability before and after myocardial infarction in conscious dogs at high and low risk of sudden death. J Am Coll Cardiol 1990;16:978.
51. Ewing D. Heart rate variability: an important new risk factor in patients following myocardial infraction. Clin Cardiol 1991;14:683.
52. Odemuyiwa O, Malik M, Farrell T, Bashir Y, Poloniecki J, Camm J. Comparison of the predictive characteristics of heart rate variability index and left ventricular ejection fraction for all-cause mortality, arrhythmic events and sudden death after acute myocardial infarction. Am J Cardiol 1991;68:434.

53. Chen CT. One-dimensional digital signal processing. New York, Marcel Dekker, 1979.
54. Kay SM. Modern spectral estimation: theory and application. Englewood Cliffs, NJ, Prentice Hall, 1988.
55. Hayano J, Taylor JA, Yamada A, et al. Continuous assessment of hemodynamic control by complex demodulation of cardiovascular variability. Am J Physiol 1993;264:H1229.
56. Pincus SM, Gladstone IM, Ehrenkranz RA. A regularity statistic for medical data analysis. J Clin Monit 1991;7:335.
57. Fleisher LA, Pincus SM, Rosenbaum H. Approximate entropy of heart rate as a correlate of postoperative ventricular dysfunction. Anesthesiology 1993;78:683.
58. Pincus SM, Huang WM. Approximate entropy: statistical properties and applications. Commun Statis Theory Methods 1992;21:3061.
59. Pincus SM, Viscarello RR. Approximate entropy: a regularity measure for fetal heart rate analysis. Obstet Gynecol 1992;79:249.
60. Kaplan DT, Furman MI, Pincus SM, Ryan SM, Lipsitz LA, Goldberger AL. Aging and the complexity of cardiovascular dynamic. Biophys J 1991;59:945.
61. Estafanous FG, Brum JM, Ribeiro MP, et al. Analysis of heart rate variability to assess hemodynamic alterations following induction of anesthesia. J Cardiothorac Vasc Anesth 199;6:651.
62. Muzi M, Ebert TJ, Kampine JP. Power spectral analysis (PSA) of heart rate variability during clonidine administration (abstract). Anesthesiology 1992;77(3A):A528.
63. McCabe PM, Younge BG, Ackles PK, Porges SW. Changes in heart period, heart-period variability, and a spectral analysis estimate of respiratory sinus arrhythmia in response to pharmacological manipulation of the baroreceptor reflex in cats. Psychophysiology 1985;22:195.
64. Saul JP, Yutaka A, Berger RD, Lilly LS, Colucci WS, Cohen RJ. Assessment of autonomic regulation in chronic congestive heart failure by heart rate spectral analysis. Am J Cardiol 1988;61:1292.
65. Guzzetti S, Piccaluga E, Casati R, et al. Sympathetic predominance in essential hypertension: a study employing spectral analysis of heart rate variability. J Hypertens 1988;6:711.
66. Lombardi F, Sandrone G, Pernpruner S, et al. Heart rate variability as an index of sympathovagal interaction after acute myocardial infarction. Am J Cardiol 1987;60:1239.
67. Grossman P, Karemaker J, Wieling W. Prediction of tonic parasympathetic cardiac control using respiratory sinus arrhythmia: the need for respiratory control. Psychophysiology 1991;28:201.
68. Pagani M, Somers V, Furlan R, et al. Changes in autonomic regulation induced by physical training in mild hypertension. Hypertension 1998;12:600.
69. Pagani M, Mazzuero G, Ferrari A, et al. Sympathovagal interaction during mental stress: a study employing spectral analysis of heart rate variability in healthy controls and in patients with a prior myocardial infarction. Circulation 1991;(suppl II):II-43.
70. Inoue K, Miyake S, Kumashiro M, Ogata H, Yoshimura O. Power spectral analysis of heart rate variability in traumatic quadriplegic humans. Am J Physiol 1990;258:H1722.

71. Pagani M, Rimoldi O, Pizzinelli P, et at. Assessment of the neural control of the circulation during psychological stress. J Auton Nerv Syst 1991;35:33.

72. Rimoldi O, Pagani M, Pagani MR, Baselli G, Malliani A. Sympathetic activation during treadmill exercise in the conscious dog: assessment with spectral analysis of heart period and systolic pressure variabilities. J Auton Nerv Syst 1990;30:S129.

73a. Saul JP, Rea RF, Berger RD, Cohen RJ. Heart rate and muscle sympathetic nerve variability during reflex changes of autonomic activity. Am J Physiol 1990;258:H713.

73. Latson TW, Mirhej A, McCarroll SM, et al. Effects of three anesthetic techniques on heart rate variability. J Clin Anesth 1992;4:265.

74. Mancia G, Parati G, Pomidossi G, Casadel R, Di Rienzo M, Zanchetti A. Arterial baroreflexes and blood pressure and heart rate variabilities in humans. Hypertension 1986;8:147.

75. Pagani M, Malfatto G, Pierini S, et al. Spectral analysis of heart rate variability in the assessment of autonomic diabetic neuropathy. J Auton Nerv Syst 1988;23:143.

76. Comi G, Sora MGN, Bianchi A, et al. Spectral analysis of short-term heart rate variability in diabetic patients. J Auton Nerv Syst 1990;30:S45.

77. Latson TW, Salwar P, Ashmore TH. Correlation between measurements of heart rate variability and other standard autonomic tests (abstract). Anesthesiology 1993;79:A1240.

77a. Latson TW, Ashmore TH, Rinehart DJ, Klein KW, Giesecke AH. Autonomic reflex dysfunction in patients presenting for elective surgery is associated with hypotension after anesthesia induction. Anesthesiology 1994;80 (in press February 1994).

77b. Freeman R, Saul JP, Roberts MS, Berger RD, Broadbridge C, Cohen RJ. Spectral analysis of heart rate in diabetic autonomic neuropathy. Arch Neurol 1991;48:185.

78. Eckberg DL, Brabinsky M, Braunwald E. Defective cardiac parasympathetic control in patients with heart disease. N Engl J Med 1971;285:877.

79. Schwartz PJ, Zaza A, Pala M, Grassi G, et al. Transient impairment in baroreceptor reflexes in the first year post myocardial infarction: a prospective study. Circulation 1984;70(suppl II):874.

80. LaRovere MT, Specchia G, Mazzoleni C, Mortara A, Schwartz PJ. Baroflex sensitivity in post-myocardial infarction patients. Correlation with physical training and prognosis. Circulation 1986;74(suppl II):514.

81. Ryan C, Hollenberg M, Harvey DB, Gwynn R. Impaired parasympathetic responses in patients after myocardial infarction. Am J Cardiol 1976;37:1013.

82. Higgins CB, Vatner SF, Eckberg DL, Braunwald E. Alterations in the baroreceptor reflex in conscious dogs with heart failure. J Clin Invest 1972;51:715.

83. Eckberg DW. Parasympathetic cardiovascular control in human disease: a critical review of methods and results. Am J Physiol 1980;239:H581.

84. Ferrari AU, Daffonchio A, Gerosa S, Mancia G. Alterations in cardiac parasympathetic function in aged rats. Am J Physiol 1991;260:H647.

85. Ferrari AU, Daffonchio A, Franzelli C, Mancia G. Cardiac parasympathetic hyperresponsiveness in spontaneously hypertensive rats. Hypertension 1992;19:653.

86. Zipes D. Influence of myocardial ischemia and infarction on autonomic innervation of the heart. Circulation 1990;82:1095.
87. Ellenbogen KA, Mohanty PK, Szentpetery S, Thames MD. Baroreflex abnormalities in heart failure: reversal after orthotopic cardiac transplantation. Circulation 1989;79:51.
88. Smith ML, Ellenbogen KA, Eckberg DL, Szentpetery S, Thames MD. Subnormal heart period variability in heart failure: Effect of cardiac transplantation. J Am Coll Cardiol 1989;14:106.
89. Rea RF, Berg WJ. Abnormal baroreflex mechanisms in congestive heart failure. Circulation 1990;81:2026.
90. Cohn JN. Abnormalities of peripheral sympathetic nervous system control in congestive heart failure. Circulation 1990;82(suppl I):I-59.
91. Schwartz PJ, Zaza A, Pala M, Locati E, Beriaq G, Zanchetti A. Baroreflex sensitivity and its evolution during the first year after myocardial infarction. J Am Coll Cardiol 1988;12:629.
92. Wolf MM, Varigos GA, Hunt D. Sinus arrhythmia in acute myocardial infarction. Med J Aust 1978;2:52.
93. Billman GE, Schwartz PJ, Stone HL. Baroreceptor reflex control of heart rate: a predictor of sudden cardiac death. Circulation 1982;66:874.
94. Kent KM, Smith ER, Redwood DR, Epstein SE. Electrical stability of acutely ischemic myocardium. Influence of heart rate and vagal stimulation. Circulation 1973;47:291.
95. Myers RW, Pearlman AS, Hyman RM, et al. Beneficial effects of vagal stimulation and bradycardia during experimental acute myocardial ischemia. Circulation 1974;49:943.
96. Kolman B, Verrier RL, Lown B. The effect of vagus nerve stimulation upon vulnerability of the canine ventricle: role of sympathetic-parasympathetic interactions. Am J Cardiol 1976;37:1041.
97. Zuanetti G, De Derrari GM, Priori SG, Schwartz PJ. Protective effect of vagal stimulation on reperfusion arrhythmias in cats. Circ Res 1987; 61:429.
98. Farrell TG, Paul V, Cripps TR, et al. Baroreflex sensitivity and electrophysiological correlates in patients after acute myocardial infarction. Circulation 1991;83:945.
99. Lown B, Verrier RL. Neural activity and ventricular fibrillation. N Engl J Med 1976;294:1165.
100. Vanoli E, Schwartz PJ. Sympathetic-parasympathetic interaction and sudden death. Basic Res Cardiol 1990;85:305.
101. Kleiger RE, Miller JP, Bigger JT, Moss AM. Heart rate variability: a variable predicting mortality following acute myocardial infarction. J Am Coll Cardiol 1984;3:2.
102. La Rovere MT, Specchia G, Mortara A, Schwartz PJ. Baroreflex sensitivity, clinical correlates, and cardiovascular mortality among patients with a first myocardial infarction—a prospective study. Circulation 1998;78:816.
103. Cripps TR, Malik M, Farrell TG, Camm AJ. Prognostic value of reduced heart rate variability after myocardial infarction: clinical evaluation of a new analysis method. Br Heart J 1991;65:14.
104. Hayano J, Sakakibara Y, Yamada M, et al. Decreased magnitude of heart rate spectral components in coronary artery disease—its relation to angiographic severity. Circulation 1990;81:1217.

105. Rich MW, Saini JS, Kleiger RE, Carney RM, teVelde A, Freedland KE. Correlation of heart rate variability with clinical and angiographic variables and late mortality after coronary angiography. Am J Cardiol 1988;62:714.

106. Latson TW, Sakai T, Whitten CW, Lipton JM. Reduced heart rate variability is associated with an altered norepinephrine response to head-up tilt (abstract). Circulation 1992;86:A2620.

107. Latson TW, Sakai T, Whitten CW, Lipton JM. Altered neurohumoral response to head-up tilt is associated with post-induction hypotension in cardiac surgical patients (abstract). Anesthesiology 1992;77:A102.

108. Latson T, Sakai T, Whitten C. Autonomic reflex dysfunction in cardiac surgery patients may predict hypotension at induction (abstract). Anesth Analg 1992;74:S174.

109. Bernardi L, Lumina C, Ferrari MR, et al. Relationship between fluctuations in heart rate and asymptomatic nocturnal ischemia. Int J Cardiol 1988;20:39.

110. Huikuri HV, Valkama JO, Airaksinen KEJ, et al. Frequency domain measures of heart rate variability before the onset of nonsustained and sustained ventricular tachycardia in patients with coronary artery disease. Circulation 1993;87:1220.

111. Muller JE, Tofler GH, Stone PH. Circadian variation and triggers of onset of acute cardiovascular disease. Circulation 1989;79:733.

112. Coumel P, Hermida J-S, Wennerblom B, Leenhardt A, Maison-Blanche P, Cauchemez B. Heart rate variability in left ventricular hypertrophy and heart failure, and the effects of beta-blockade. Eur Heart J 1991;12:412.

113. Casolo G, Balli E, Taddei T, Amuhasi J, Gori C. Decreased spontaneous heart rate variability in congestive heart failure. Am J Cardiol 1989;64:1162.

114. Casolo G, Balli E, Fazi A, Gori C, Freni A, Gensini G. Twenty-four-hour spectral analysis of heart rate variability in congestive heart failure secondary to coronary artery disease. Am J Cardiol 1991;67:1154.

115. Van Hoogenhuyze D, Weinstein N, Martin GJ, et al. Reproducibility and relation to mean heart rate of heart rate variability in normal subjects and in patients with congestive heart failure secondary to coronary artery disease. Am J Cardiol 1991;68:1668.

116. Binkley PF, Nunziata E, Haas GJ, Nelson SD, Cody RJ. Parasympathetic withdrawal is an integral component of autonomic imbalance in congestive heart failure: demonstration in human subjects and verification in a paced canine model of ventricular failure. J Am Coll Cardiol 1991; 18:464.

117. Hrushesky WJM, Fader DJ, Berestka JS, Sommer M, Hayes J, Cope FO. Diminishment of respiratory sinus arrhythmia foreshadows doxorubicin-induced cardiomyopathy. Circulation 1991;84:697.

118. Binkley PF, Cody RJ. Measurement of the autonomic profile in congestive heart failure by spectral analysis of heart rate variability. Heart Failure 1992;8:154.

119. Parati G, Castiglioni P, Di Rienzo M, Omboni S, Pedotti A, Mancia G. Sequential spectral analysis of 24-hour blood pressure and pulse interval in humans. Hypertension 1990;16:414.

120. Seals DR, Chase PB. Influence of physical training on heart rate variability and baroreflex circulatory control. J Appl Physiol 1989;66:1886.

121. Goldsmith RL, Bigger JT, Steinman RC, Fleiss JL. Comparison of 24-hour parasympathetic activity in endurance-trained and untrained young men. J Am Coll Cardiol 1992;20:552.

122. Mazzuero G, Lanfranchi P, Colombo R, Giannuzzi P, Giordano A. Long-term adaptation of 24-h heart rate variability after myocardial infarction. The EAMI Study Group. Exercise Training in Anterior Myocardial Infarction. Chest 1992;101:304S.

123. Cook JR, Bigger JT, Kleiger RE, Fleiss JL, Steinman RC, Rolnitzky LA. Effect of atenolol and diltiazem on heart period variability in normal persons. J Am Coll Cardiol 1991;17:480.

124. Binkley PF, Kurkemeyer JJ, Nunziata E, et al. Demonstration of resting beta adrenergic hypersensitivity following propranolol withdrawal via analysis of heart rate variability. Clin Res 1990;38:254A.

125. Schwartz JB. In vivo evidence for parasympathetic inhibition by verapamil (abstract). Circulation 1989;80:2377

126. Vybiral T, Bryg RJ, Maddens ME, et al. Effects of transdermal scopolamine on heart rate variability in normal subjects. Am J Cardiol 1990; 65:604.

127. Muzi M, Ebert TJ, Kampine JP. Power spectral analysis (PSA) of heart rate variability during clonidine administration (abstract). Anesthesiology 1992;77:A528.

128. Binkley PF, Nunziata E, Haas GJ, Cody RJ. Sustained augmentation of parasympathetic tone with chronic angiotensin converting enzyme inhibition in congestive heart failure (abstract). J Am Coll Cardiol 1991; 17:23A.

129. Binkley PF, Nunziata E, Cody RJ. Digoxin mediated reduction of sympathetic tone contributes to vasodilation in dilated cardiomyopathy (abstract). Circulation 1990;82:111.

130. Binkley PF, Nunziata E, Hatton P, et al: Flosequinan augments parasympathetic tone and attenuates sympathetic drive in congestive heart failure: Demonstration by analysis of heart rate variability (abstract). J Am Coll Cardiol 1992;19:147A.

131. Sands KEF, Appel ML, Lilly LS, Schoen FJ, Mudge GH, Cohen RJ. Power spectrum analysis of heart rate variability in human cardiac transplant recipients. Circulation 1989;79:76.

132. Luciano B, Keller F, Sanders M, et al. Respiratory sinus arrhythmia in the denervated human heart. J Appl Physiol 1989;67:1447.

133. Zbilut JP, Murdock DK, Lawson L, Lawless CE, Von Dreele MM, Porges SW. Use of power spectral analysis of respiratory sinus arrhythmia to detect graft rejection. J Heart Transplant 1998;7:280.

134. Zeuzem S, Olbrich HG, Seeger C, Kober G, Schoffling K, Caspary WF. Beat-to-beat variations of heart rate in diabetic patients with autonomic neuropathy and in completely cardiac denervated patients following orthotopic heart transplantation. Int J Cardiol 1991;33:105.

135. Fallen EL, Kamath MV, Ghista DN, Fitchett D. Spectral analysis of heart rate variability following human heart transplantation: evidence for functional reinnervation. J Auton Nerv Syst 1988;23:199.

136. Wilson RF, Christensen BV, Olivari MT, Simon A, White CW, Laxson DD. Evidence for structural sympathetic reinnervation after orthotopic cardiac transplantation in humans. Circulation 1991;83:1210.

137. Lishner M, Akselrod S, Avi VM, Oz O, Divon M, Ravid M. Spectral analysis of heart rate fluctuations. A non-invasive, sensitive method for the early diagnosis of autonomic neuropathy in diabetes mellitus. J Auton Nerv Syst 1987;19:119.

138. Freeman R, Saul JP, Roberts MS, Berger RD, Broadbridge C, Cohen RJ. Spectral analysis of heart rate in diabetic autonomic neuropathy—a comparison with standard tests of autonomic function. Arch Neurol 1991;48:185.

139. Lipsitz LA, Mietus J, Moody GB, Goldberger AL. Spectral characteristics of heart rate variability before and during postural tilt-relations to aging and risk of syncope. Circulation 1990;81:1803.

140. Shannon DC, Carley DW, Benson H. Aging of modulation of heart rate. Am J Physiol 1987;253:H874.

141. Schwartz JB, Gibb WJ, Tran T. Aging effects on heart rate variation. J Gerontol 1991;46:M99

142. Korkushko OV, Shatilo VB, Plachinda YI, Shatilo TV. Autonomic control of cardiac chronotropic function in man as a function of age-assessment by power spectral analysis of heart rate variability. J Auton Nerv Syst 1991;32:191.

143. Shnider SM, Levinson G. Anesthesia for obstetrics. Baltimore, Williams & Wilkins, 1987.

144. Gordon D, Cohen RJ, Kelly D, Akselrod S, Shannon DC. Sudden infant death syndrome: abnormalities in short term fluctuations in heart rate and respiratory activity. Pediatr Res 1984;18:921.

145. Gordon D, Herrera VL, McAlpine L, et al. Heart-rate spectral analysis: a noninvasive probe of cardiovascular regulation in critically ill children with heart disease. Pediatr Cardiol 1988;9:69.

146. Lowenshoh RI, Weiss M, Hon EH. Heart rate variability in brain damaged adults. Lancet 1977;192:626.

147. Donchin Y, Feld JM, Porges SW. Respiratory sinus arrhythmia during recovery from isoflurane-nitrous oxide anesthesia. Anesth Analg 1985; 648:811.

148. Latson T, Martin D, Isaac P. Intraoperative changes in respiratory sinus arrhythmia: ? an indicator of anesthetic depth (abstract). Anesth Analg 1988;67:S170

149. Pfeifer BL, Sernaker HL, Porges SW. Respiratory sinus arrhythmia: an index of anesthetic depth? (abstract). Anesth Analg 1988;67:S170.

150. Ebert TJ, Muzi M. Can the amount of heart rate variability be used as a monitor of anesthetic depth. Anesthesiology 1993;79:A1241.

151. Komatsu T, Kimura T, Sanchafa V, et al. Effects of fentanyl-diazepam-pancuronium anesthesia on heart rate variability: a spectral analysis. J Cardiothorac Vasc Anesth 1992;6:444.

152. Latson TW, Martin DC, Whitten CW, Hyndman VA, Elmore J. Changes in heart rate variability correlate with decreases in blood pressure (abstract). Anesthesiology 1990;73:A1207.

153. Latson TW, Hyndman VA, Mirhej MA. Rapid changes in autonomic reflexes induced by Pentothal: assessment by a new method (abstract). Anesth Analg 1990;70:S226.

154. Galletly DC, Corfiatis T, Westenbert AM, Robinson BJ. Heart rate periodicities during induction of propofol-nitrous oxide-isoflurane anesthesia. Br J Anaesth 1992;68:360.

155. Latson TW, O'Flaherty D. Effects of surgical stimulation on autonomic reflex function: assessment by changes in heart rate variability. Br J Anaesth 1993;70:301.

156. Ishikawa T, Kimura T, Kemmotsu O. Spectral analysis of heart rate in sevoflurane/nitrous oxide anesthesia (abstract). Anesthesiology 1989; 70:A97.

157. Kato M, Komatsu T, Kimura T, Sugiyama F, Nakashima K, Shimada Y. Spectral analysis of heart rate variability during isoflurane anesthesia. Anesthesiology 1992;77:669.

158. Vinik HR, Bradley EL, Kissin I. Propofol-midazolam-alfentanil combination: is hypnotic synergism present? (abstract). Anesth Analg 1993;76:S450.

159. Thomson IR, Bergstrom RG, Rosenbloom M, Meatherall RC. Premedication and high-dose fentanyl anesthesia for myocardial revascularization: a comparison of lorazepam versus morphine-scopolamine. Anesthesiology 1988;68:194.

160. Marty J, Reves JG. Cardiovascular control mechanisms during anesthesia. Anesth Analg 1989;69:273.

161. Takeshima R, Dohi S. Comparison of arterial baroreflex function in humans anesthetized with enflurane or isoflurane. Anesth Analg 1989; 69:284.

162. Pruett JK, Yolowski EH, Introna RPS, Buggay DS, Crumrine RS. The influence of spinal anesthetics on heart rate variations. Pharmacol (Life Sci Adv) 1991;10:51.

163. Burgos LG, Ebert TJ, Asiddao C, et al. Increased intraoperative cardiovascular morbidity in diabetics with autonomic neuropathy. Anesthesiology 1989;70:591.

164. Latson TW, Ashmore TH, Reinhart DJ, Klein KW, Giesecke AH. Autonomic reflex dysfunction is associated with post-induction hypotension in both diabetic and non-diabetic patients (abstract). Anesthesiology 1992;77:A103.

165. Latson TW, Liu J, Kellin K. Increased incidence of hypotension after propofol narcotic induction in day surgery patients with autonomic reflex dysfunction (abstract). Anesth Analg 1992;76:S210.

166. Ebert TJ. Preoperative evaluation of the autonomic nervous system. Adv Anesth 1993;10:49.

167. Donchin J, Constantini S, Szold A, Byrne EA, Porges SW. Cardiac vagal tone predicts outcome in neurosurgical patients. Crit Care Med 1992;20:942.

168. Ebert TJ. Autonomic balance and cardiac function. Curr Opin Anaesthesiol 1992;5:3.

169. Nakashima K, Komatsu T, Kimura T, Fujiwara Y, Shimada Y. Rapid recovery of autonomic nervous activity after propofol anesthesia (abstract). Anesth Analg 1993;76:S285.

170. Kimura T, Komatsu T, Takezawa J, Shimada Y. Absence of heart rate variability in esophagectomied patients: probability of autonomic denervation of the heart (abstract). Anesthesiology 1991;75:A89.

171. Komatsu T, Kimura T, Sakakibara Y, Ito M, Shimada Y. Depression of heart rate variability after cardiac surgery in patients (abstract). Anesthesiology 1992;77(3A):A94.

Index

Page numbers followed by (*t*) indicate tables; page numbers followed by (*f*) indicate figures.

ISBN 0-397-51405-0

90000